Introduction to Finite Element Analysis

using I-DEAS 9®

Randy H. Shih
Oregon Institute of Technology

SDC
PUBLICATIONS

Mission, Kansas

Schroff Development Corporation
P.O. Box 1334, Mission, KS 66222
(913) 262-2664
www.schroff.com

Shih, Randy H.
    Introduction to finite element analysis using I-DEAS 9/
Randy H. Shih

    ISBN 1-58503-085-6

Printed and bound in the United States of America

Preface

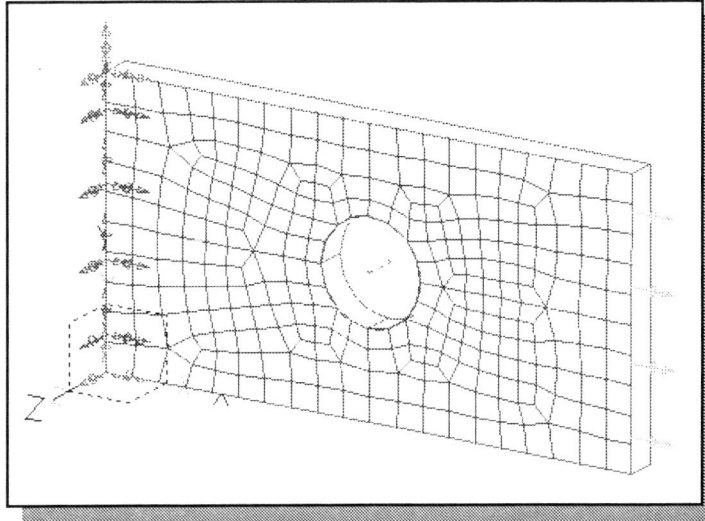

The primary goal of ***Introduction to Finite Element Analysis using I-DEAS 9*** is to introduce the aspects of finite element analysis that are important to the software users. Theoretical aspects are also introduced as they are needed to help better understanding of the operation. Emphasis is on the practical concepts and procedures to using I-DEAS in performing *Linear Static Stress Analysis*. This text is intended to be used as a training guide for students and professionals. This text covers *I-DEAS 9* and the lessons proceed in a pedagogical fashion to guide you from constructing basic truss elements to generating three-dimensional solid elements from solid models. This text takes a hands-on, exercise-intensive approach to all the important *Linear Static Stress Analysis* techniques and concepts. This textbook contains a series of ten tutorial style lessons designed to introduce beginning FEA users to ***I-DEAS***. This text is also helpful to *I-DEAS* users upgrading from a previous release of the software. The finite element analysis techniques and concepts discussed in this text are also applicable to other FEA packages. The basic premise of this book is that the more designs you create using *I-DEAS*, the better you learn the software. With this in mind, each lesson introduces a new set of commands and concepts, building on previous lessons. This book does not attempt to cover all of *the I-DEAS'* features, only to provide an introduction to the software. It is intended to help you establish a good basis for exploring and growing in the exciting field of **computer aided engineering**.

Acknowledgments

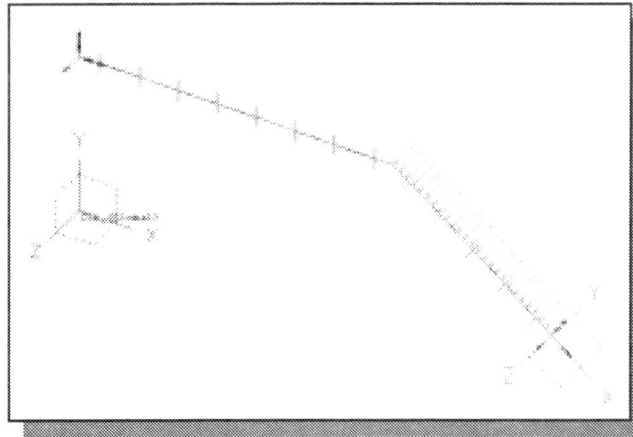

This book would not have been possible without a great deal of support. First, special thanks to two great teachers, Prof. George R. Schade of University of Nebraska-Lincoln and Mr. Denwu Lee, who taught me the fundamentals, the intrigue, and the sheer fun of Computer Aided Engineering.

The effort and support of the editorial and production staff of Schroff Development Corporation is gratefully acknowledged. I would especially like to thank Stephen Schroff and Mary Schmidt for their support and helpful suggestions during this project.

I am grateful that the Mechanical Engineering Technology Department of Oregon Institute of Technology has provided me with an excellent environment in which to pursue my interests in teaching and research.

I also want to thank Prof. J. E. Akin of Rice University for helpful comments and suggestions. I would also like to thank Mr. Mark H. Lawry of EDS for his suggestions and encouragement.

Finally, truly unbounded thanks are due to my wife Hsiu-Ling and our daughter Casandra for their understanding and encouragement throughout this project.

Randy H. Shih
Klamath Falls, Oregon

Table of Contents

Preface
Acknowledgments

Chapter 1
Introduction

Chapter 2
The Direct Stiffness Method

Chapter 3
Two-Dimensional Truss Element

Chapter 4
I-DEAS Two-Dimensional Truss Analysis

Chapter 5
Three-Dimensional Truss Analysis

Chapter 6
Basic Beam Analysis

Chapter 7
Beam Analysis Tools

Chapter 8
Statically Indeterminate Structures

Chapter 9
Two-Dimensional Solid Elements

Chapter 10
Three-Dimensional Solid Elements

Index

Chapter One
Introduction

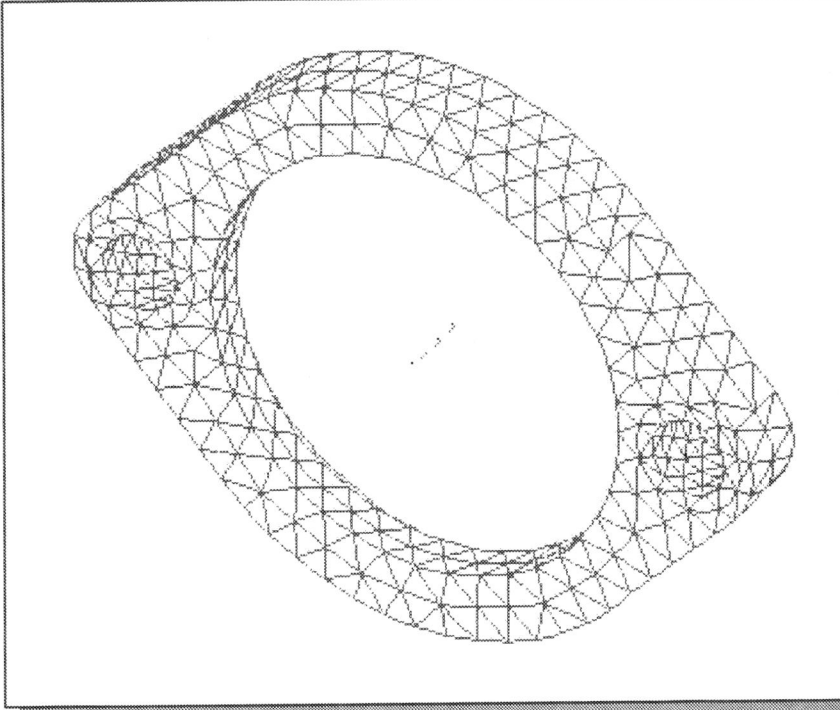

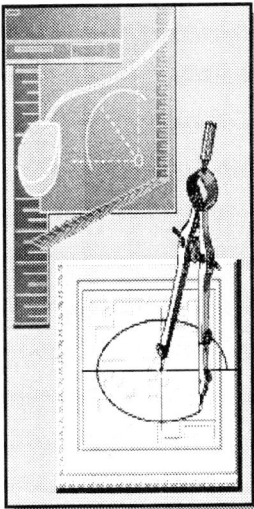

- ◆ Development of Finite Element Analysis
- ◆ FEA Modeling Considerations
- ◆ Finite Element Analysis Procedure
- ◆ Getting Started with I-DEAS
- ◆ The I-DEAS Startup Window and Units Setup
- ◆ I-DEAS Screen Layout
- ◆ Mouse Buttons
- ◆ I-DEAS Data management concepts

1.1 Introduction

Design includes all activities involved from the original concept to the finished product. Design is the process by which products are created and modified. For many years, designers sought ways to describe and analyze three-dimensional designs without building physical models. With the advancements in computer technology, the creation of three-dimensional models on computers offers a wide range of benefits. Computer models are easier to interpret and easily altered. Simulations of real-life loads can be applied to computer models and the results graphically displayed.

Finite Element Analysis (FEA) is a numerical method for solving engineering problems by simulating real-life-operating situations on computers. Typical problems solved by Finite Element Analysis include structural analysis, heat transfer, fluid flow, soil mechanics, acoustics, and electromagnetism. *I-DEAS* (Integrated Design Engineering Analysis Software) is an integrated package of mechanical computer aided engineering software tools developed by Structural Dynamic Research Corporation (**SDRC**). *I-DEAS* is a suite of programs, including *Finite Element Analysis*, that is used to facilitate a concurrent engineering approach to the design, analysis, and manufacturing of mechanical engineering products. This text focuses on basic structural analysis using the *Finite Element Analysis* application of *I-DEAS 9*.

1.2 Development of Finite Element Analysis

Finite element analysis procedures evolved gradually from the work of many people in the fields of engineering, physics, and applied mathematics. The finite element analysis procedure was first applied to problems of stress analysis. The essential ideas began to appear in publications during the 1940s. In 1941, Hrenikoff proposed that the elastic behavior of a physically continuous plate would be similar to a framework of one-dimensional rods and beams, connected together at discrete points. The problem could then be handled by familiar methods for trusses and frameworks. In 1943, Courant's paper detailed an approach to solving the torsion problem in elasticity. Courant described the use of piecewise linear polynomials over a triangularized region. Courant's work was not noticed and soon forgotten, since the procedure was impractical to solve by hand.

In the early 1950s, with the developments in digital computers, Argyris and Kelsey converted the well-established "framework-analysis" procedure into matrix format. In 1956, Turner, Clough, Matin, and Topp derived stiffness matrices for truss elements, beam elements and two-dimensional triangular and rectangular elements in plane stress. Clough introduced the first use of the phrase "finite element" in 1960. In 1961, Melosh developed a flat, rectangular-plate bending-element, followed by development of the curved-shell bending-element by Grafton and Strome in 1963. Martin developed the first three-dimensional element in 1961. Followed by Gallagher, Padlog and Bijlaard in 1962 and Melosh in 1964.

From the mid-1960s to the end of the 1970s, finite element analysis procedures spread beyond structural analysis to many other fields of application. Large general purpose FEA software began to appear. By the late 1980s, FEA software became available on microcomputers, complete with automatic mesh-generation, interactive graphics, and pre-processing and post-processing capabilities.

In this text, we will follow a logical order, parallel to the historical development of the finite element analysis procedures, in learning the fundamental concepts and commands for performing finite element analysis using *I-DEAS*. We will begin with the one-dimensional truss element, beam element, and move toward the more advanced features of *I-DEAS*. This text also covers the general procedures of performing two-dimensional and three-dimensional solid FE analyses. The concepts and techniques presented in this text are also applicable to other FEA packages.

1.3 FEA Modeling Considerations

The analysis of an engineering problem requires the idealization of the problem into a mathematical model. It is clear that we can only analyze the selected mathematical model, and that all the assumptions in this model will be reflected in the predicted results. We cannot expect any more information in the prediction than the information contained in the model. Therefore it is crucial to select an appropriate mathematical model that will most closely represent the actual situation. It is also important to realize that we cannot predict the response exactly, because it is impossible to formulate a mathematical model that will represent all the information contained in an actual system.

As a general rule, finite element modeling should start with a simple model. Once a mathematical model has been solved accurately and the results have been interpreted, it is feasible to consider a more refined model in order to increase the accuracy of the prediction of the actual system. For example, in a structural analysis, the formulation of the actual loads into appropriate models can drastically change the results of the analysis. The results from the simple model, combined with an understanding of the behavior of the system, will assist us in deciding whether and at which part of the model we want to use further refinements. Clearly, the more complicated model will include more complex response effects, but it will also be more costly and sometimes more difficult to interpret the solutions.

Modeling requires that the physical action of the problem be understood well enough to choose suitable kinds of analyses. We want to avoid the waste of time and computer resources caused by over-refinement and badly shaped elements. Once the results have been calculated, we must check them to see if they are reasonable. Checking is very important because it is easy to make mistakes when we rely upon the FEA software to solve complicated systems.

1.4 Types of Finite Elements

The finite element analysis method is a numerical solution technique that finds an approximate solution by dividing a region into small sub-regions. The solution within each sub-region that satisfies the governing equations can be reached much more simply than that required for the entire region. The sub-regions are called *elements* and the elements are assembled through interconnecting a finite number of points on each element called *Nodes*. Numerous types of finite elements can be found in commercial FEA software and new types of elements are being developed as research is done worldwide. Depending on the dimensions, finite elements can be divided into three categories:

1. **One-dimensional** line elements: Truss, beam, and boundary elements.

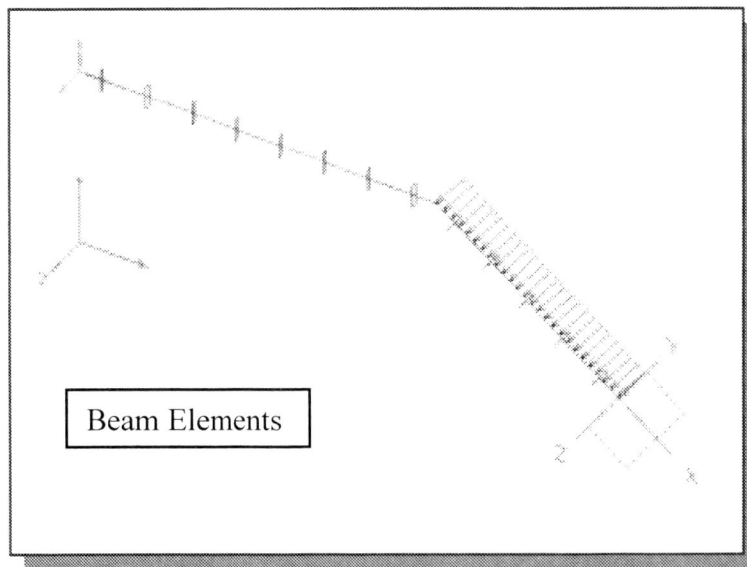

Beam Elements

2. **Two-dimensional** plane elements: Plane stress, plane strain, axisymmetric, membrane, plate, and shell elements.

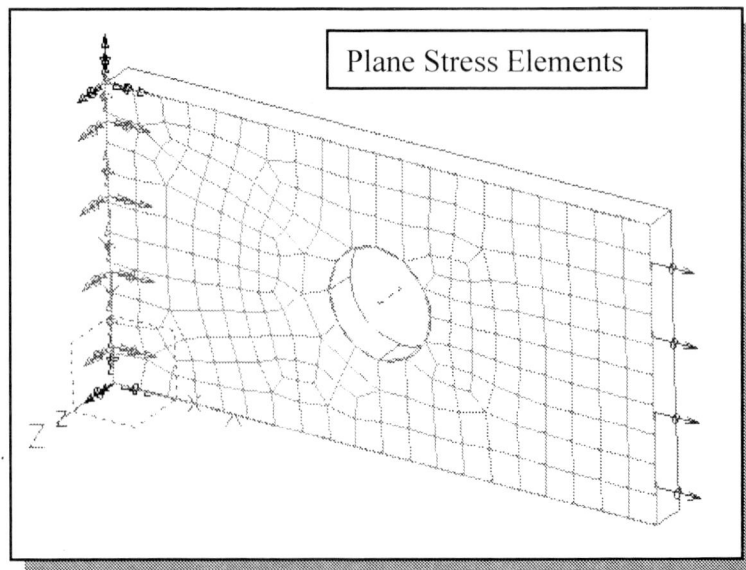

Plane Stress Elements

3. **Three-dimensional** volume elements: Tetrahedral, hexahedral, and brick elements.

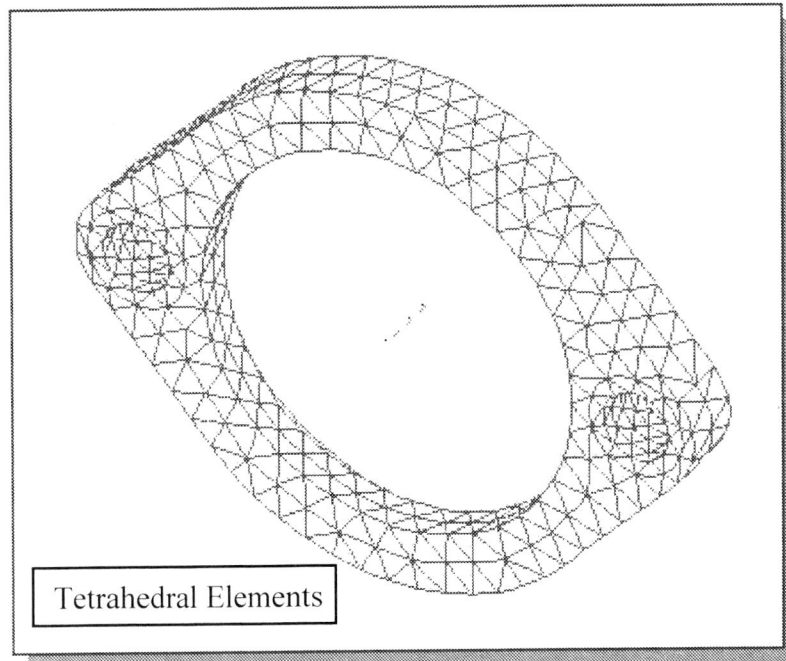

Tetrahedral Elements

Typically, finite element solutions using one-dimensional line elements are as accurate as solutions obtained using conventional truss and beam theories. It is usually easier to get FEA results than doing hand calculations using conventional theories. However, very few closed form solutions exist for two-dimensional elements and almost none exist for three-dimensional solid elements.

In theory, all designs could be modeled with three-dimensional volume elements. However, this is not practical since many designs can be simplified with reasonable assumptions to obtain adequate FEA results without any loss of accuracy. Using simplified models greatly reduces the time and efforts in reaching FEA solutions.

1.5 Finite Element Analysis Procedure

Prior to carrying out the finite element analysis, it is important to do an approximate preliminary analysis to gain some insights into the problem and as a means of checking the finite element analysis results.

For a typical linear static analysis problem, the finite element analysis requires the following steps:

1. Preliminary Analysis.

2. Preparation of the finite element model:
   a. Model the problem into finite elements.
   b. Prescribe the geometric and material information of the system.
   c. Prescribe how the system is supported.
   d. Prescribe how the loads are applied to the system.

3. Perform calculations:
   a. Generate a stiffness matrix of each element
   b. Assemble the individual stiffness matrices to obtain the overall, or global, stiffness matrix.
   c. Solve the global equations and compute displacements, strains, and stresses.

4. Post-processing of the results:
   a. Viewing the stress contours and the displaced shape.
   b. Checking any discrepancy between the preliminary analysis results and the FEA results.

1.6 Matrix Definitions

The use of vectors and matrices is of fundamental importance in engineering analysis because it is with the use of these quantities that complex procedures can be expressed in a compact and elegant manner. One need not understand vectors or matrices in order to use FEA software. However, by studying the matrix structural analysis, we develop an understanding of the procedures common to the implementation of structural analysis as well as the general finite element analysis. The objective of this section is to present the fundamentals of matrices, with emphasis on those aspects that are important in finite element analysis. In chapter two, we will introduce the derivation of matrix structural analysis, the stiffness matrix method. *MATRIX Algebra* is a powerful tool for use in programming the FEA methods for electronic digital computers. Matrix notation represents a simple and easy-to-use notation for writing and solving sets of simultaneous algebraic equations.

A **matrix** is a rectangular array of elements arranged in rows and columns. Applications in this text deal only with matrices whose elements are real numbers.

For example,

$$[A] = \begin{bmatrix} A_{11} & A_{12} & A_{13} \\ A_{21} & A_{22} & A_{23} \end{bmatrix}$$

[A] is a rectangular array of two rows and three columns, thus called a 2 x 3 matrix. The element A_{ij} is the element in the ith row and j^{th} column.

- **Column Matrix (Row Matrix)**   A column (row) matrix is a matrix having one column (row). A single-column array, is commonly called a column matrix or *vector*. For example:

$$\{F\} = \begin{Bmatrix} F_1 \\ F_2 \\ F_3 \end{Bmatrix}$$

- **Square Matrix**   A square matrix is a matrix having equal numbers of rows and columns.

- **Diagonal Matrix**   A diagonal matrix is a square matrix with nonzero elements only along the diagonal of the matrix.

- **Addition**   The addition of matrices involves the summation of elements having the same "address" in each matrix. The matrices to be summed must have identical dimensions. The addition of matrices of different dimensions is not defined. Matrix addition is associative and commutative.

For example:

$$[A] + [B] = \begin{bmatrix} \textcircled{1} & 2 \\ 3 & \boxed{4} \end{bmatrix} + \begin{bmatrix} \textcircled{2} & 4 \\ 6 & \boxed{8} \end{bmatrix} = \begin{bmatrix} \textcircled{3} & 6 \\ 9 & \boxed{12} \end{bmatrix}$$

- **Multiplication by a Constant**   If a matrix is to be multiplied by a constant, every element in the matrix is multiplied by that constant. Also, if a constant is factored out of a matrix, it is factored out of each element. For example:

$$3 \times [A] = 3 \times \begin{bmatrix} 1 & 2 \\ 3 & 4 \end{bmatrix} \qquad \begin{bmatrix} 3 & 6 \\ 9 & 12 \end{bmatrix}$$

- **Multiplication of Two Matrices** Assume that $[C] = [A][B]$, where $[A]$, $[B]$, and $[C]$ are matrices. Element C_{ij} in matrix $[C]$ is defined as follows:

$$C_{ij} = A_{i1} \times B_{1j} + A_{i2} \times B_{2j} + \cdots + A_{ik} \times B_{kj}$$

For example:

$$[C] = [A][B] = \begin{bmatrix} 1 & 2 \\ 3 & 4 \end{bmatrix} \begin{bmatrix} 2 & 4 \\ 6 & 8 \end{bmatrix} = \begin{bmatrix} 14 & 20 \\ 30 & 44 \end{bmatrix}$$

$$C_{11} = 1 \times 2 + 2 \times 6 = 14, \quad C_{12} = 1 \times 4 + 2 \times 8 = 20$$
$$C_{21} = 3 \times 2 + 4 \times 6 = 30, \quad C_{22} = 3 \times 4 + 4 \times 8 = 44$$

- **Identity Matrix** An identity matrix is a diagonal matrix with each diagonal element equal to unity.
  For example:

$$[I] = \begin{bmatrix} 1 & 0 & 0 & 0 \\ 0 & 1 & 0 & 0 \\ 0 & 0 & 1 & 0 \\ 0 & 0 & 0 & 1 \end{bmatrix}$$

- **Transpose of a Matrix** The transpose of a matrix is a matrix obtained by interchanging rows and columns. Every matrix has a transpose. The transpose of a column matrix (vector) is a row matrix; the transpose of a row matrix is a column matrix.
  For example:

$$[A] = \begin{bmatrix} 1 & 2 & 3 \\ 4 & 5 & 6 \end{bmatrix} \quad [A]^T = \begin{bmatrix} 1 & 4 \\ 2 & 5 \\ 3 & 6 \end{bmatrix}$$

- **Inverse of a Square Matrix** A square matrix *may* have an inverse. The product of a matrix and its inverse matrix yields the identity matrix.

$$[A][A]^{-1} = [A]^{-1}[A] = [I]$$

The reader is referred to the following techniques for matrix inversion:

  1. Gauss-Jordan elimination method.
  2. Gauss-Seidel iteration method.

These techniques are popular and are discussed in most texts on numerical techniques.

Getting Started with *I-DEAS*

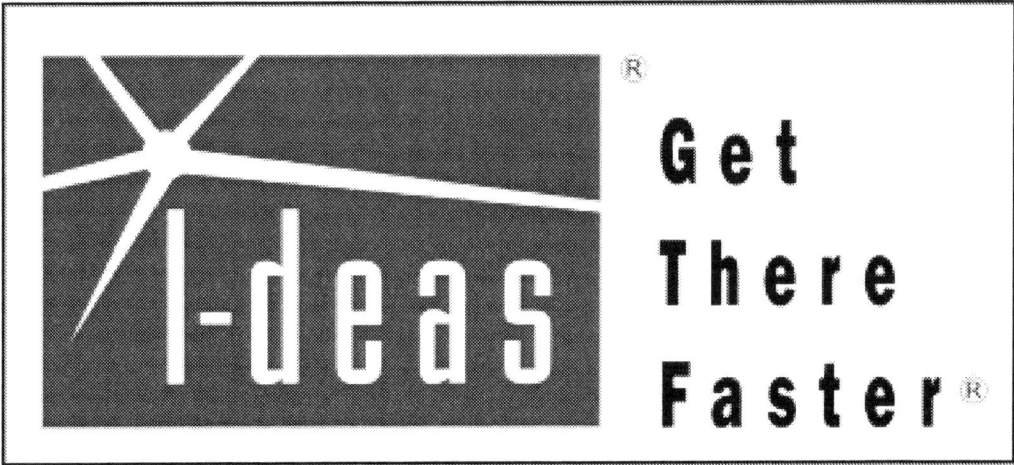

- *I-DEAS* is composed of a number of application software modules (called *applications* and *tasks*), all sharing a common database. In this text, we will be dealing only with the solid modeling modules used for part design, the tools necessary for the creation of models, and engineering drawings. You can select an application or a task when you start *I-DEAS*, or change applications/tasks using the pull-down menu within *I-DEAS*.

I-DEAS Applications

Design
Simulation
Test
Manufacturing
Management
Open Data/PCB

→

I-DEAS Design Tasks

Master Modeler
Master Assembly
Master Drafting
Mechanism Design
Harness Design

◆ I-DEAS Data Management Concepts

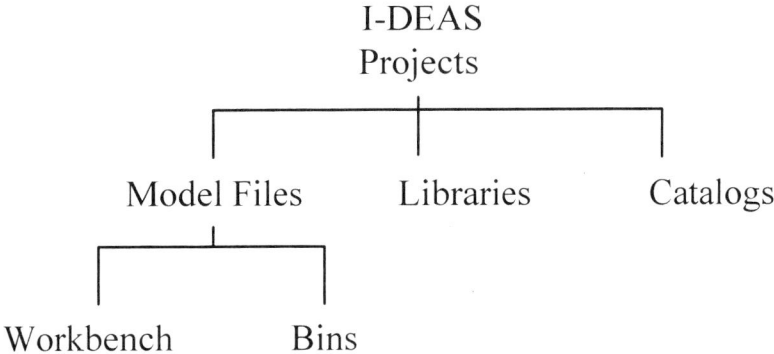

I-DEAS
Projects

Model Files   Libraries   Catalogs

Workbench   Bins

- **Model Files**

  A model file is made up of a workbench and any number of bins. Think of a model file as a personal workspace, like a desk. Parts that are visible on the screen are described as being on the workbench. Parts in the model file can also be placed into bins in the model file, much like storing items in the drawers of a desk.

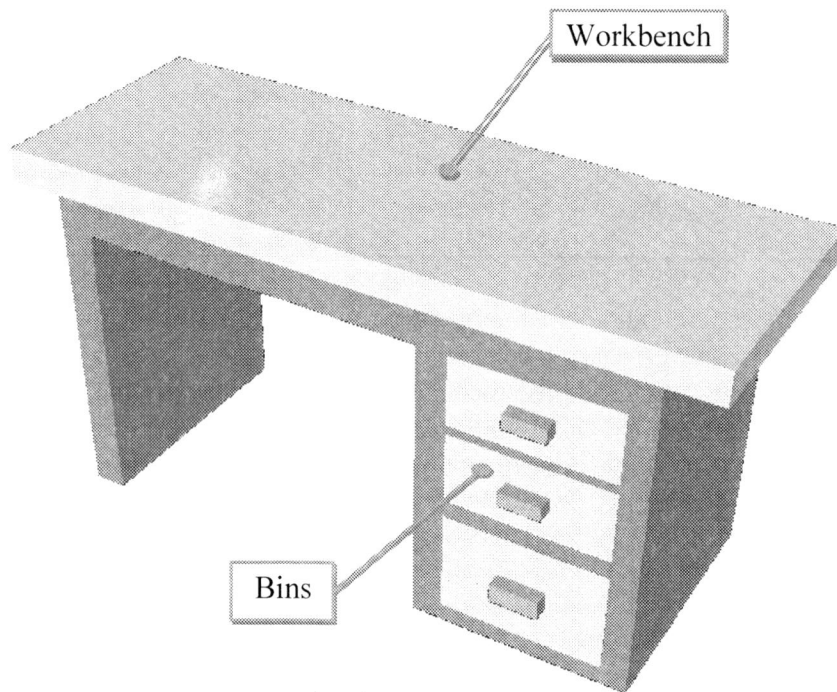

  Workbench

  Bins

- **Libraries**

  Libraries are used to share information among a project team. The libraries allow a project team to concurrently access parts. Libraries can also be used to automatically update your work, provide version control, and provide a central location to store models.

- **Catalogs**

  Catalogs are used for standardized parts and features that would be referenced by the entire company. Information placed in catalogs is typically information that does not change very often.

Starting *I-DEAS*

- To start *I-DEAS*, select the **I-DEAS** icon or type "*ideas*" at your system prompt. Depending upon the type of workstation you are using, you may choose to use different display device options. Typical display device options are *X3D* and *OGL*. Choosing X3D will only allow real-time rotation of wireframe models using the X11 protocol. Choosing OGL will use the Open-Graphics protocol, which allows real-time rotation of both the shaded solid models and wireframe models. Several parameters are available at start-up; you can type "*ideas -h*" at the system prompt to find out what other parameters are available. The program takes a while to load, so be patient. Eventually the *I-DEAS Start* window will appear on the screen, and you will be asked for a project name, a model file name, and the application and task you want to use.

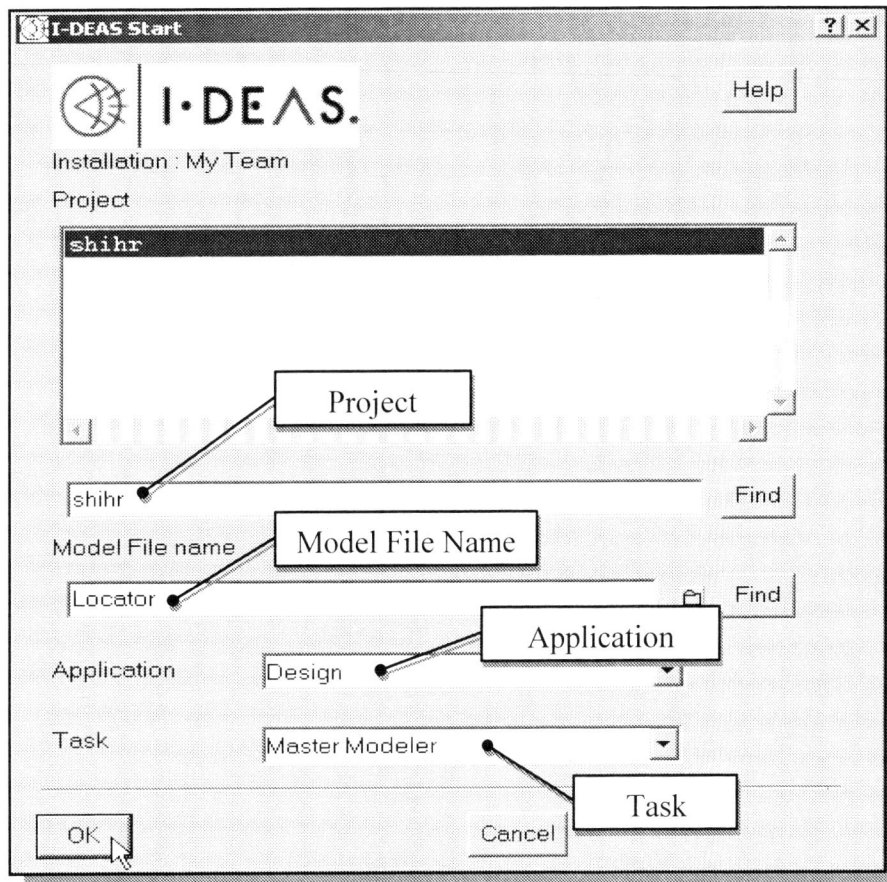

- **Project Name**
  The project name identifies the files associated with the project, allowing team members to easily locate the project files. Enter a project name by clicking with the left-mouse-button in the space provided and typing in a name, or select an existing project out of the list. The first time a new project name is entered, *I-DEAS* will alert you that a new project will be created. For a user not working on a team project, the user account name may be entered for simplicity.

- **Model File Name**
  The model file name identifies the part file that you want to work on. Enter a model file name by clicking with the left-mouse-button in the space provided and typing in a name, or select an existing model file name out of the folder. On some systems, the file name may not contain spaces or special characters, and the number of characters may be limited to 10 characters. The first time a new model file name is entered, *I-DEAS* will alert you that it is a new model file. A model file can contain many parts or assemblies, and actually consists of two files with the extension .MF1 and .MF2. One file contains model data and the other contains graphical data.

- **Application Menu**
  The *application* menu allows you to select the specific application you want. Click on the icon with the left-mouse-button to display the menu of choices. The *Design* application will be the first application we use in this text.

- **Task Menu**
  After selecting the application, select the particular *task* from the task menu. *Master Modeler* is the basic task for creating solid models. Click and drag the left mouse button to make the selection from the menu of choices.

1. In the *Start* dialog box, use your account name as the *Project* name as shown.

2. Enter **Locator** as the *Model File name*.

3. Select **Design** in the *Application* menu.

4. Select **Master Modeler** in the *Task* menu.

5. Click on the **OK** button to accept the settings and launch the selected application and task.

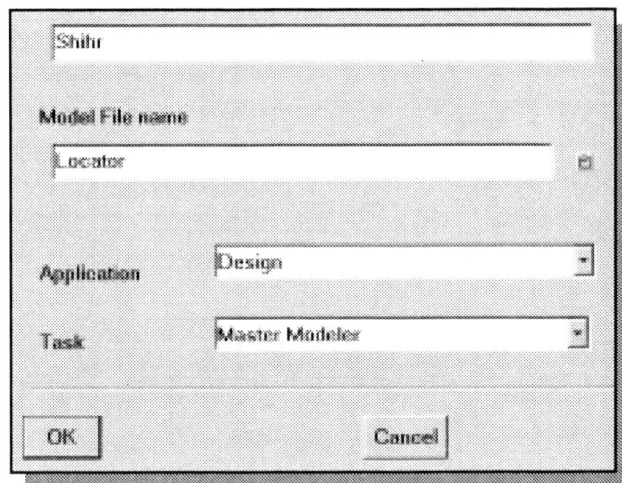

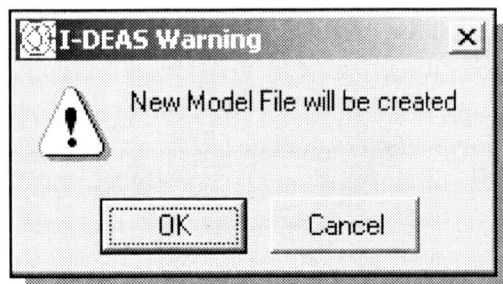

6. In the *I-DEAS warning* dialogue box, click on the **OK** button to acknowledge the creation of the new model file.

I-DEAS Screen Layout

- The four windows are the *graphics window*, the *prompt window*, the *list window*, and the *icon panel*. A line of *quick help* text appears at the bottom of the graphics window as you move the mouse cursor over the icons. You may resize by click and drag on the edges of the window or relocate these windows by click and drag on the window title area. If your computer system provides an icon or menu choice to close the window, DO NOT use it while in *I-DEAS*. If you close a window used for input or output, you run the risk of corrupting the *I-DEAS* model files. To leave *I-DEAS*, select **Exit** in the icon panel's **File** pull-down menu.

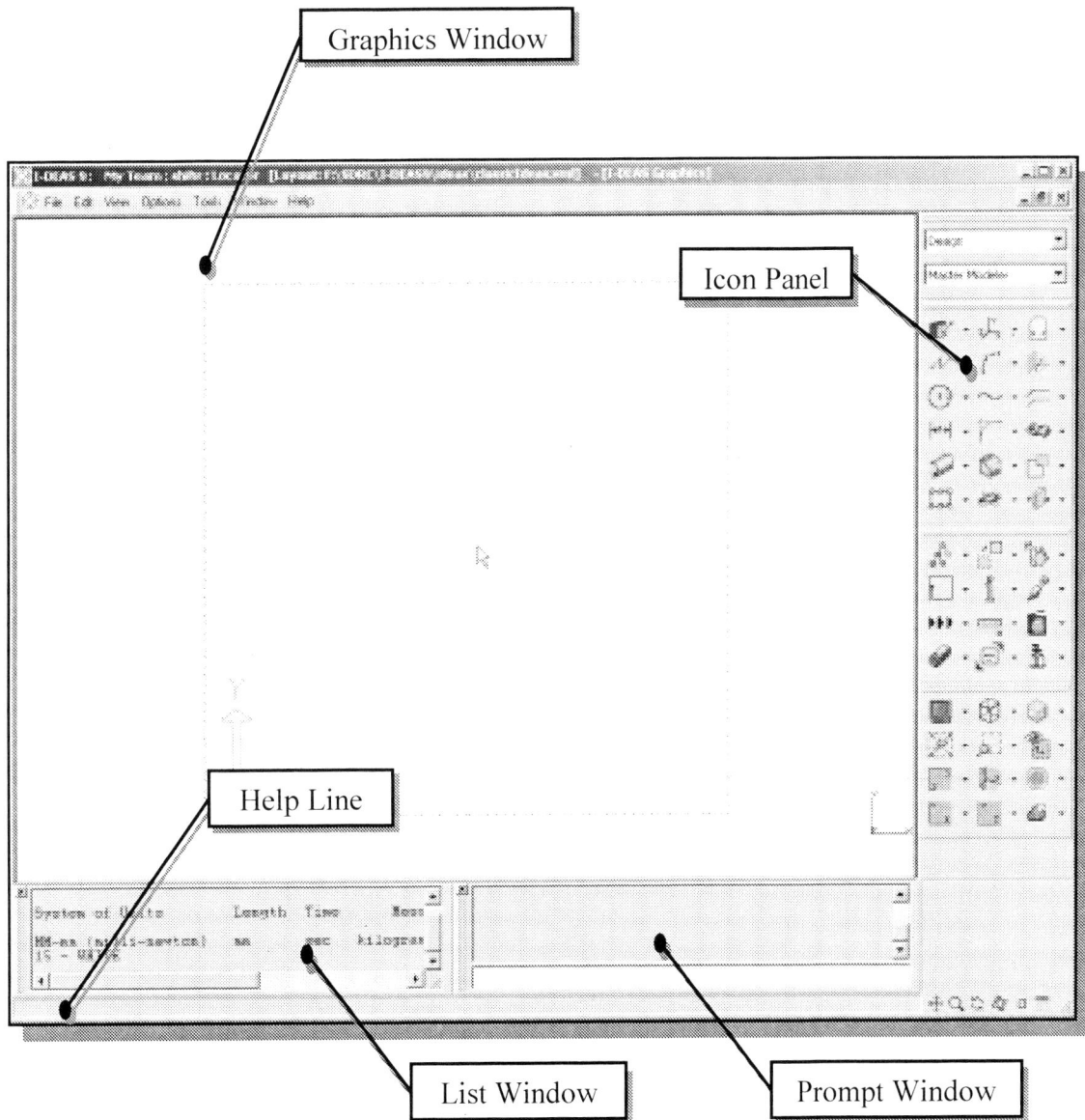

Graphics Window

Icon Panel

Help Line

List Window

Prompt Window

Mouse Buttons

- *I-DEAS* utilizes the mouse buttons extensively. In learning *I-DEAS*' interactive environment, it is important to understand the basic functions of the mouse buttons. It is highly recommended that you use a three-button mouse with *I-DEAS* since the package uses all three buttons for various functions.

- The left-mouse-button is used for most operations, such as selecting menus and icons, or picking graphic entities. *One click* of the button is used to select icons, menus and window entries, and to pick a graphic item. To pick multiple graphic items, hold down the **SHIFT** key and click on each item. The *press, hold, and drag* operation is used to bring up more pull-down menus and icon choices. Use the *double-click* operation in windows to open a listed item. *Multiple clicks* of the left button are used to access the part's *History Tree* hierarchy. The first click picks the *edge* or *face* (selection is highlighted), the second click picks the whole *part* (a white bounding box around the part), and the third click picks the *feature* (a yellow bounding box around the feature).

- The software utilizes the middle-mouse-button the same as the **ENTER** key, and is often used to accept the default setting to a prompt or to end a process. If you are using a two-button mouse, you can always hit the **ENTER** key to get the same result as clicking the middle button.

- The right-mouse-button brings up a pop-up option menu with different choices available. Press and hold down the right button, then slide up and down to select the desired option.

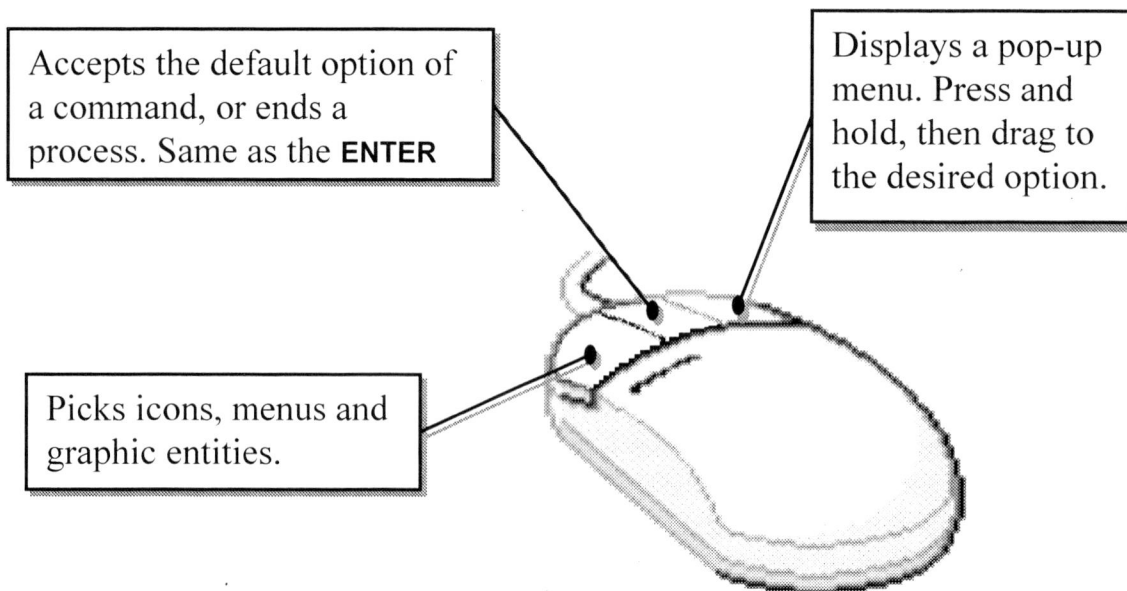

Accepts the default option of a command, or ends a process. Same as the **ENTER**

Displays a pop-up menu. Press and hold, then drag to the desired option.

Picks icons, menus and graphic entities.

Icon Panel

Most of the command input in *I-DEAS* is made by picking icons in the icon panel. The icon panel is arranged in three icon sections, plus a set of pull down menu bars at the top.

◆ Top Menu Bars

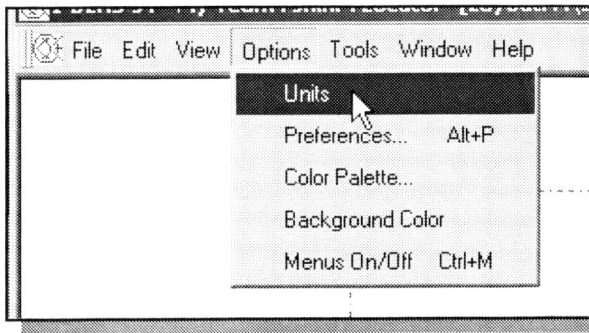

The top menu bars contain the pull-down menus: **File**, **Edit**, **View**, **Options**, **Tools**, **Window** and **Help**. The **File** menu contains the *I-DEAS* input/output options such as **Open**, **Save**, **Plotting**, etc. The **Options** menu contains option settings that allow you to set up units and other personal preferences.

◆ Application and Task menus

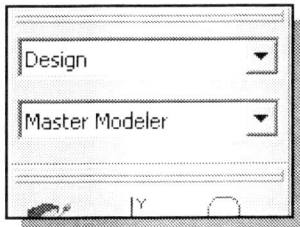

The **Application** and **Task** menus, located above the icon panels, let us change to other *I-DEAS* applications and tasks.

◆ Task Specific Icon Panel

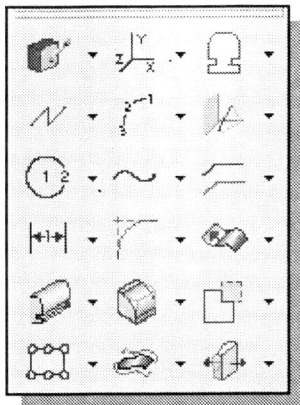

The top icon section contains icons that are specific to the *task* you have chosen. The icons will change as different tasks are selected. The icons shown at the left are from the ***Master Modeler***.

◆ Application Specific Icon Panel

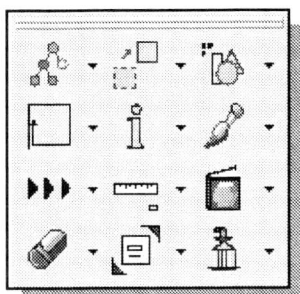

The middle section contains icons that are specific to the application you have chosen. The icons will change as different applications are selected. The icons shown at the left are from the ***Design*** application.

♦ **Display Icon Panel**

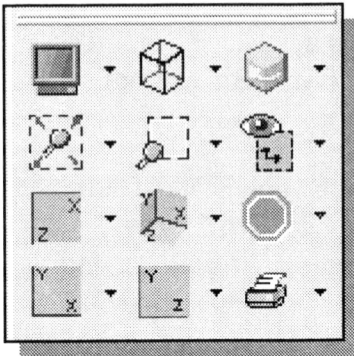

The bottom icon section contains icons that handle the various display operations. These icons control the screen display, such as the view scale, the view angle, redisplay, and shaded and hidden line displays.

Lists

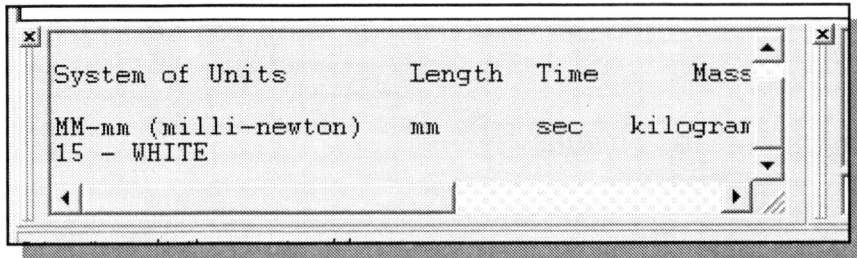

The *List* window is used by *I-DEAS* to display the current status of the system and inquiry information requested by the user. For example, in the above figure, *I-DEAS* displayed the default system units as we first entered *I-DEAS*. Other geometric information, such as the coordinates, of a selected geometric entity can be displayed in this window.

Prompts

Some commands will require keyboard input in the prompt window. Prompts for keyboard input will usually start with the word "*Enter*" as opposed to "*Pick*," which implies picking model entities, or "*Select*," which is used to select an icon or menu command. If a default answer to a prompt is available, it is shown in parentheses *( )* at the end of the prompt. To abort a prompt, use the right-mouse-button and select **Cancel** from the pop-up menu, or type "**$**" on the keyboard.

Quick Help

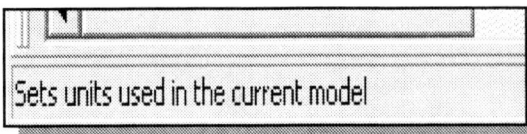

A line of *quick help* text appears at the bottom of the graphics window as the mouse cursor is moved over the icons.

Icon Operation

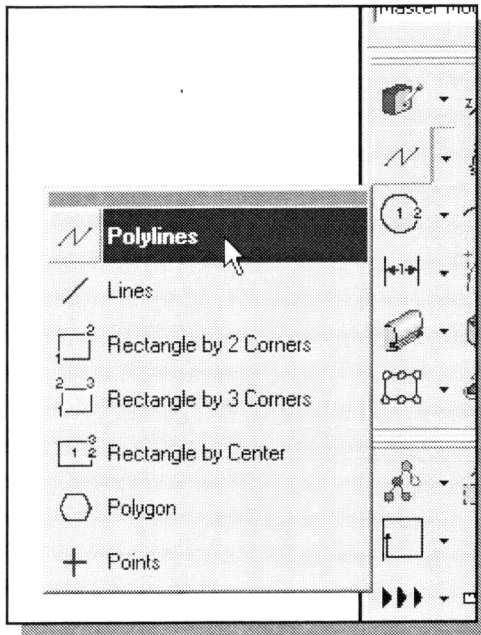

Most of the icons shown are actually a "stack" of related icons. The inverted triangle to the right of the displayed icon indicates there are pull down selections. To select a different icon, hold down the left-mouse-button on the displayed icon to show the available list of icons. Slide up and down with the mouse cursor to switch to other choices in the pull down menu. The icon you used last will stay at the top of the "stack" of icons.

Leaving *I-DEAS*

To leave *I-DEAS*, use the left mouse button, click on **File** in the toolbar menu, and choose **Exit** from the pull-down menu. Note that the quick key combination **[Ctrl] [E]** can also be used for this option.

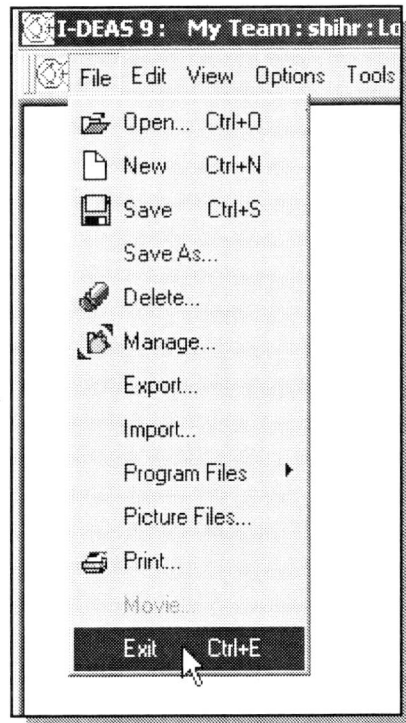

Questions:

1. Can we expect the finite element analysis to predict the response of the system EXACTLY?

2. What is the advantage of starting with a simple model in doing finite element modeling vs. a more complicated model?

3. List the three basic types of element commonly used in FEA.

4. Describe the importance of doing an approximate preliminary analysis prior to performing the finite element analysis?

5.
$$[\mathbf{A}] + [\mathbf{B}] = \begin{bmatrix} 1 & 2 \\ 3 & 4 \end{bmatrix} + \begin{bmatrix} 6 & 2 \\ 4 & 5 \end{bmatrix} = \begin{bmatrix} ? \end{bmatrix}$$

6.
$$5 \times [\mathbf{A}] = 5 \times \begin{bmatrix} 6 & 2 \\ 3 & 7 \end{bmatrix} = \begin{bmatrix} ? \end{bmatrix}$$

7.
$$[\mathbf{C}] = [\mathbf{A}] \, [\mathbf{B}] = \begin{bmatrix} 1 & 2 \\ 3 & 4 \end{bmatrix} \begin{bmatrix} 6 & 7 \\ 3 & 8 \end{bmatrix} = \begin{bmatrix} ? \end{bmatrix}$$

8. What is an identity matrix?

9. The product of a matrix and its inverse matrix equals to what kind of matrix?

10. The transpose of a column matrix (vector) is what kind of matrix?

Chapter 2
The Direct Stiffness Method

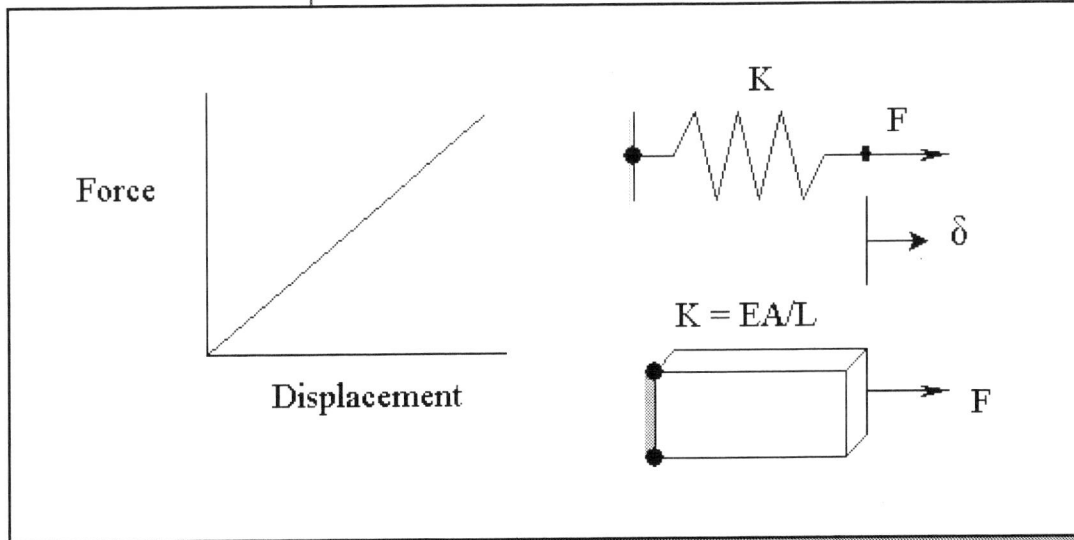

Force

Displacement

$$K$$

$$F$$

$$\delta$$

$$K = EA/L$$

$$F$$

Learning Objectives

When you have completed this lesson, you will be able to:
- ◆ **Understand system equations for truss elements.**
- ◆ **Understand the setup of a Stiffness Matrix.**
- ◆ **Apply the Direct Stiffness Method.**
- ◆ **Create an Extruded solid model using I-DEAS.**
- ◆ **Use the Display Viewing commands.**
- ◆ **Use the Sketch In Place command.**
- ◆ **Create Cutout features.**

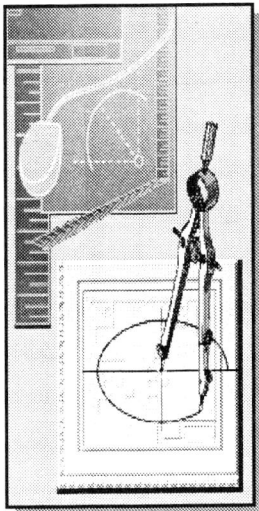

2.1 Introduction

The **direct stiffness method** is used mostly for Linear Static analysis. The development of the direct stiffness method originated in the 1940s and is generally considered the fundamental of finite element analysis. Linear Static analysis is appropriate if deflections are small and vary only slowly. Linear Static analysis omits time as a variable. It also excludes plastic action and deflections that change the way loads are applied. The direct stiffness method for Linear Static analysis follows the *laws of Statics* and the *laws of Strength of Materials*.

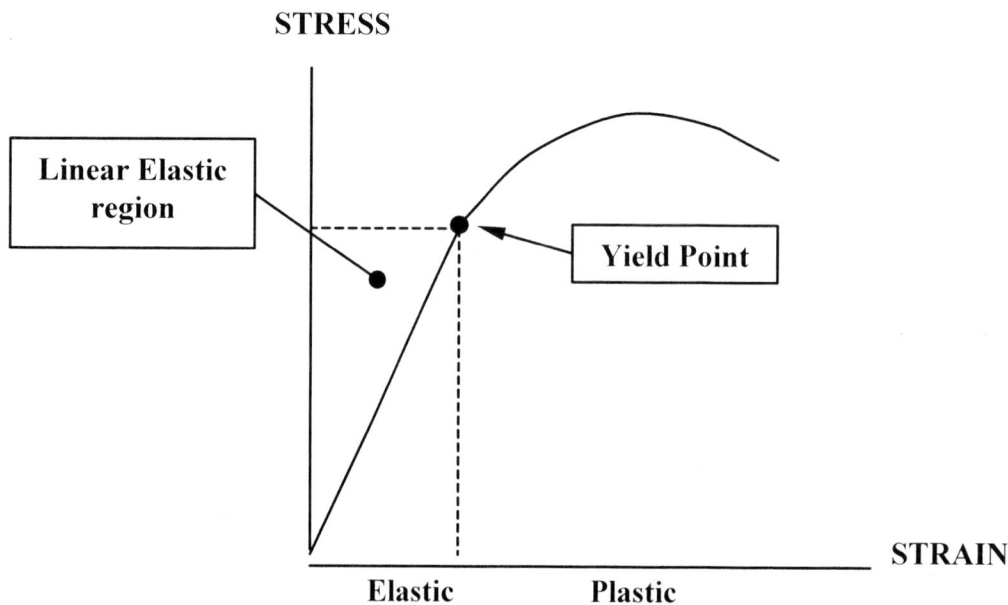

Stress-Strain diagram of typical ductile material

This chapter introduces the fundamentals of finite element analysis by illustrating an analysis of a one-dimensional truss system using the direct stiffness method. The main objective of this chapter is to present the classical procedure common to the implementation of structural analysis. The direct stiffness method utilizes *matrices* and *matrix algebra* to organize and solve the governing system equations. Matrices, which are ordered arrays of numbers that are subjected to specific rules, can be used to assist the solution process in a compact and elegant manner. Of course, only a limited discussion of the direct stiffness method is given here, but we hope that the focused practical treatment will provide a strong basis for understanding the procedure to perform finite element analysis with *I-DEAS*.

The later sections of this chapter demonstrate the procedure to create a solid model using I-DEAS *Master Modeler*. The step-by-step tutorial introduces the *I-DEAS* user interface and serves as a preview to some of the basic modeling techniques demonstrated in the later chapters.

2.2 One-dimensional Truss Element

The simplest type of engineering structure is the truss structure. A truss member is a slender (the length is much larger than the cross section dimensions) **two-force** member. Members are joined by pins and only have the capability to support tensile or compressive loads axially along the length. Consider a uniform slender prismatic bar (shown below) of length L, cross-sectional area A, and elastic modulus E. The ends of the bar are identified as nodes. The nodes are the points of attachment to other elements. The nodes are also the points for which displacements are calculated. The truss element is a two-force member element; forces are applied to the nodes only, and the displacements of all nodes are confined to the axes of elements.

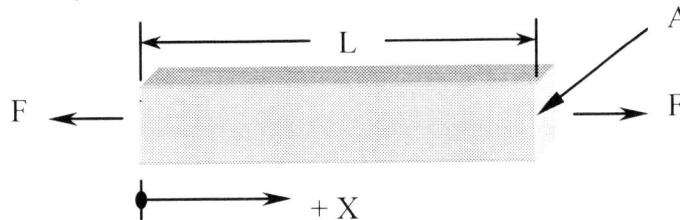

In this initial discussion of the truss element, we will consider the motion of the element to be restricted to the horizontal axis (one-dimensional). Forces are applied along the X-axis and displacements of all nodes will be along the X-axis.

For the analysis, we will establish the following sign conventions:

1. Forces and displacements are defined as positive when they are acting in the positive X direction as shown in the above figure.

2. The position of a node in the undeformed condition is the finite element position for that node.

If equal and opposite forces of magnitude F are applied to the end nodes, from the elementary strength of materials, the member will undergo a change in length according to the equation:

$$\delta = \frac{FL}{EA}$$

This equation can also be written as $\delta = F/K$, which is similar to *Hooke's Law* used in a linear spring. In a linear spring, the symbol K is called the **spring constant** or **stiffness** of the spring. For a truss element, we can see that an equivalent spring element can be used to simplify the representation of the model, where the spring constant is calculated as **K=EA/L.**

Force-Displacement Curve of a Linear Spring

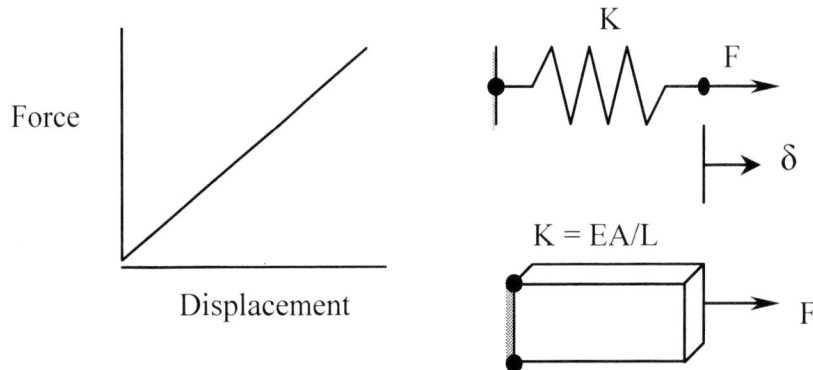

We will use the general equations of a single one-dimensional truss element to illustrate the formulation of the stiffness matrix method:

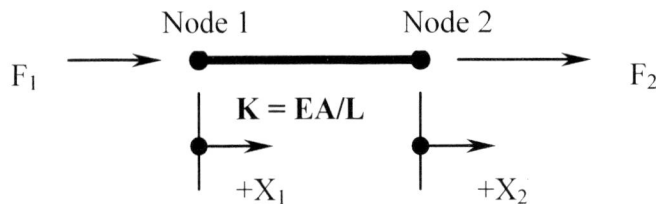

By using the *Relative Motion Analysis* method, we can derive the general expressions of the applied forces (F_1 and F_2) in terms of the displacements of the nodes (X_1 and X_2) and the stiffness constant (K).

   1. Let $X_1 = 0$,

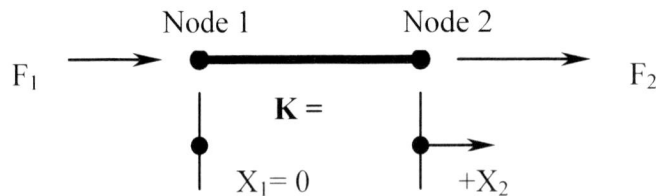

Based on Hooke's law and equilibrium equation:

$$\begin{cases} F_2 = K\, X_2 \\ F_1 = -\, F_2 = -\, K\, X_2 \end{cases}$$

2. Let $X_2 = 0$,

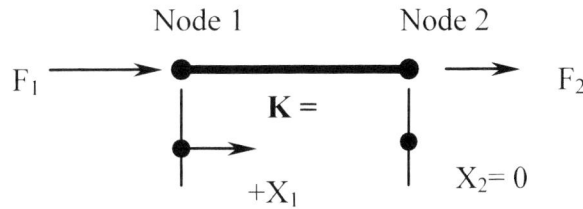

Based on Hooke's law and equilibrium:

$$\begin{cases} F_1 = K\,X_1 \\ F_2 = -F_1 = -K\,X_1 \end{cases}$$

Using the *Method of Superposition,* the two sets of equations can be combined:

$$F_1 = K\,X_1 - K\,X_2$$
$$F_2 = -K\,X_1 + K\,X_2$$

The two equations can be put into matrix form as follows:

$$\begin{Bmatrix} F_1 \\ F_2 \end{Bmatrix} = \begin{bmatrix} +K & -K \\ -K & +K \end{bmatrix} \begin{Bmatrix} X_1 \\ X_2 \end{Bmatrix}$$

This is the general force-displacement relation for a two-force member element, and the equations can be applied to all members in an assemblage of elements. The following example illustrates a system with three elements.

Example 2.1:

Consider an assemblage of three of these two-force member elements. (Motion is restricted to one-dimension, along the X-axis.)

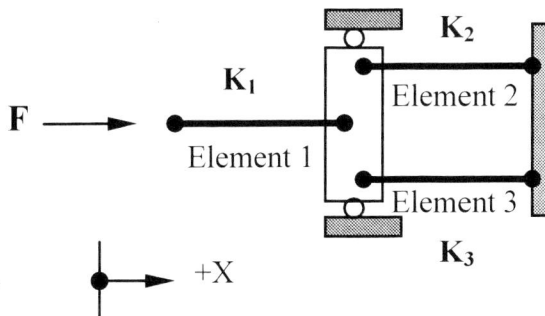

The assemblage consists of three elements and four nodes. The *Free Body Diagram* of the system with node numbers and element numbers labeled:

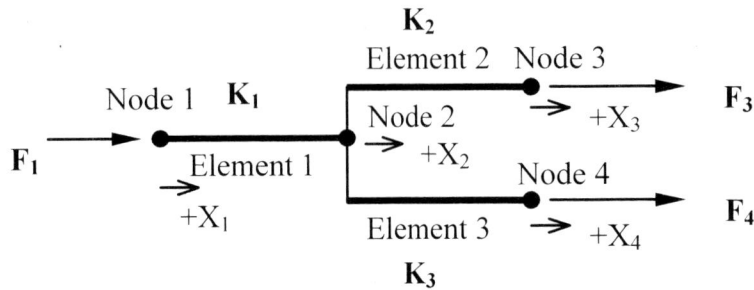

Consider now the application of the general force-displacement relation equations to the assemblage of the elements.

Element 1:

$$\left\{ \begin{array}{c} F_1 \\ F_{21} \end{array} \right\} = \left[\begin{array}{cc} +K_1 & -K_1 \\ -K_1 & +K_1 \end{array} \right] \left\{ \begin{array}{c} X_1 \\ X_2 \end{array} \right\}$$

Element 2:

$$\left\{ \begin{array}{c} F_{22} \\ F_3 \end{array} \right\} = \left[\begin{array}{cc} +K_2 & -K_2 \\ -K_2 & +K_2 \end{array} \right] \left\{ \begin{array}{c} X_2 \\ X_3 \end{array} \right\}$$

Element 3:

$$\left\{ \begin{array}{c} F_{23} \\ F_4 \end{array} \right\} = \left[\begin{array}{cc} +K_3 & -K_3 \\ -K_3 & +K_3 \end{array} \right] \left\{ \begin{array}{c} X_2 \\ X_4 \end{array} \right\}$$

Expanding the general force-displacement relation equations into an Overall Global Matrix (containing all nodal displacements):

Element 1:

$$\left\{ \begin{array}{c} F_1 \\ F_{21} \\ 0 \\ 0 \end{array} \right\} = \left[\begin{array}{cccc} +K_1 & -K_1 & 0 & 0 \\ -K_1 & +K_1 & 0 & 0 \\ 0 & 0 & 0 & 0 \\ 0 & 0 & 0 & 0 \end{array} \right] \left\{ \begin{array}{c} X_1 \\ X_2 \\ X_3 \\ X_4 \end{array} \right\}$$

Element 2:

$$\begin{Bmatrix} 0 \\ F_{22} \\ F_3 \\ 0 \end{Bmatrix} = \begin{bmatrix} 0 & 0 & 0 & 0 \\ 0 & +K_2 & -K_2 & 0 \\ 0 & -K_2 & +K_2 & 0 \\ 0 & 0 & 0 & 0 \end{bmatrix} \begin{Bmatrix} X_1 \\ X_2 \\ X_3 \\ X_4 \end{Bmatrix}$$

Element 3:

$$\begin{Bmatrix} 0 \\ F_{23} \\ 0 \\ F_4 \end{Bmatrix} = \begin{bmatrix} 0 & 0 & 0 & 0 \\ 0 & +K_3 & 0 & -K_3 \\ 0 & 0 & 0 & 0 \\ 0 & -K_3 & 0 & +K_3 \end{bmatrix} \begin{Bmatrix} X_1 \\ X_2 \\ X_3 \\ X_4 \end{Bmatrix}$$

Summing the three sets of general equation: (Note $F_2 = F_{21} + F_{22} + F_{32}$)

$$\begin{Bmatrix} F_1 \\ F_2 \\ F_3 \\ F_4 \end{Bmatrix} = \begin{bmatrix} K_1 & -K_1 & 0 & 0 \\ -K_1 & (K_1+K_2+K_3) & -K_2 & -K_3 \\ 0 & -K_2 & K_2 & 0 \\ 0 & -K_3 & 0 & +K_3 \end{bmatrix} \begin{Bmatrix} X_1 \\ X_2 \\ X_3 \\ X_4 \end{Bmatrix}$$

Overall Global Stiffness Matrix

Once the Overall Global Stiffness Matrix is developed for the structure, the next step is to substitute boundary conditions and solve for the unknown displacements. At every node in the structure, either the externally applied load or the nodal displacement is needed as a boundary condition. We will demonstrate this procedure with the following example.

Example 2.2:

Given:

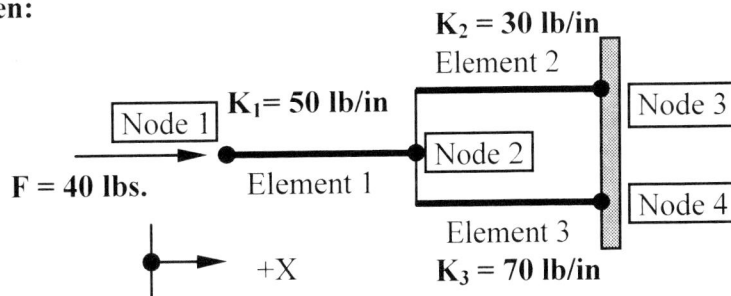

Find: Nodal displacements and reaction forces.
Solution:

From example 2.1, the overall global force-displacement equation set:

$$\begin{Bmatrix} F_1 \\ F_2 \\ F_3 \\ F_4 \end{Bmatrix} = \begin{bmatrix} 50 & -50 & 0 & 0 \\ -50 & (50+30+70) & -30 & -70 \\ 0 & -30 & 30 & 0 \\ 0 & -70 & 0 & 70 \end{bmatrix} \begin{Bmatrix} X_1 \\ X_2 \\ X_3 \\ X_4 \end{Bmatrix}$$

Next, apply the known boundary conditions to the system: the right-end of element 2 and element 3 are attached to the vertical wall; therefore, these two nodal displacements (X_3 and X_4) are zero.

$$\begin{Bmatrix} F_1 \\ F_2 \\ F_3 \\ F_4 \end{Bmatrix} = \begin{bmatrix} 50 & -50 & 0 & 0 \\ -50 & (50+30+70) & -30 & -70 \\ 0 & -30 & 30 & 0 \\ 0 & -70 & 0 & 70 \end{bmatrix} \begin{Bmatrix} X_1 \\ X_2 \\ 0 \\ 0 \end{Bmatrix}$$

The two displacements we need to solve the system are X_1 and X_2. Remove any unnecessary columns in the matrix:

$$\begin{Bmatrix} F_1 \\ F_2 \\ F_3 \\ F_4 \end{Bmatrix} = \begin{bmatrix} 50 & -50 \\ -50 & 150 \\ 0 & -30 \\ 0 & -70 \end{bmatrix} \begin{Bmatrix} X_1 \\ X_2 \end{Bmatrix}$$

Next, include the applied loads into the equations. The external load at *Node 1* is 40 lbs. and there is no external load at *Node 2*.

$$\begin{Bmatrix} 40 \\ 0 \\ F_3 \\ F_4 \end{Bmatrix} = \begin{bmatrix} 50 & -50 \\ -50 & 150 \\ 0 & -30 \\ 0 & -70 \end{bmatrix} \begin{Bmatrix} X_1 \\ X_2 \end{Bmatrix}$$

The Matrix represents the following four simultaneous system equations:

$$\begin{aligned} 40 &= 50\,X_1 - 50\,X_2 \\ 0 &= -50\,X_1 + 150\,X_2 \\ F_3 &= 0\,X_1 - 30\,X_2 \\ F_4 &= 0\,X_1 - 70\,X_2 \end{aligned}$$

From the first two equations, we can solve for X_1 and X_2:

$$X_1 = 1.2 \text{ in.}$$
$$X_2 = 0.4 \text{ in.}$$

Substituting these known values into the last two equations, we can now solve for F_3 and F_4:

$$F_3 = 0 X_1 - 30 X_2 = -30 \times 0.4 = 12 \text{ lbs.}$$
$$F_4 = 0 X_1 - 70 X_2 = -70 \times 0.4 = 28 \text{ lbs.}$$

From the above analysis, we can now reconstruct the *Free Body Diagram* (FBD) of the system:

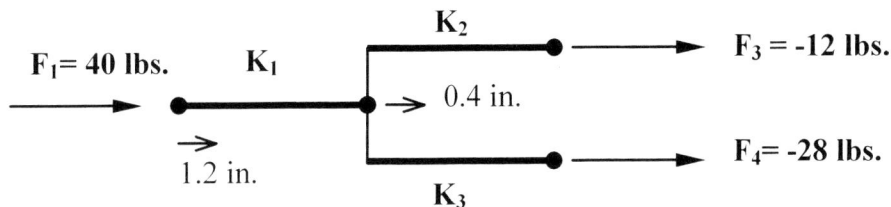

➤ The above sections illustrated the fundamental operation of the direct stiffness method, the classical finite element analysis procedure. As can be seen, the formulation of the global force-displacement relation equations is based on the general force-displacement equations of a single one-dimensional truss element. The two-force-member element (truss element) is the simplest type of element used in FEA. The procedure to formulate and solve the global force-displacement equations is straightforward, but somewhat tedious. In real-life application, the use of a truss element in one-dimensional space is rare and very limited. In the next chapter, we will expand the procedure to solve two-dimensional truss frameworks.

The following sections illustrate the procedure to create a solid model using *I-DEAS Master Modeler*. The step-by-step tutorial introduces the basic *I-DEAS* user-interface and the tutorial serves as a preview to some of the basic modeling techniques demonstrated in the later chapters.

2.3 Basic Solid Modeling using I-DEAS Master Modeler

One of the methods to create solid models in *I-DEAS Master Modeler* is to create a two-dimensional shape and then *extrude* the two dimensional shape to define a volume in the third dimension. This is an effective way to construct three-dimensional solid models since many designs are in fact the same shape in one direction. Computer input and output devices used today are largely two-dimensional in nature, which makes this modeling technique quite practical. This method also conforms to the design process that helps the designer with conceptual design along with the capability to capture the design intent. *I-DEAS Master Modeler* provides many powerful modeling tools and there are many different approaches available to accomplish modeling tasks. We will start by introducing the basic two-dimensional sketching and parametric modeling tools.

The *Adjuster Block* design

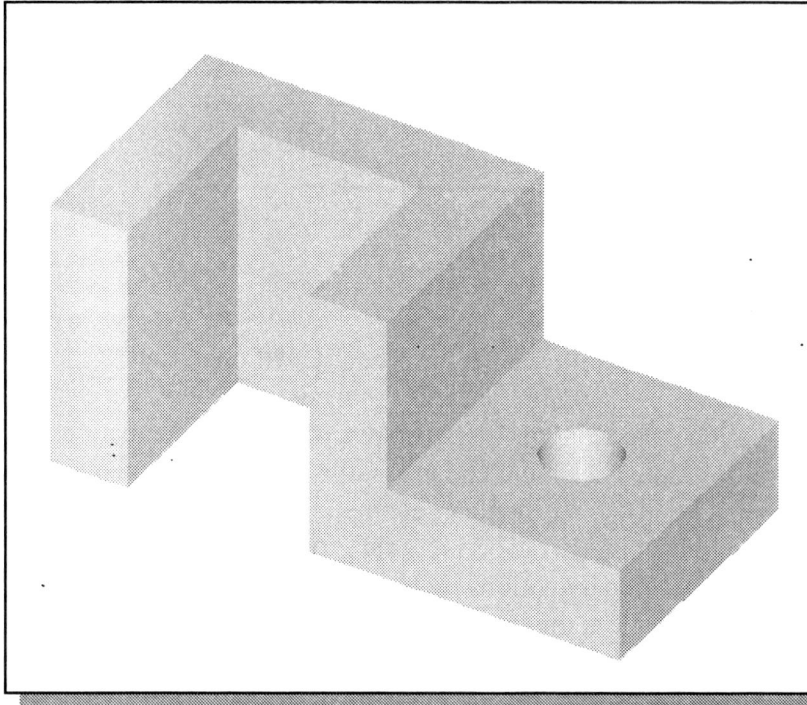

Starting *I-DEAS*

1. Select the **I-DEAS** icon or type "*ideas*" at your system prompt to start *I-DEAS*. The *I-DEAS Start* window will appear on the screen.

2. Fill in and select the items as shown below:

> Project Name: **(Your account name)**
> Model File Name: **Adjuster**
> Application: **Design**
> Task: **Master Modeler**

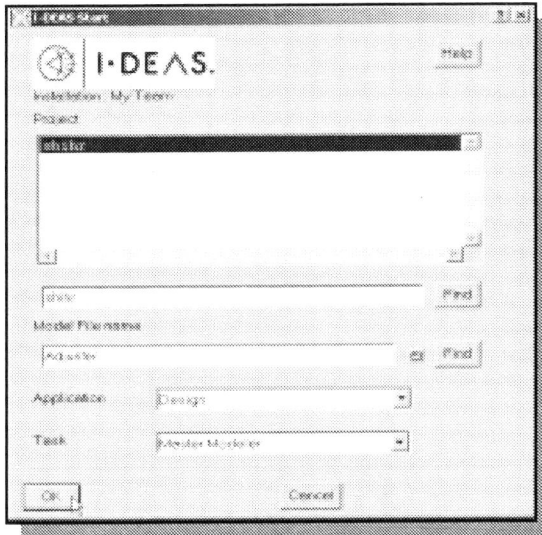

3. After you click **OK**, two *warning windows* will appear to tell you that a new model file will be created. Click **OK** on both windows as they come up.

> **I-DEAS Warning**
> **! New Model File will be created**
> **OK**

4. Next, *I-DEAS* will display the main screen layout, which includes the *graphics window*, the *prompt window*, the *list window* and the *icon panel*.

Units Setup

❖ When starting a new model, the first thing we should do is to determine the set of units we would like to use. *I-DEAS* displays the default set of units in the list window.

1. Use the left-mouse-button and select the **Options** menu in the icon panel as shown.

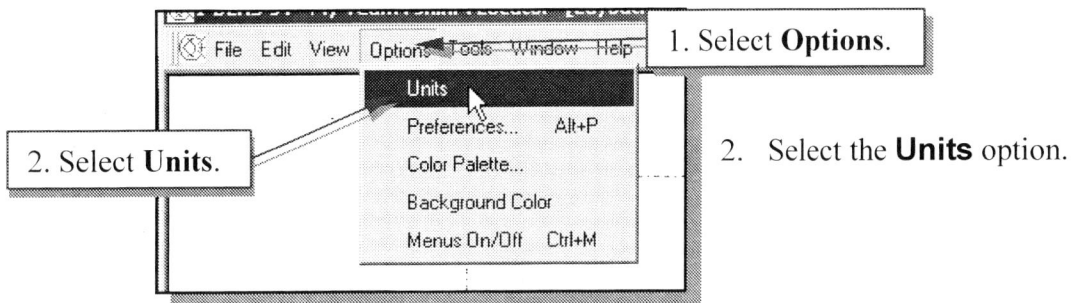

1. Select **Options**.

2. Select **Units**.

2. Select the **Units** option.

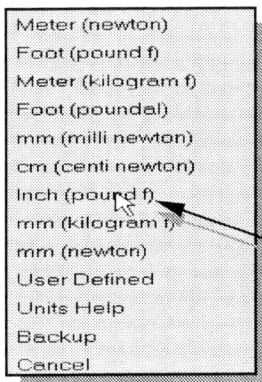

3. Inside the graphics window, pick **Inch (pound f)** from the pop-up menu. The set of units is stored with the model file when you save.

3. Select **Inch (pound f)**.

Step 1: Creating a rough sketch

❖ In this lesson we will begin by building a 2D sketch, as shown in the figure below.

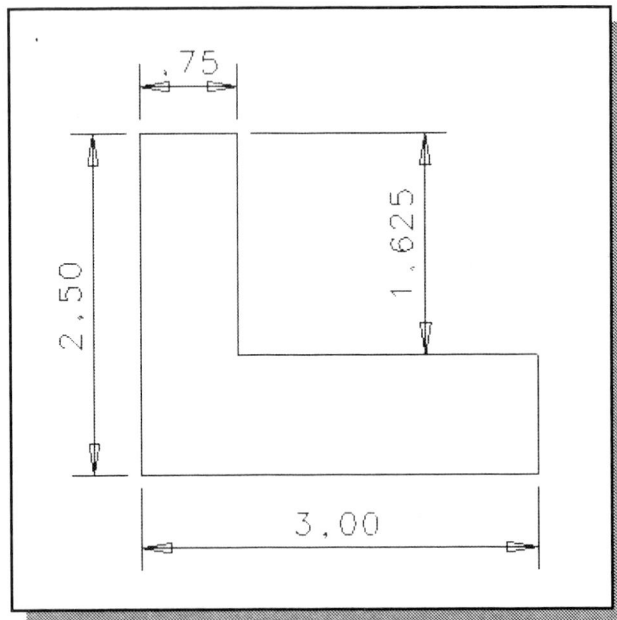

I-DEAS provides many powerful tools for sketching 2D shapes. In the previous generation CAD programs, exact dimensional values were needed during construction, and adjustments to dimensional values were quite difficult once the model is built. In *I-DEAS*, we can now treat the sketch as if it is being done on a piece of napkin, and it is the general shape of the design that we are more interested in defining. The *I-DEAS* part model contains more than just the final geometry, it also contains the **design intent** that governs what will happen when geometry changes. The design philosophy of "shape before size" is implemented through the use of *I-DEAS'* **Variational Geometry**. This allows the designer to construct solid models in a higher level and leave all the geometric details to *I-DEAS*. We will first create a rough sketch, by using some of the visual aids available, and then update the design through the associated control parameters.

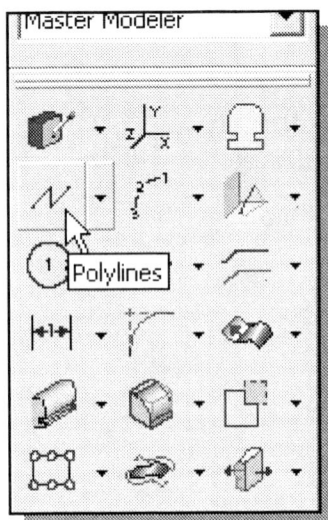

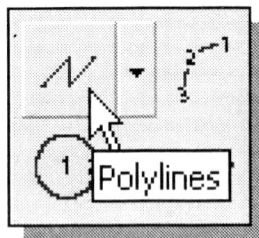

1. Pick **Polylines** in the icon panel. (The icon is located in the second row of the task specific icon panel. If the icon is not on top of the stack, press and hold down the left-mouse-button on the displayed icon to display all the choices. Select the desired icon by clicking with the left-mouse-button when the icon is highlighted.)

Graphics Cursors

Notice the cursor changes from an arrow to a crosshair when graphical input is expected. Look in the prompt window for a description of what you are to choose. The cursor will change to a *double crosshair* when there is a possibly ambiguous choice. When the *double crosshair* appears, you can press the middle-mouse-button to accept the highlighted pick or choose a different item.

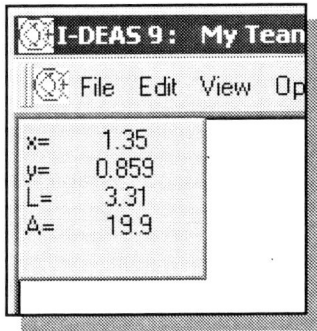

```
I-DEAS 9:  My Team
 File  Edit  View  Op

x=    1.35
y=    0.859
L=    3.31
A=    19.9
```

2. The message "*Locate start*" is displayed in the *prompt window*. Left-click a starting point of the shape, roughly at the center of the graphics window; it could be inside or outside of the displayed grids. In *I-DEAS*, the sketch plane actually extends into infinity. As you move the graphics cursor, you will see a digital readout in the upper left corner of the graphics window. The readout gives you the cursor location, the line length, and the angle of the line measured from horizontal. Move the cursor around and you will also notice different symbols appear along the line as it occupies different positions.

Dynamic Navigator

I-DEAS provides you with visual clues as the cursor is moved across the screen; this is the *I-DEAS Dynamic Navigator*. The *Dynamic Navigator* displays different symbols to show you alignments, perpendicularities, tangencies, etc. The *Dynamic Navigator* is also used to capture the *design intent* by creating constraints where they are recognized. The *Dynamic Navigator* displays the governing geometric rules as models are built.

⫽	**Vertical**	indicates a line is vertical
⟋⟋	**Horizontal**	indicates a line is horizontal
- - - - - -	**Alignment**	indicates the alignment to the center point or endpoint of an entity
⟍⟍	**Parallel**	indicates a line is parallel to other entities
⟋⟍	**Perpendicular**	indicates a line is perpendicular to other entities

Endpoint   indicates the cursor is at the endpoint of an entity

Intersection   indicates the cursor is at the intersection point of
                             two entities

Center   indicates the cursor is at the centers or midpoints of
                             entities

Tangent   indicates the cursor is at tangency points to curves

3. Move the graphics cursor directly below *point 1*. Pick the second point when
    the *vertical constraint* is displayed and the length of the line is about 2 inches.

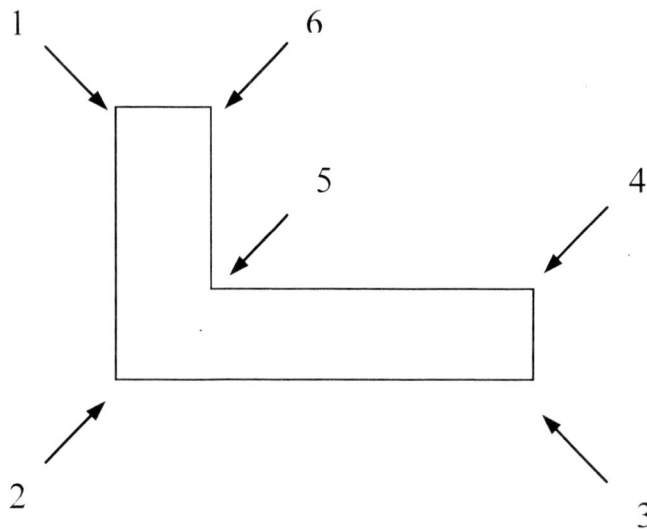

4. Move the graphics cursor horizontally to the right of *point 2*. The
    perpendicular symbol indicates when the line from *point 2* to *point 3* is
    perpendicular to the vertical line. Left-click to select the third point. Notice
    that dimensions are automatically created as you sketch the shape. These
    dimensions are also constraints, which are used to control the geometry.
    Different dimensions are added depending upon how the shape is sketched.
    Do not worry about the values not being exactly what we want. We will
    modify the dimensions later.

5. Move the graphics cursor directly above *point 3*. Do not place this point in
    alignment with the midpoint of the other vertical line. An additional constraint

will be added if they are aligned. Left-click the fourth point directly above *point 3*.

6. Move the graphics cursor to the left of *point 4*. Again, watch the displayed symbol to apply the proper geometric rule that will match the design intent. A good rule of thumb is to exaggerate the features during the initial stage of sketching. For example, if you want to construct a line that is five degrees from horizontal, it would be easier to sketch a line that is 20 to 30 degrees from horizontal. We will be able to adjust the actual angle later. Left-click once to locate the fifth point horizontally from *point 4*.

7. Move the graphics cursor directly above the last point. Watch the different symbols displayed and place the point in alignment with *point 1*. Left-click the sixth point directly above *point 5*.

8. Move the graphics cursor near the starting point of the sketch. Notice the *Dynamic Navigator* will jump to the endpoints of entities. Left-click *point 1* again to end the sketch.

9. In the prompt window, you will see the message "*Locate start.*" By default, *I-DEAS* remains in the **Polylines** command and expects you to start a new sequence of lines.

10. Press the **ENTER** key or click once with the middle-mouse-button to end the **Polylines** command.

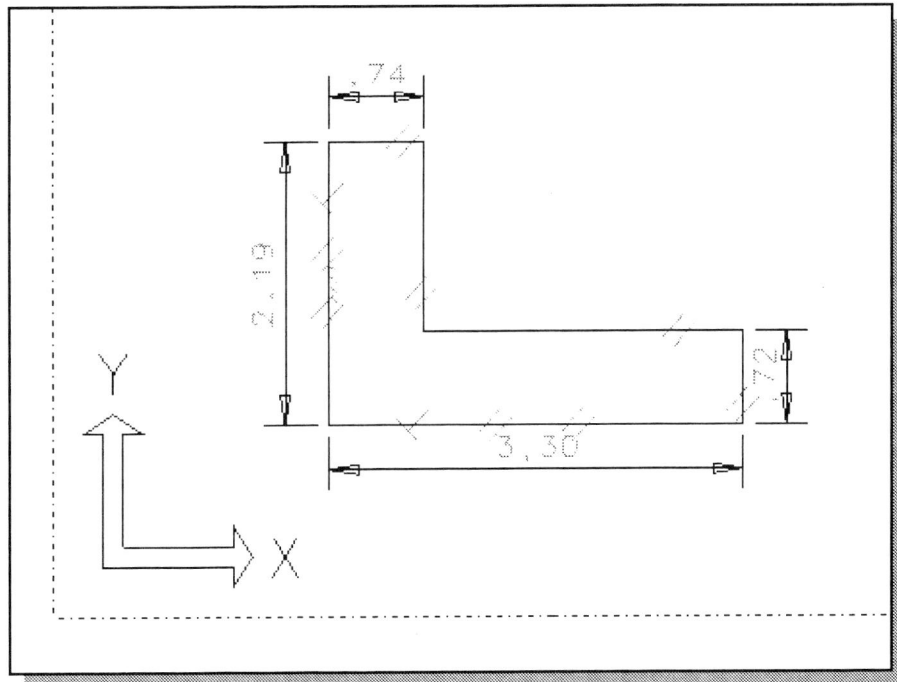

◆ Your sketch should appear similar to the figure above. Note that the displayed dimension values may be different on your screen. In the following sections, we will discuss the procedure to adjust the dimensions. At this point in time, our main concern is the SHAPE of the sketch.

Dynamic Viewing Functions

I-DEAS provides a special user interface called *Dynamic Viewing* that enables convenient viewing of the entities in the graphics window. The *Dynamic Viewing* functions are controlled with the function keys on the keyboard and the mouse.

❖ Panning – F1 and the mouse

Hold the **F1** function key down, and move the mouse to pan the display. This allows you to reposition the display while maintaining the same scale factor of the display. This function acts as if you are using a video camera. You control the display by moving the mouse.

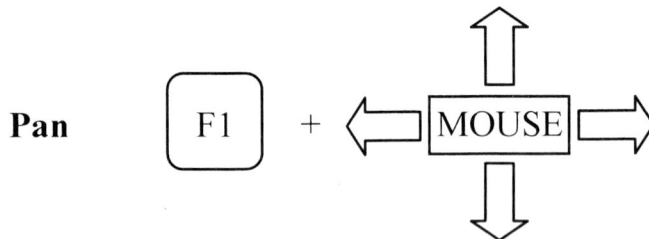

Pan   F1   +   ⇦ MOUSE ⇨ (⇧⇩)

❖ Zooming – F2 and the mouse

Hold the **F2** function key down, and move the mouse vertically on the screen to adjust the scale of the display. Moving upward will reduce the scale of the display, making the entities display smaller on the screen. Moving downward will magnify the scale of the display.

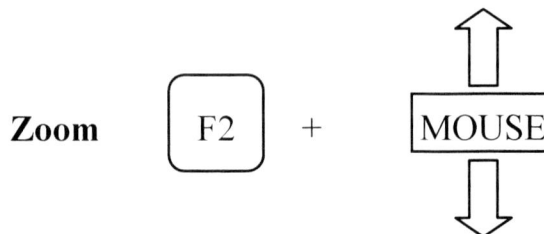

Zoom   F2   +   MOUSE (⇧⇩)

♦ On your own, experiment with the two *Dynamic Viewing* functions. Adjust the display so that your sketch is near the center of the graphics window and adjust the scale of your sketch so that it is occupies about two-thirds of the graphics window.

Basic Editing – Using the Eraser

One of the advantages of using a CAD system is the ability to remove entities without leaving any marks. We will delete one of the lines using the *Delete* command.

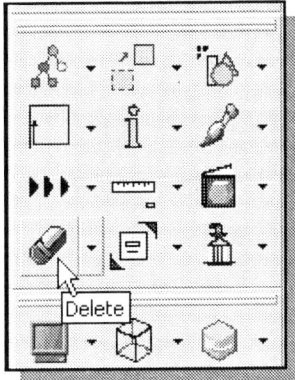

1. Pick **Delete** in the icon panel. (The icon is located in the last row of the *application icon panel*. The icon is a picture of an eraser at the end of a pencil.)

2. In the prompt window, the message *"Pick entity to delete"* appears. Pick the line as shown in the figure below.

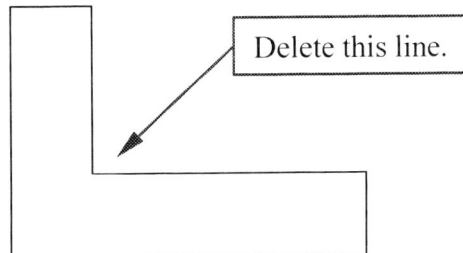

Delete this line.

3. The prompt window now reads *"Pick entity to delete (done)."* Press the **ENTER** key or the **middle-mouse-button** to indicate you are done picking entities to be deleted.

4. In the prompt window, the message *"OK to delete 1 curve, 1 constraint and 1 dimension? (Yes)"* will appear. The *"1 constraint"* is the *parallel constraint* created by the **Dynamic Navigator**.

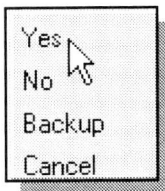

5. Press **ENTER**, or pick **Yes** in the pop-up menu to delete the selected line. The constraints and dimensions are used as geometric control variables. When the geometry is deleted, the associated control features are also removed.

6. In the prompt window, you will see the message "*Pick entity to delete.*" By default, *I-DEAS* remains in the **Delete** command and expects you to select additional entities to be erased.

7. Press the **ENTER** key or the **middle-mouse-button** to end the **Delete** command.

Creating a Single Line

Now we will create a line at the same location by using the **Lines** command.

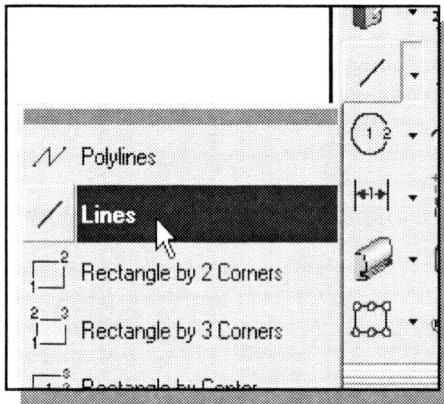

1. Pick **Lines** in the icon panel. (The icon is located in the same stack as the *Polylines* icon.) Press and hold down the **left-mouse-button** on the *Polylines* icon to display the available choices. Select the **Lines** command with the **left-mouse-button** when the option is highlighted.

2. The message "*Locate start*" is displayed in the prompt window. Move the graphics cursor near *point 1* and, as the *endpoint* symbol is displayed, pick with the **left-mouse-button**.

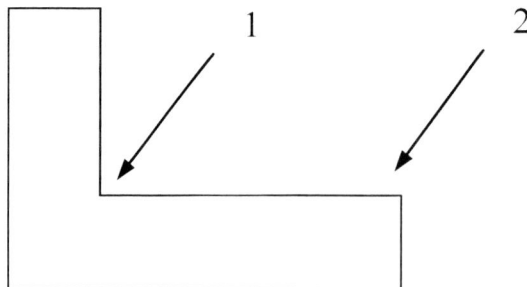

3. Move the graphics cursor near *point 2* and click the **left-mouse-button** when the *endpoint* symbol is displayed.

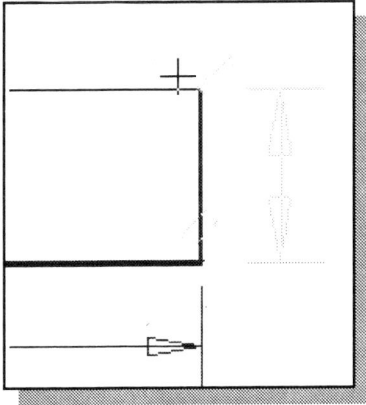

> ➤ Notice the *Dynamic Navigator* creates the parallel constraint and the dimension as the geometry is constructed.

4. The message "*Locate start*" is displayed in the prompt window. Press the **ENTER** key or use the **middle-mouse-button** to end the **Lines** command.

Consideration of Design Intent

While creating the sketch, it is very important to keep in mind the design intent. Always consider functionality of the part and key features of the design. Using *I-DEAS*, we can accomplish and maintain the design intent at all levels of the design process.

The dimensions automatically created by *I-DEAS* might not always match with the designer's intent. For example, in our current design, we may want to use the vertical distance between the top two horizontal lines as a key dimension. Even though it is a very simple calculation to figure out the corresponding length of the vertical dimension at the far right, for more complex designs it might not be as simple, and to do additional calculations is definitely not desirable. The next section describes re-dimensioning the sketch.

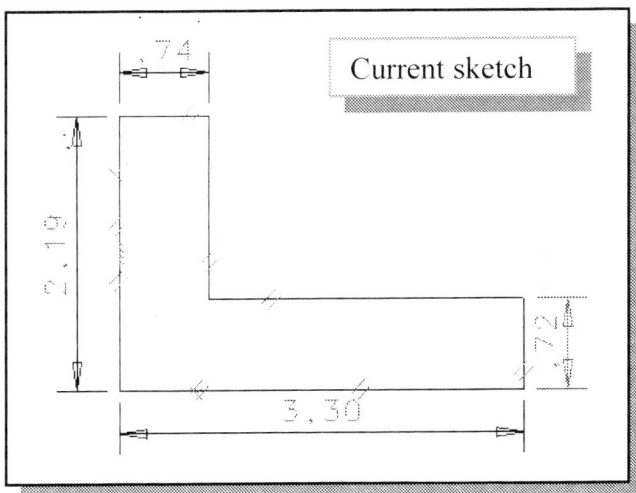

Current sketch

.75

2.50

1.625

The design we
have in mind

3.00

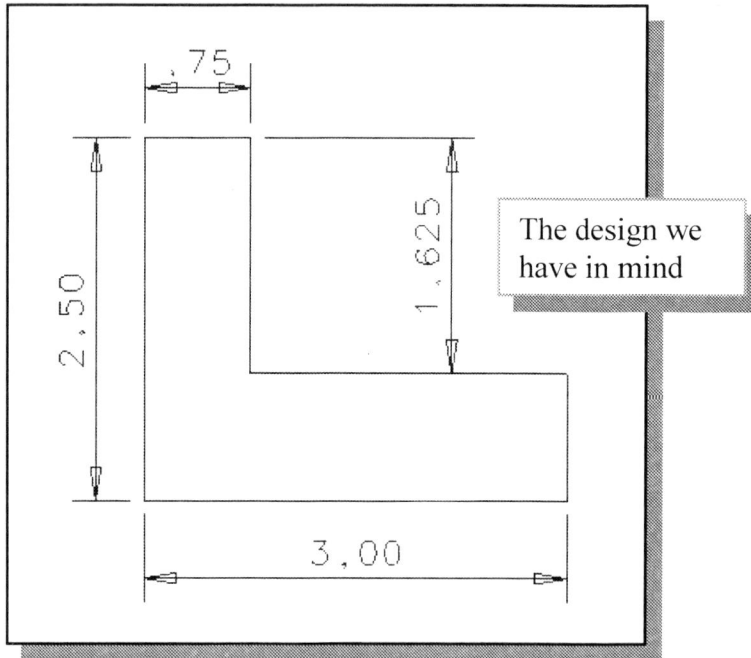

Step 2: Apply/Delete/Modify constraints and dimensions

As the sketch is made, *I-DEAS* automatically applies some of the geometric constraints (such as *horizontal*, *parallel* and *perpendicular*) to the sketched geometry. We can continue to modify the geometry, apply additional constraints, and/or define the size of the existing geometry. In this example, we will illustrate deleting existing dimensions and add new dimensions to describe the sketched entities.

To maintain our design intent, we will first remove the unwanted dimension and then create the desired dimension.

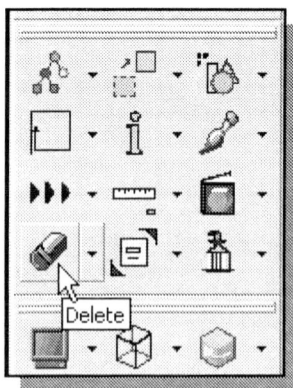

1. Pick **Delete** in the icon panel. (The icon is located in the last row of the application icon panel.)

2. Pick the dimension as shown.

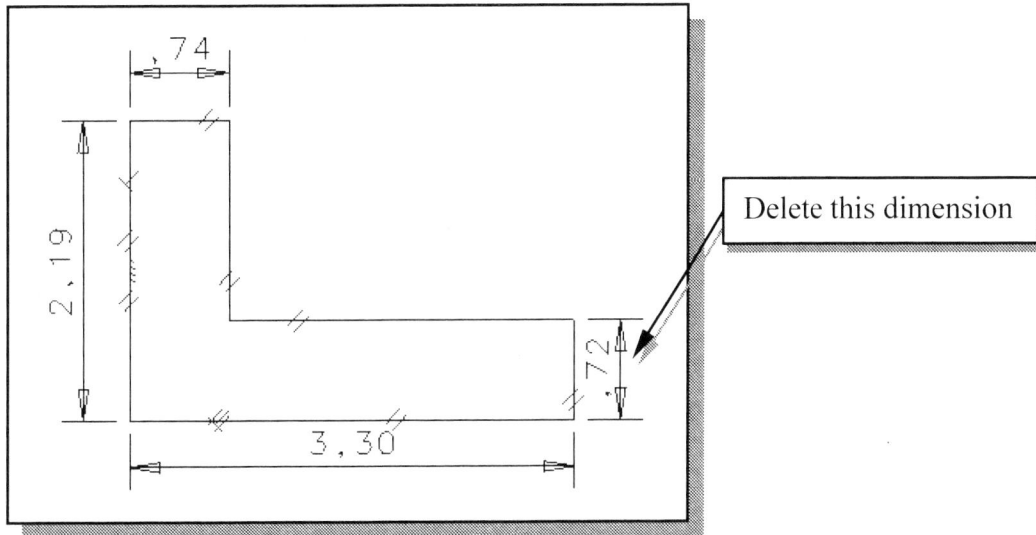

Delete this dimension

3. Press the **ENTER** key or the middle-mouse-button to accept the selection.

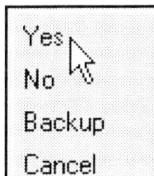

4. In the prompt window, the message "*OK to delete 1 dimension?*" is displayed. Pick **Yes** in the popup menu, or press the **ENTER** key or the middle-mouse-button to delete the selected dimension. End the ***Delete*** command by hitting the middle-mouse-button again.

Creating Desired Dimensions

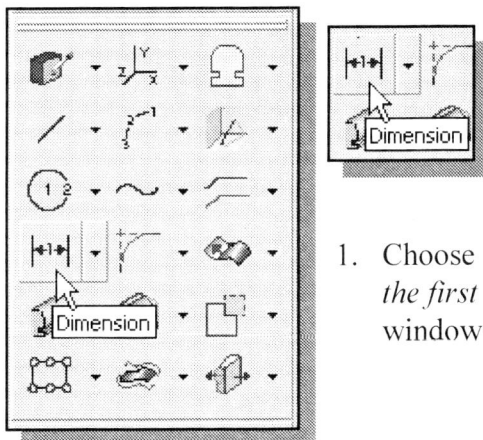

1. Choose ***Dimension*** in the icon panel. The message "*Pick the first entity to dimension*" is displayed in the prompt window.

2. Pick the **top horizontal line** as shown in the figure below.

3. Pick the **second horizontal line** as shown.

4. Place the text to the right of the model.

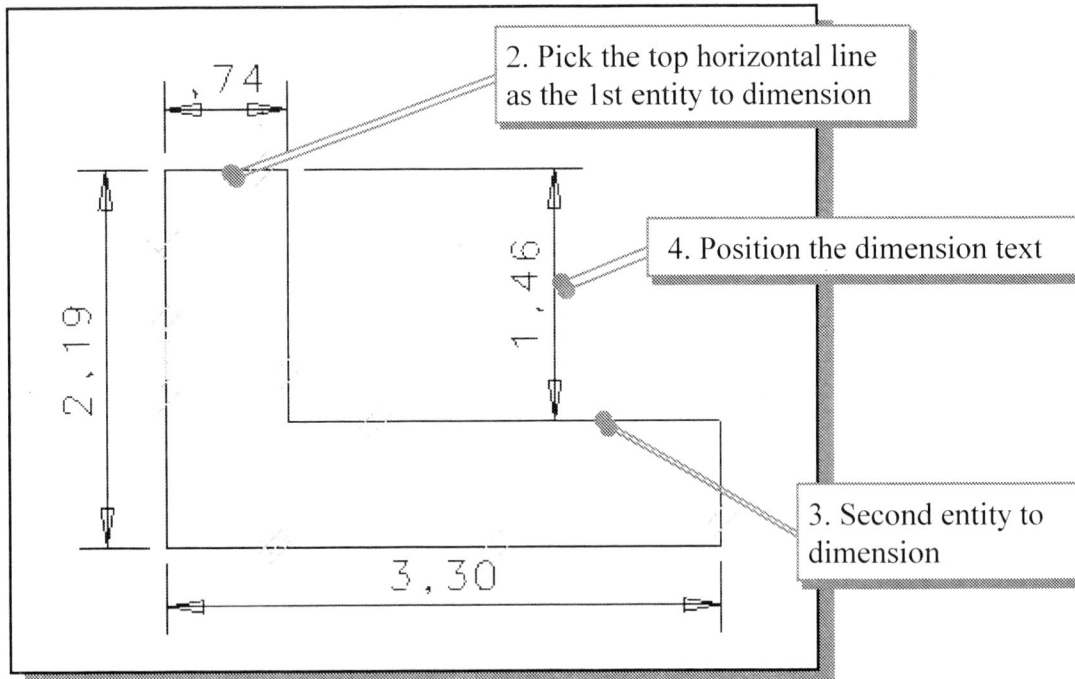

5. Press the **ENTER** key or the middle-mouse-button to end the **Dimension** command.

➢ In *I-DEAS*, the *Dimension* command will create a linear dimension if two parallel lines are selected (distance in between the two lines). Selecting two lines that are not parallel will create an angular dimension (angle in between the two lines).

Modifying Dimensional Values

Next we will adjust the dimensional values to the desired values. One of the main advantages of using a feature-based parametric solid modeler, such as *I-DEAS*, is the ability to easily modify existing entities. The operation of modifying dimensional values will demonstrate implementation of the design philosophy of "shape before size." In *I-DEAS*, several options are available to modify dimensional values. In this lesson, we will demonstrate two of the options using the **Modify** command. The *Modify* command icon is located in the second row of the application icon panel; the icon is a picture of an arrowhead with a long tail.

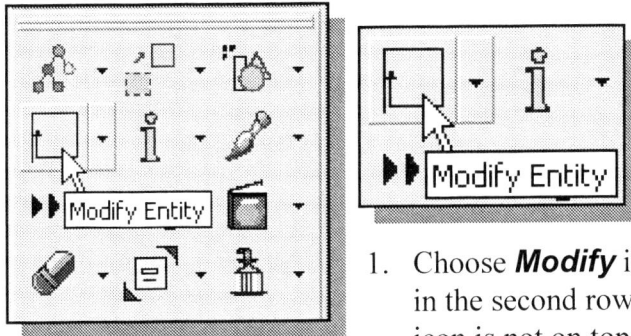

1. Choose **Modify** in the icon panel. (The icon is located in the second row of the application icon panel. If the icon is not on top of the stack, press and hold down the left-mouse-button on the displayed icon, then select the **Modify** icon.) The message "*Pick entity to modify*" is displayed in the prompt window.

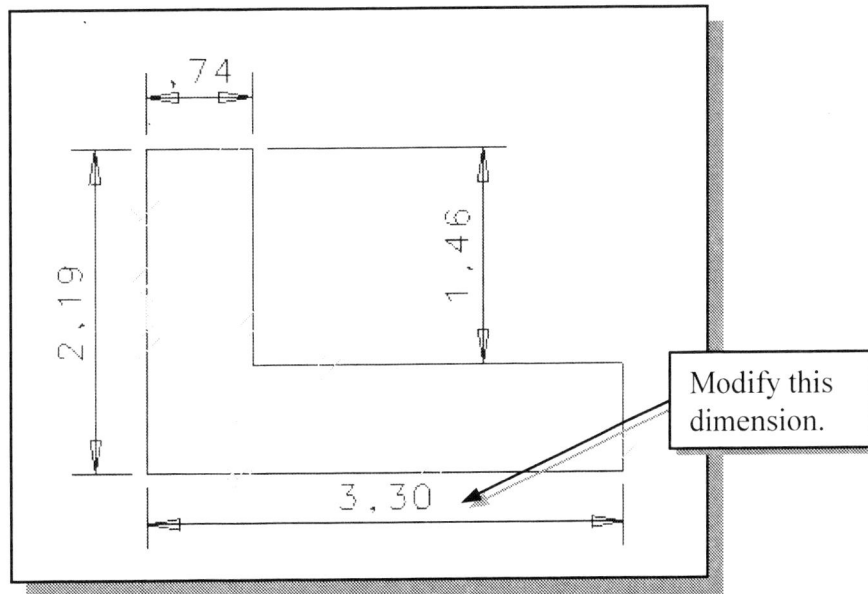

Modify this dimension.

1. Pick the dimension as shown (the number might be different than displayed). The selected dimension will be highlighted. The *Modify Dimension* window appears.

❖ In the *Modify Dimension* window, the value of the selected dimension is displayed and also identified by a *name* in the format of "Dxx," where the "D" indicates it is a dimension and the "xx" is a number incremented automatically as dimensions are added. You can change both the name and the value of the dimension by clicking and typing in the appropriate boxes.

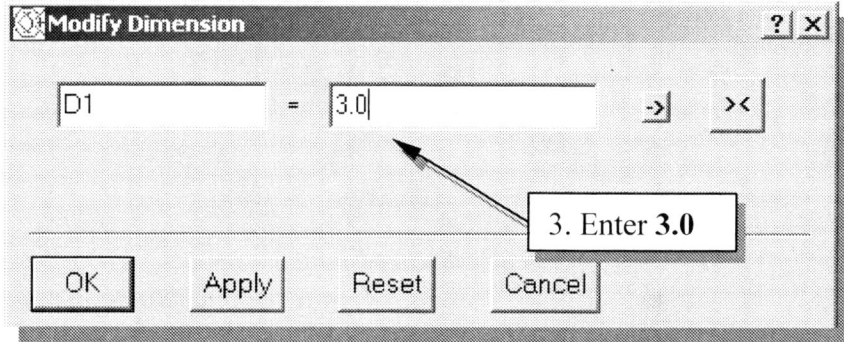

2. Type in **3.0** to modify the dimensional value as shown in the figure above.

3. Click on the **OK** button to accept the value you have entered.

➢ *I-DEAS* will adjust the size of the object based on the new value entered.

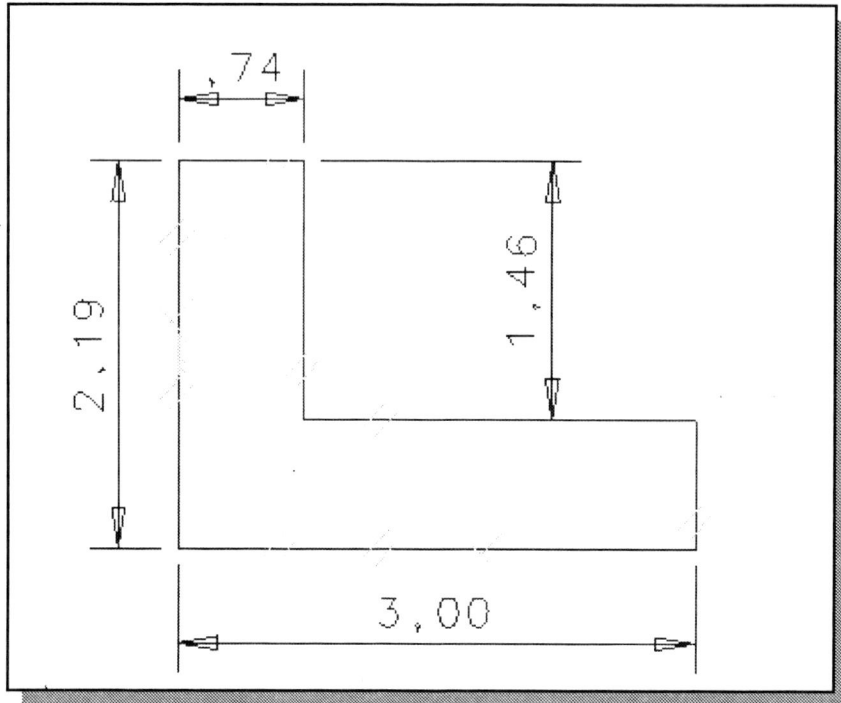

4. On your own, click on the top horizontal dimension and adjust the dimensional value to **0.75**.

5. Press the **ENTER** key or the middle-mouse-button to end the **Modify** command.

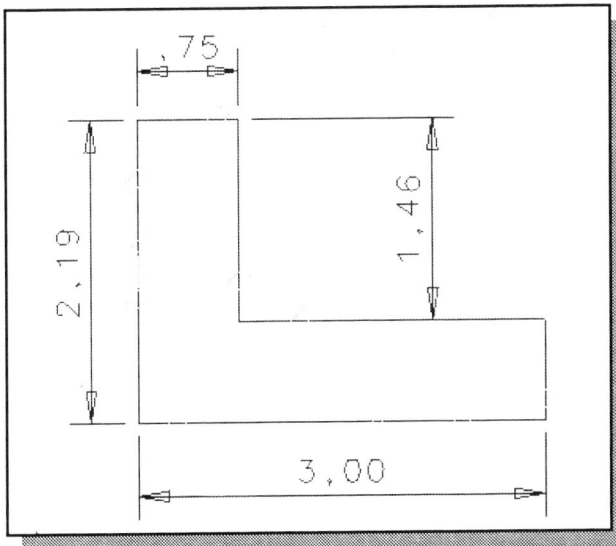

❖ The size of our design is automatically adjusted by *I-DEAS* based on the dimensions we have entered. *I-DEAS* uses the dimensional values as control variables and the geometric entities are modified accordingly. This approach of rough sketching the shape of the design first then finalizing the size of the design is known as the "**shape before size**" approach.

Pre-selection of Entities

I-DEAS provides a flexible graphical user interface that allows users to select graphical entities BEFORE the command is selected (*pre-selection*), or AFTER the command is selected (*post-selection*). The procedure we have used so far is the *post-selection* option. To pre-select one or more items to process, hold down the **SHIFT** key while you pick. Selected items will stay highlighted. You can *deselect* an item by selecting the item again. The item will be toggled on and off by each click. Another convenient feature of pre-selection is that the selected items remain selected after the command is executed.

1. Pre-select all of the dimensions by holding down the **SHIFT** key and clicking the **left-mouse-button** on each dimension value.

PRE-SELECT | SHIFT | + | LEFT-mouse-button |

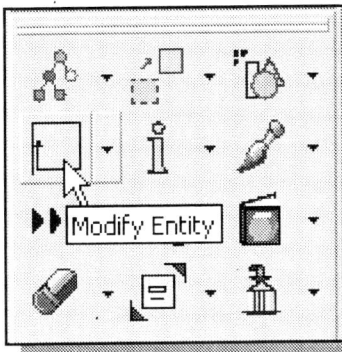

2. Select the **Modify** icon. The *Dimensions* window appears.

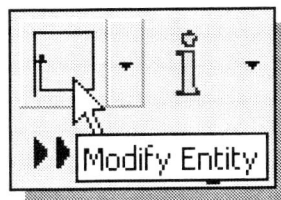

3. Move the *Dimensions* window around so that it does not overlap the part drawing. Do this by "clicking and dragging" the window's title area with the left-mouse-button. You can also use the *Dynamic Viewing* functions (activate the graphics window first) to adjust the scale and location of the entities displayed in the graphics window (**F1** and the mouse, **F2** and the mouse).

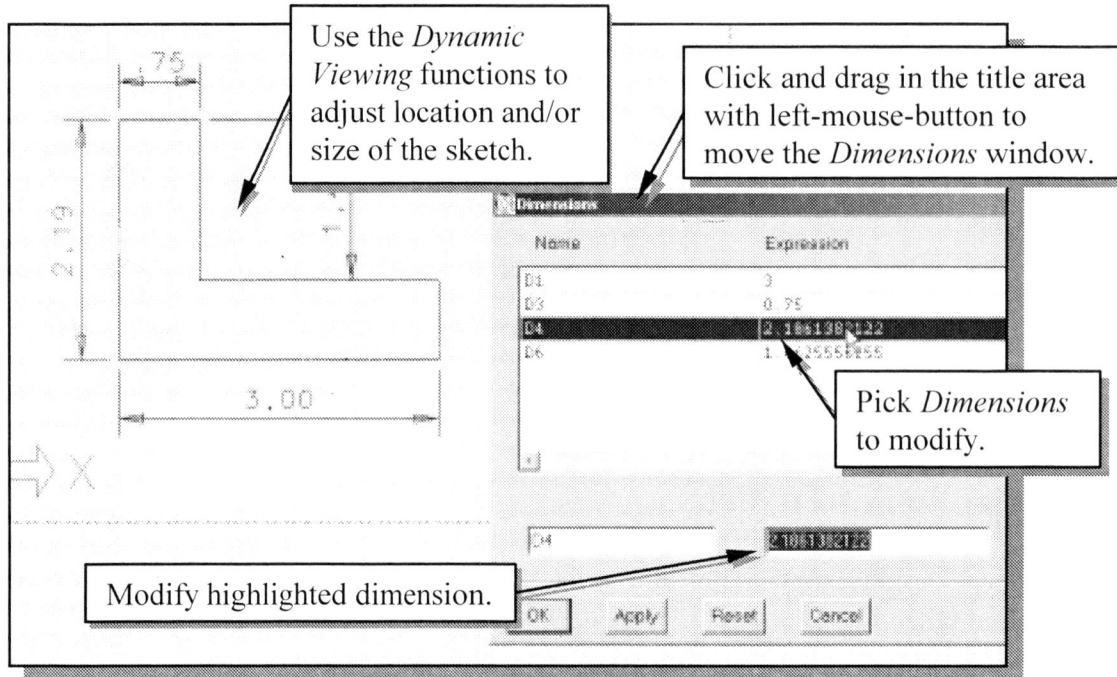

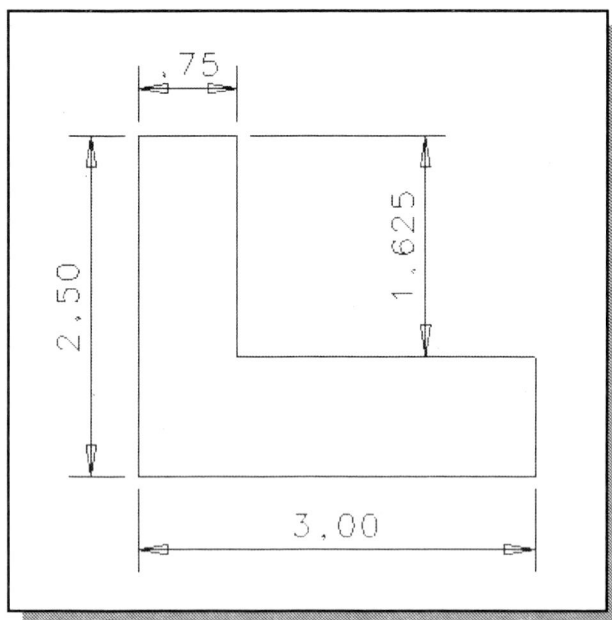

Use the *Dynamic Viewing* functions to adjust location and/or size of the sketch.

Click and drag in the title area with left-mouse-button to move the *Dimensions* window.

Pick *Dimensions* to modify.

Modify highlighted dimension.

4. Click on one of the dimensions in the pop-up window. The selected dimension will be highlighted in the graphics window. Type in the desired value for the selected dimension. **DO NOT** hit the **ENTER** key. Select another dimension from the list to continue modifying. Modify all of the dimensional values to the values as shown.

5. Click the **OK** button to accept the values you have entered and close the *Dimensions* window.

> *I-DEAS* will now adjust the size of the shape to the desired dimensions. The design philosophy of "shape before size" is implemented quite easily. The geometric details are taken care of by *I-DEAS*.

Step 3: Completing the Base Solid Feature

♦ Now that the 2D sketch is completed, we will proceed to the next step: create a 3D feature from the 2D profile. Extruding a 2D profile is one of the common methods that can be used to create 3D parts. We can extrude planar faces along a path.

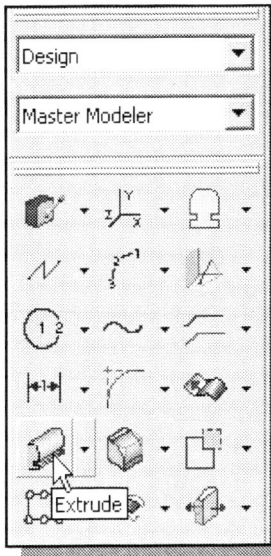

1. Choose **Extrude** in the icon panel. The **Extrude** icon is located in the fifth row of the task specific icon panel. Press and hold down the left-mouse-button on the icon to display all the choices. If a different choice were to be made, you would slide the mouse up and down to switch between different options. In the prompt window, the message "*Pick curve or section*" is displayed.

2. Pick any edge of the 2D shape. By default, the **Extrude** command will automatically select all segments of the shape that form a closed region. Notice the different color signifying the selected segments.

3. Notice the *I-DEAS* prompt "*Pick curve to add or remove. (Done)*" We can select more geometric entities or deselect any entity that has been selected. Picking the same geometric entity will again toggle the selection of the entity "on" or "off" with each left-mouse-button click. Press the **ENTER** key to accept the selected entities.

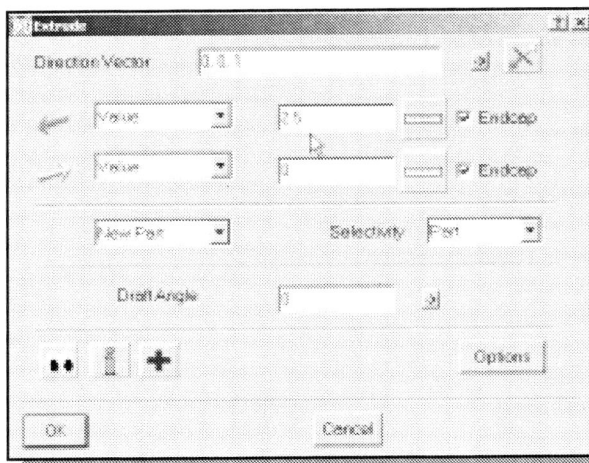

4. The *Extrude Section* window will appear on the screen. Enter **2.5**, in the first value box, as the *extrusion distance* and confirm that the **New part** option is set as shown in the figure.

5. Click on the **OK** button to accept the settings and extrude the 2D section into a 3D solid.

➤ Notice all of the dimensions disappeared from the screen. All of the dimensional values and geometric constraints are stored in the database by *I-DEAS* and they can be brought up at any time.

Display Viewing Commands

❖ 3D Dynamic Rotation – F3 and the mouse

The *I-DEAS Dynamic Viewing* feature allows users to do "*real-time*" rotation of the display. Hold the **F3** function key down and move the mouse to rotate the display. This allows you to rotate the displayed model about the screen X (horizontal), Y (vertical), and Z (perpendicular to the screen) axes. Start with the cursor near the center of the screen and hold down **F3**; moving the cursor up or down will rotate about the screen X-axis while moving the cursor left or right will control the rotation about the screen Y-axis. Start with the cursor in the corner of the screen and hold down **F3**, which will control the rotation about the screen Z-axis.

Dynamic Rotation

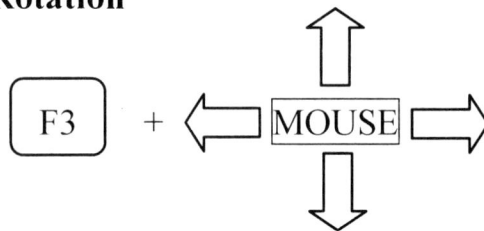

Display Icon Panel

The *Display* icon panel contains various icons to handle different viewing operations. These icons control the screen display, such as the view scale, the view angle, redisplay, and shaded and hidden line displays.

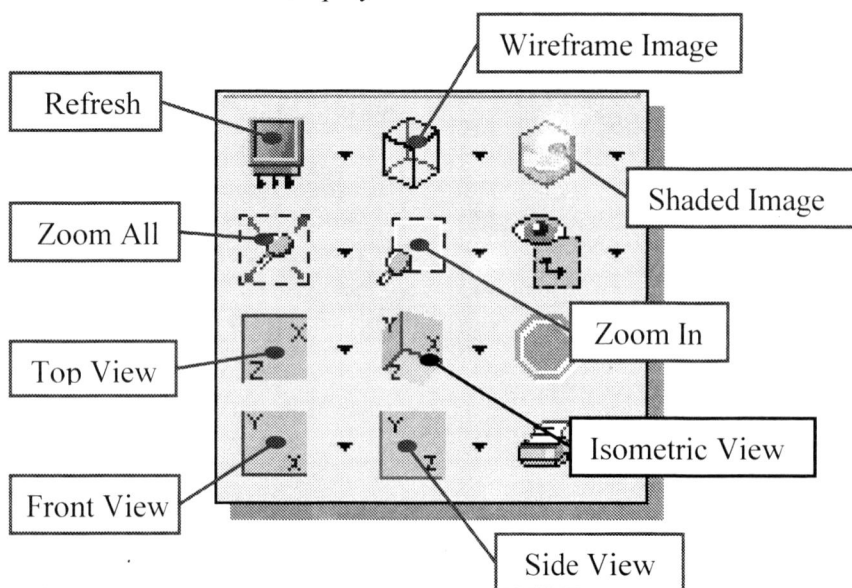

View icons:

Front, Side, Top, Bottom, Isometric, and *Perspective*: These six icons are the standard view icons. Selecting any of these icons will change the viewing angle. Try each one as you read its description below

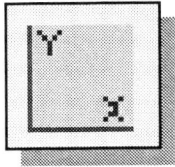

Front View (X-Y Workplane)

Right Side View

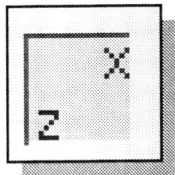

Top View

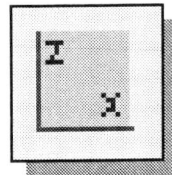

Bottom View

Isometric View

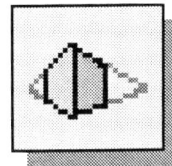

Perspective View

❖ Shaded Solids:

Depending on your display type, you will pick either *Shaded Hardware* or *Shaded Software* to get shaded images of 3D objects. **Shaded Hardware** on a workstation with **OGL** display capability allows real-time dynamic rotation (**F3** and the mouse) of the shaded 3D solids. A workstation with **X3D** display capability allows the use of the **Shaded Software** command to get the shaded image without the real-time dynamic rotation capability.

Shaded Hardware

Shaded Software

❖ **Hidden-line Removal:**
Three options are available to generate images with all the back lines removed.

Hidden Hardware   Precise Hidden   Quick Hidden

❖ **Wireframe Image:**
This icon allows the display of the 3D objects using the basic wireframe representation scheme.

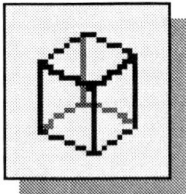   Wireframe

❖ **Refresh and Redisplay:**
Use these commands to regenerate the graphics window.

Refresh   Redisplay

❖ **Zoom-All:**
Adjust the viewing scale factor so that all objects are displayed.

   Zoom-All

❖ **Zoom-In:**
Allows the users to define a rectangular area, by selecting two diagonal corners, which will fill the graphics window.

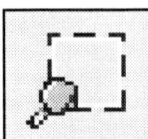   Zoom-In

Workplane – It is an XY CRT, but an XYZ World

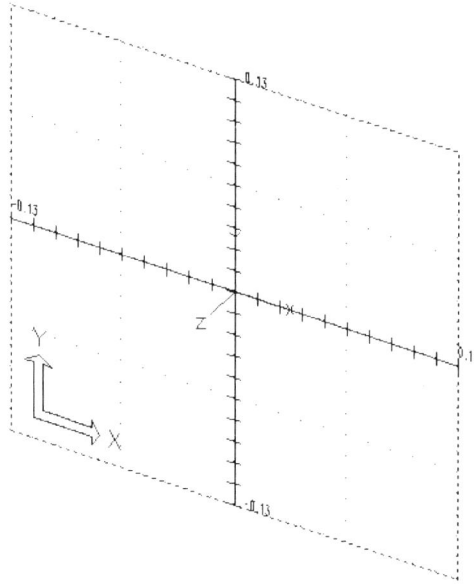

Design modeling software is becoming more powerful and user friendly, yet the system still does only what the user tells it to do. In using a geometric modeler, therefore, we need to have a good understanding of what the inherent limitations are. We should also have a good understanding of what we want to do and what results to expect based upon what is available.

In most 3D geometric modelers, 3D objects are located and defined in what is usually called **world space** or **global space**. Although a number of different coordinate systems can be used to create and manipulate objects in a 3D modeling system, the objects are typically defined and stored using the world space. The world space is usually a 3D Cartesian coordinate system that the user cannot change or manipulate.

In most engineering designs, models can be very complex; it would be tedious and confusing if only the world coordinate system were available. Practical 3D modeling systems allow the user to define **Local Coordinate Systems** or **User Coordinate Systems** relative to the world coordinate system. Once a local system is defined, we can then create geometry in terms of this more convenient system.

Although objects are created and stored in 3D space coordinates, most of the input and output is done in a 2D Cartesian system. Typical input devices such as a mouse or digitizers are two-dimensional by nature; the movement of the input device is interpreted by the system in a planar sense. The same limitation is true of common output devices, such as CRT displays and plotters. The modeling software performs a series of three-dimensional to two-dimensional transformations to correctly project 3D objects onto the 2D picture plane (monitor).

The *I-DEAS **workplane*** is a special construction tool that enables the planar nature of 2D input devices to be directly mapped into the 3D coordinate system. The workplane is a local coordinate system that can be aligned to the world coordinate system, an existing face of a part, or a reference plane. By default, the workplane is aligned to the world coordinate system.

The basic design process of creating solid features in the *I-DEAS* task is a three-step process:

1. Select and/or define the workplane.
2. Sketch and constrain 2D planar geometry.
3. Create the solid feature.

These steps can be repeated as many times as needed to add additional features to the design. The base feature of the ***Adjuster Block*** model was created following this basic design process; we used the default settings where the workplane is aligned to the world coordinate system. We will next add additional features to our design and demonstrate how to manipulate the *I-DEAS* workplane.

Workplane Appearance

The workplane is a construction tool; it is a coordinate system that can be moved in space. The size of the workplane display is only for our visual reference, since we can sketch on the entire plane, which extends to infinity.

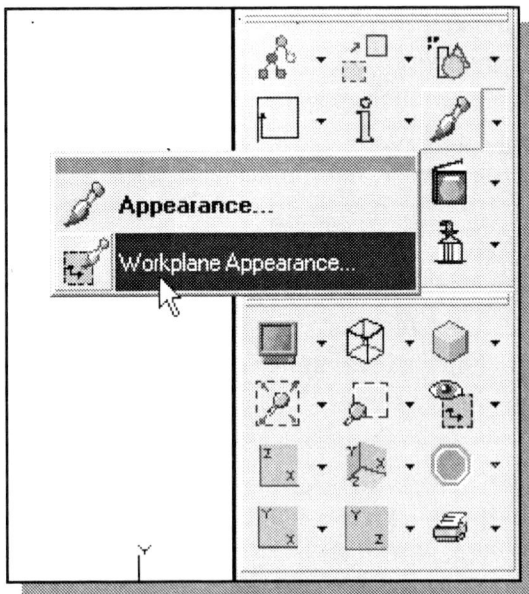

1. Choose ***Workplane Appearance*** in the icon panel. (The icon is located in the second row of the application icon panel. If the icon is not on top of the stack, press and hold down the left-mouse-button on the displayed icon to display all the choices, then select the ***Workplane Appearance*** icon.) The *Workplane Attributes* window appears.

2. Toggle **on** the three display switches as shown.

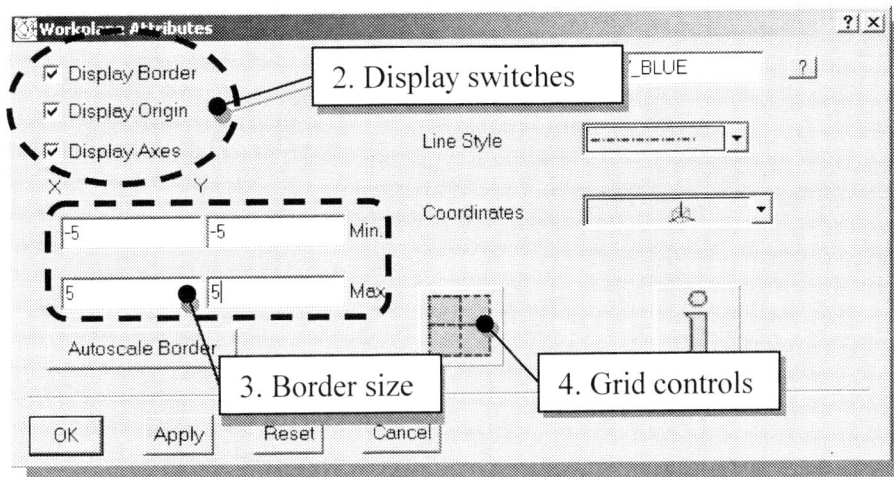

3. Adjust the **workplane border size** by entering the *Min.* and *Max.* values as shown.

4. In the *Workplane Attributes* window, click on the **Workplane Grid** button. The *Grid Attributes* window appears.

5. Change the ***Grid Size*** settings by entering the values as shown.

6. Toggle **on** the ***Display Grid*** option if it is not already switched on.

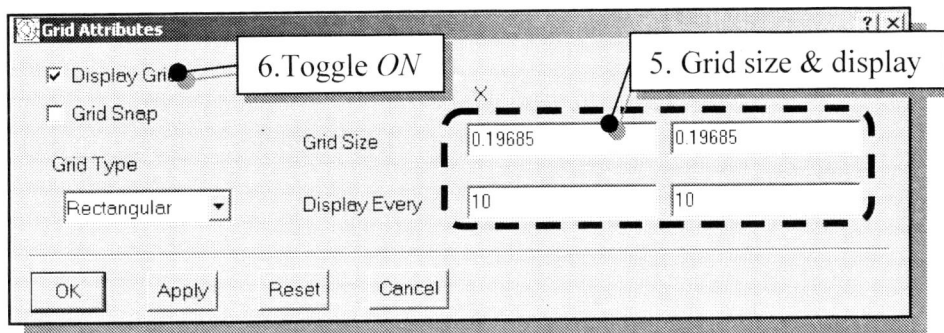

❖ Although the *Grid Snap* option is available, its usage in parametric modeling is not recommended. The *Grid Snap* concept does not conform to the *"shape before size"* philosophy and most real designs rarely have uniformly spaced dimension values.

7. Pick **Apply** to view the effects of the changes.

8. Click on the **OK** button to exit the *Grid Attributes* window.

9. Click on the **OK** button to exit the *Workplane Attributes* window.

10. On your own, use [**F3**+Mouse] to dynamically rotate the part and observe the workplane is aligned with the surface corresponding to the first sketch drawn.

Step 4: Adding additional features

➤ Sketch In Place

One option to manipulate the workplane is with the **Sketch in Place** command. The *Sketch in Place* command allows the user to sketch on an existing part face. The workplane is reoriented and is attached to the face of the part.

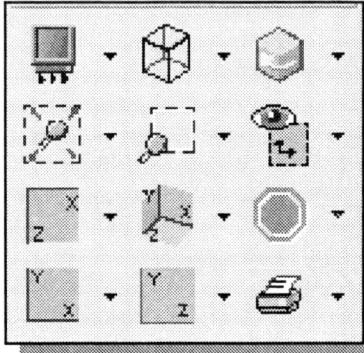

1. Choose **Isometric View** in the display viewing icon panel.

2. Choose **Zoom-All** in the display viewing icon panel.

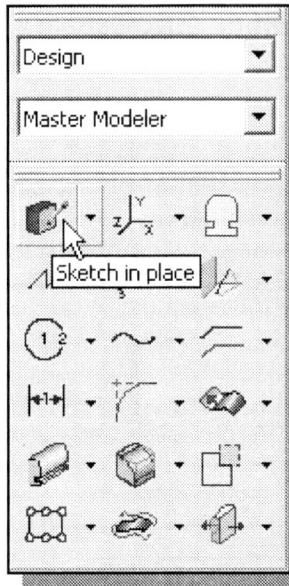

3. Choose **Sketch in Place** in the icon panel. In the prompt window, the message "*Pick plane to sketch on*" is displayed.

4. Pick the top face of the horizontal portion of the 3D object by left-clicking the surface, when it is highlighted as shown in the figure below.

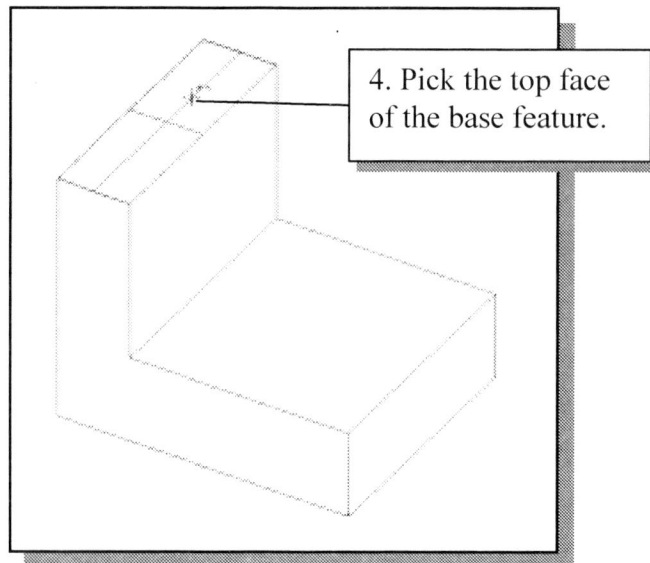

4. Pick the top face of the base feature.

❖ Notice that, as soon as the top surface is picked, *I-DEAS* automatically orients the workplane to the selected surface. The surface selected is highlighted with a different color to indicate the attachment of the workplane.

Step 4-1: Adding an extruded feature

- Next, we will create another 2D sketch, which will be used to create an extruded feature that will be added to the existing solid object.

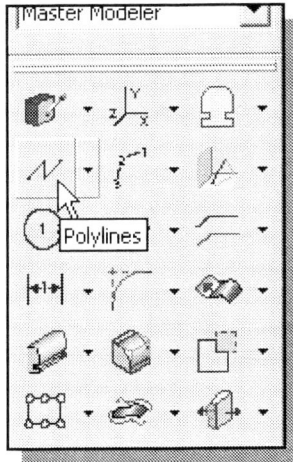

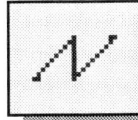

1. Pick **Polylines** in the icon panel. (The icon is located in the second row of the task specific icon panel. If the icon is not on top of the stack, press and hold down the left-mouse-button on the displayed icon to display all the choices. Select the desired icon by clicking with the left-mouse-button when the icon is highlighted.)

2. Create a sketch with segments perpendicular/parallel to the existing edges of the solid model as shown below.

- Note that the edges of the new sketch are either perpendicular or parallel to the existing edges of the solid model. Also note that none of the edges are aligned to the mid-point or corners of the existing solid model.

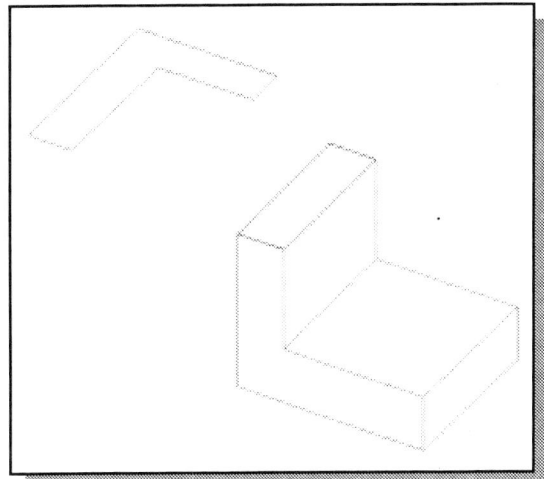

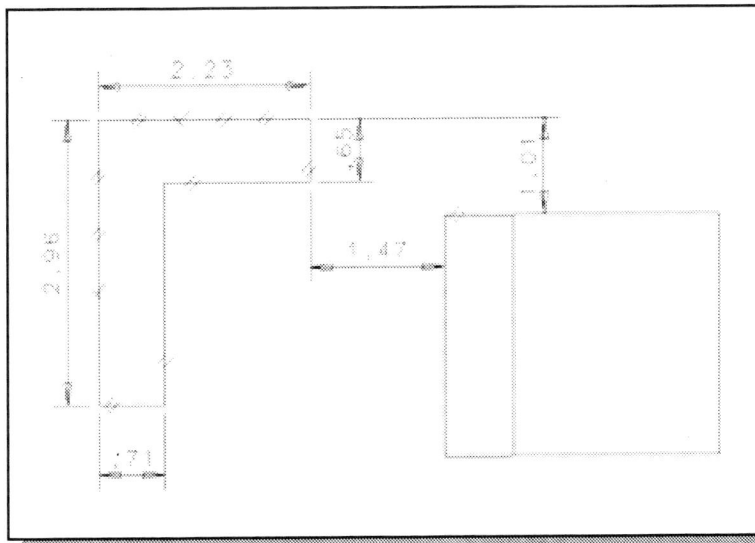

3. On your own, confirm that there are six dimensions on your screen. Create and/or delete dimensions if necessary. Do not be concerned with the actual numbers of the dimensions, which we will adjust in the next section.

4. On your own, modify the location dimensions and the size dimensions as shown in the figure below.

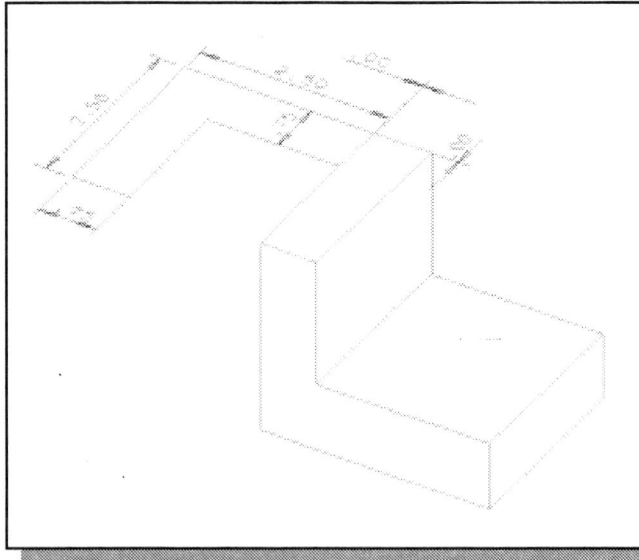

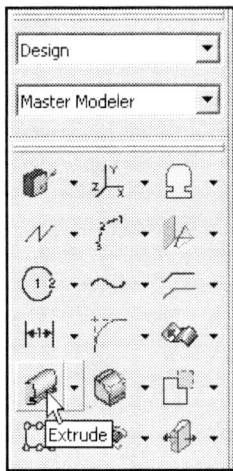

5. Choose **Extrude** in the icon panel. The *Extrude* icon is located in the fifth row of the task specific icon panel.

6. In the prompt window, the message "*Pick curve or section*" is displayed. Pick any edge of the 2D shape. By default, the *Extrude* command will automatically select all neighboring segments of the selected segment to form a closed region. Notice the different color signifying the selected segments.

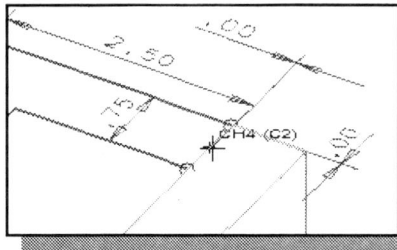

7. Pick the segment in between the displayed two small circles so that the highlighted entities form a closed region.

8. Press the **ENTER** key once, or click once with the middle-mouse-button, to accept the selected entity.

❖ Attempting to select a line where two entities lie on top of one another (i.e. coincide) causes confusion as indicated by the double line cursor ⌗ symbol and the prompt window message "*Pick curve to add or remove (Accept)***". This message indicates *I-DEAS* needs you to confirm the selected item. If the correct entity is selected, you can continue to select additional entities. To reject an erroneously selected entity, press the [**F8**] key to select a neighboring entity or press the right-mouse-button and highlight **Deselect All** from the popup menu.

9. Press the **ENTER** key once, or click once with the middle-mouse-button, to proceed with the *Extrude* command.

10. The *Extrude Section* window will appear on the screen. Enter **2.5**, in the first value box, as the *extrusion distance* and confirm that the **Join** option is set as shown in the figure.

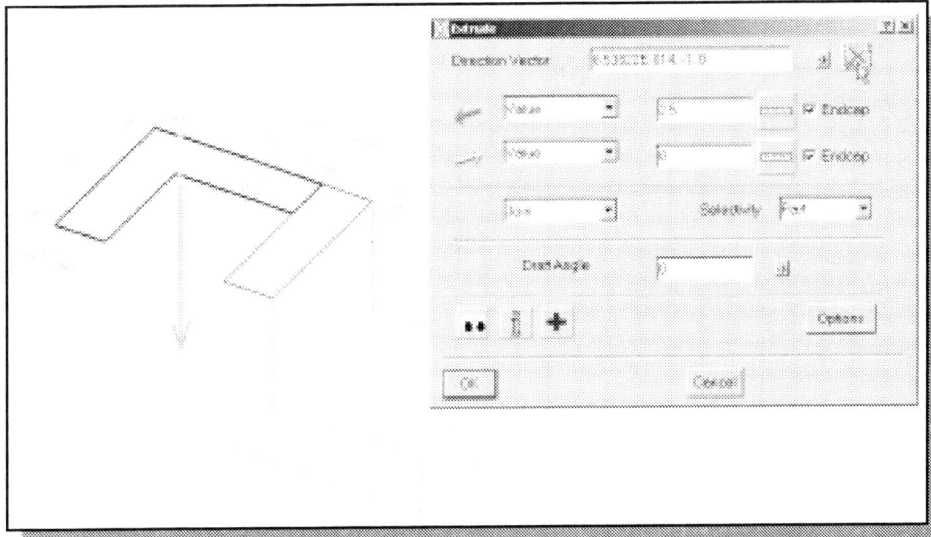

11. Click on the **Arrows** icon, near the upper-right corner of the *Extrude* window, to flip the extrusion direction so that the green arrow points downward as shown.

12. Click on the **OK** button to accept the settings and extrude the 2D section into a 3D solid feature.

Step 4-2: Adding a cut feature

• Next, we will create a circular cut feature to the existing solid object.

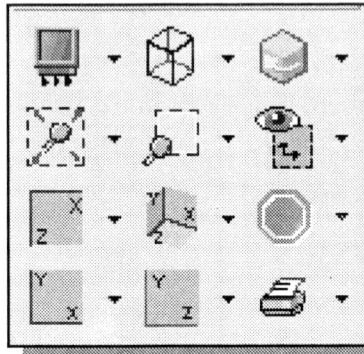

1. Choose *Isometric View* in the display viewing icon panel.

2. Choose *Zoom-All* in the display viewing icon panel.

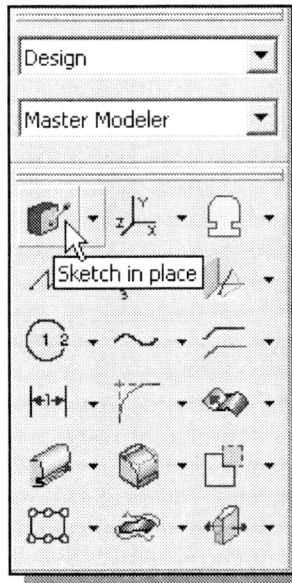

3. Choose **Sketch in Place** in the icon panel. In the prompt window, the message "*Pick plane to sketch on*" is displayed.

4. Pick the top face of the horizontal portion of the 3D object by left-clicking the surface, when it is highlighted as shown in the below figure.

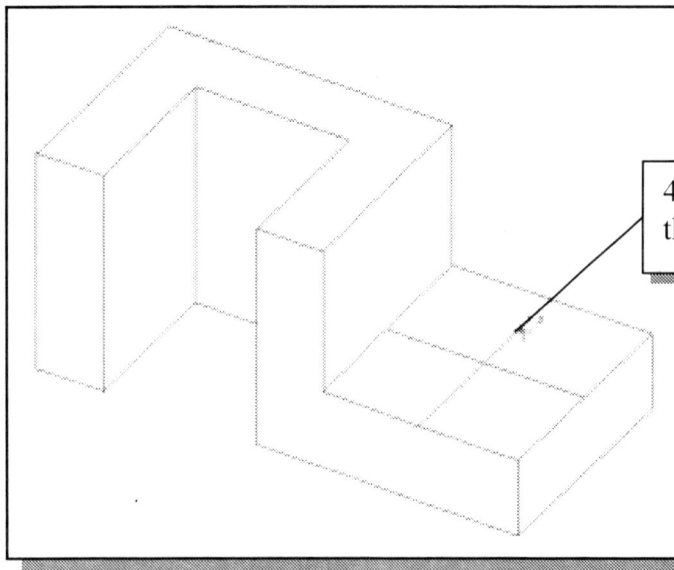

4. Pick this face of the base feature.

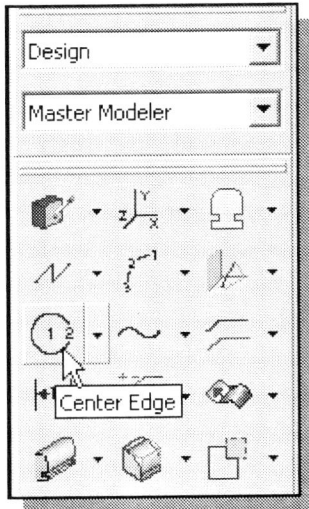

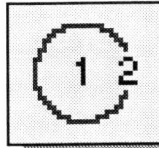

5. Choose **Circle – Center Edge** in the icon panel. This command requires the selection of two locations: first the location of the center of the circle and then a location where the circle will pass through.

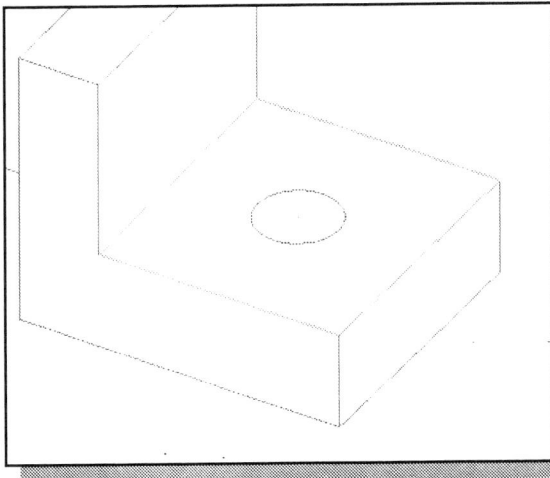

6. On your own, create a circle inside the horizontal face of the solid model as shown.

7. On your own, create and modify the three dimensions as shown.

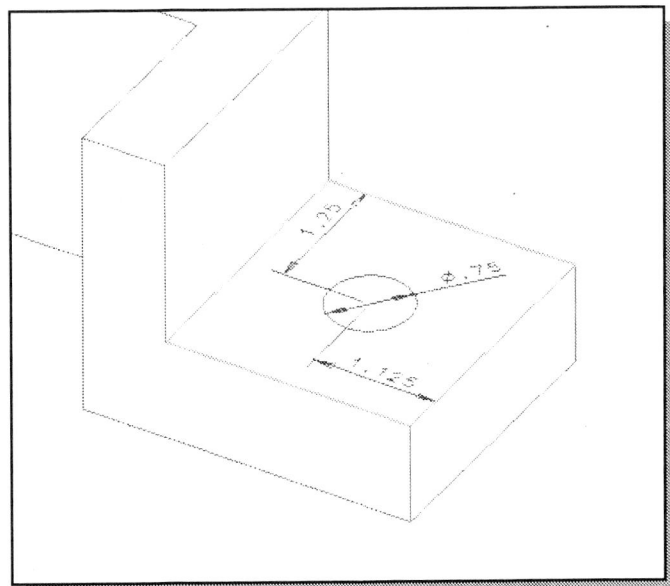

◆ Extrusion – Cut option

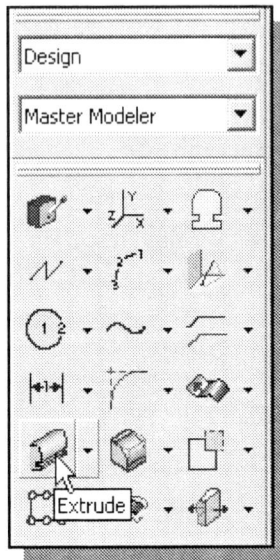

1. Choose **Extrude** in the icon panel. The *Extrude* icon is located in the fifth row of the task specific icon panel.

2. In the prompt window, the message "*Pick curve or section*" is displayed. Pick the newly sketched circle.

3. At the *I-DEAS* prompt "*Pick curve to add or remove (Done),*" press the **ENTER** key or the middle-mouse-button to accept the selection.

4. The *Extrude Section* window appears. Set the *extrude option* to **Cut**. Note the extrusion direction displayed in the graphics window.

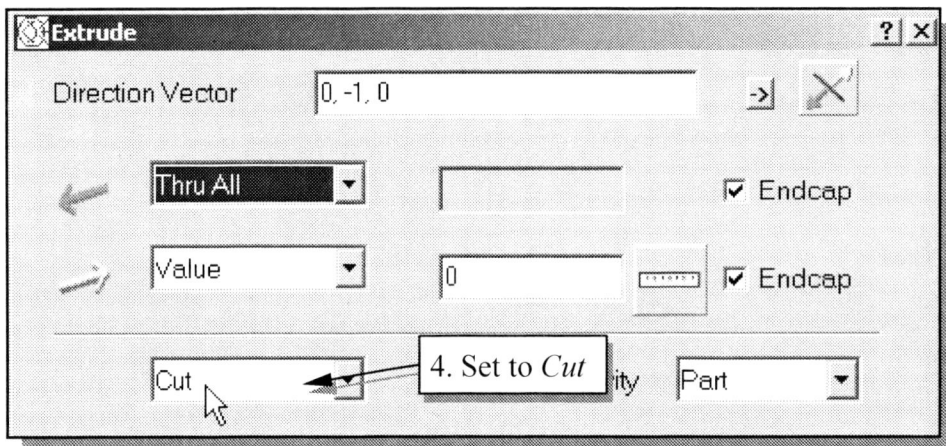

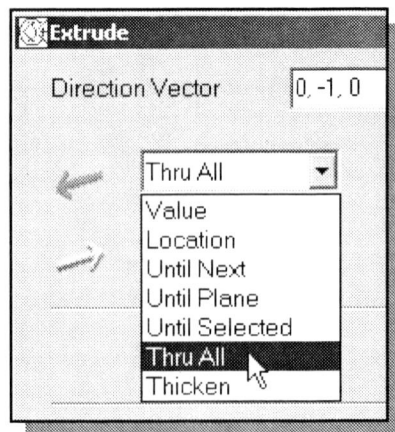

5. Click and hold down the left-mouse-button on the **depth** menu and select the **Thru All** option. *I-DEAS* will calculate the distance necessary to cut through the part.

6. Click on the **OK** button to accept the settings. The circle is extruded and the volume of the cylinder is removed.

7. On your own, generate a shaded image of the 3D object.

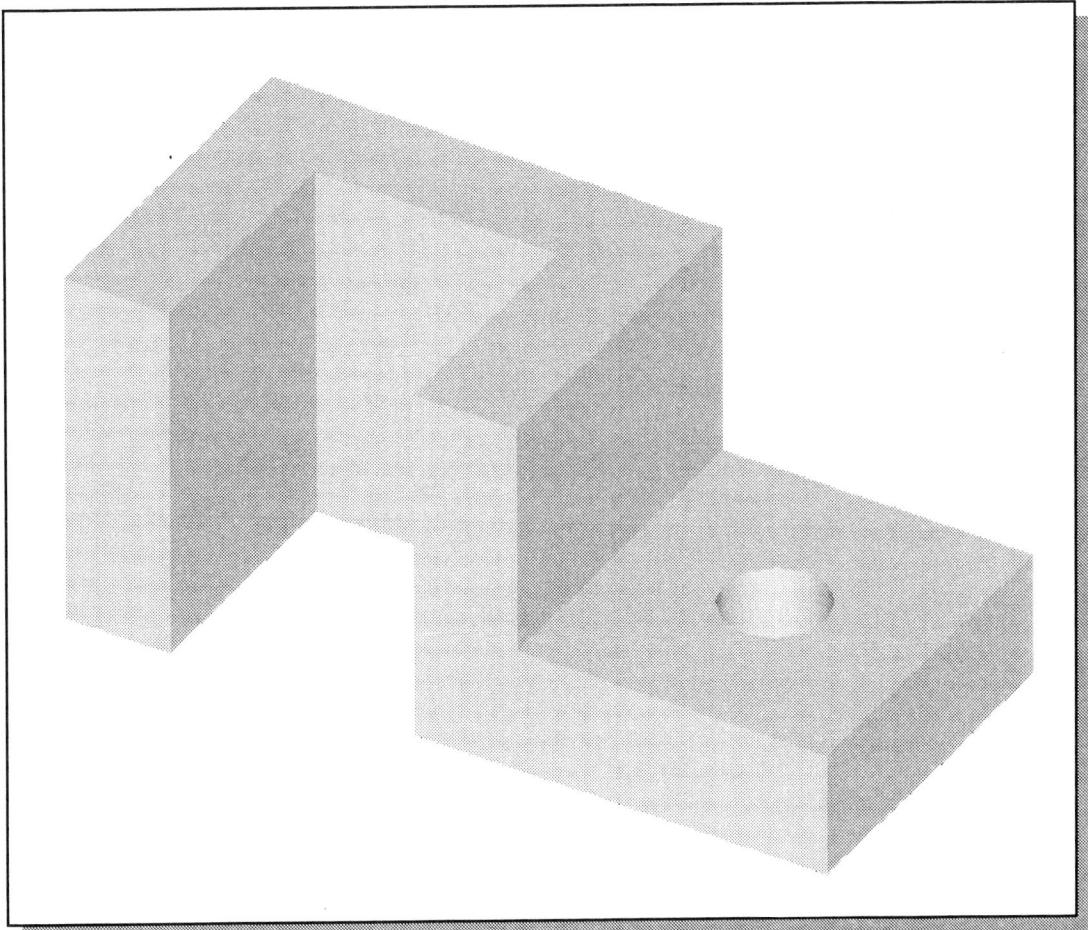

Save the Part and Exit *I-DEAS*

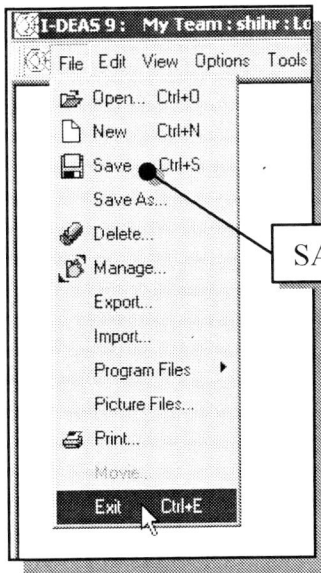

1. From the icon panel, select the **File** pull-down menu. Pick the **Save** option. Notice that you can also use the **Ctrl-S** combination (pressing down the **Ctrl** key and hitting the "**S**" key once) to save the part. A small watch appears to indicate passage of time as the part is saved.

2. Now you can leave *I-DEAS*. Use the left-mouse-button to click on **File** in the toolbar menu and select **Exit** from the pull-down menu. A pop-up window will appear with the message "*Save changes before exiting?*" Click on the **NO** button since we have saved the model already.

Questions:

1. The truss element used in finite element analysis is a two-force member element. List and describe the assumptions of a two-force member.

2. What is the size of the stiffness matrix for a single element? What is the size of the overall global stiffness matrix in example 2.2?

3. What is the first thing we should setup when starting a new CAD model in *I-DEAS*?

4. How does the *I-DEAS Dynamic Navigator* assist us in sketching?

5. How do we remove the dimensions created by the *Dynamic Navigator*?

6. How do we modify more than one dimension at a time?

7. What is the difference between *Distance* and *Thru All* when extruding?

8. Identify and describe the following commands:

    (a)

    (b)

    (c)

    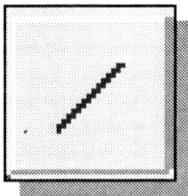

    (d)

    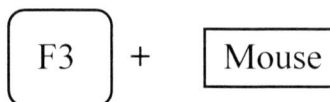

Exercises:

1. Determine the nodal displacements and reaction forces using the direct stiffness method.

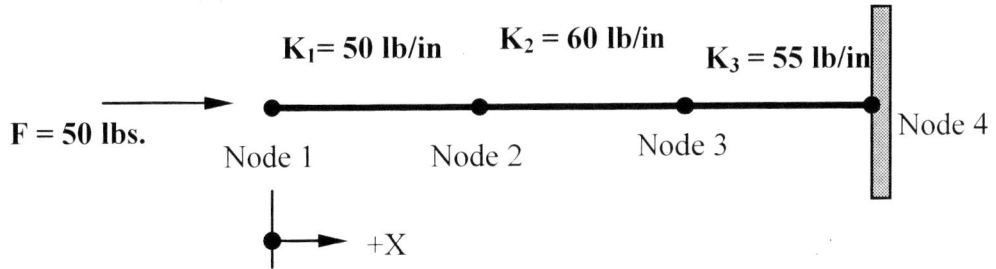

2.

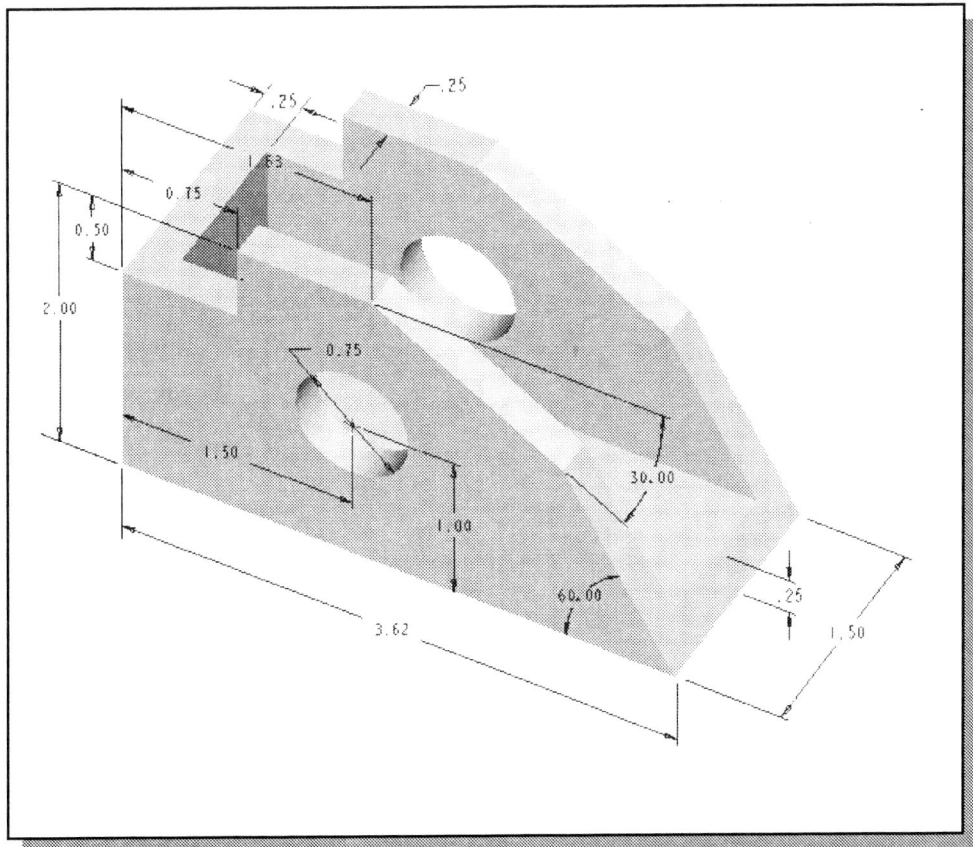

NOTES:

Chapter 3
Truss Elements In Two-Dimensional Spaces

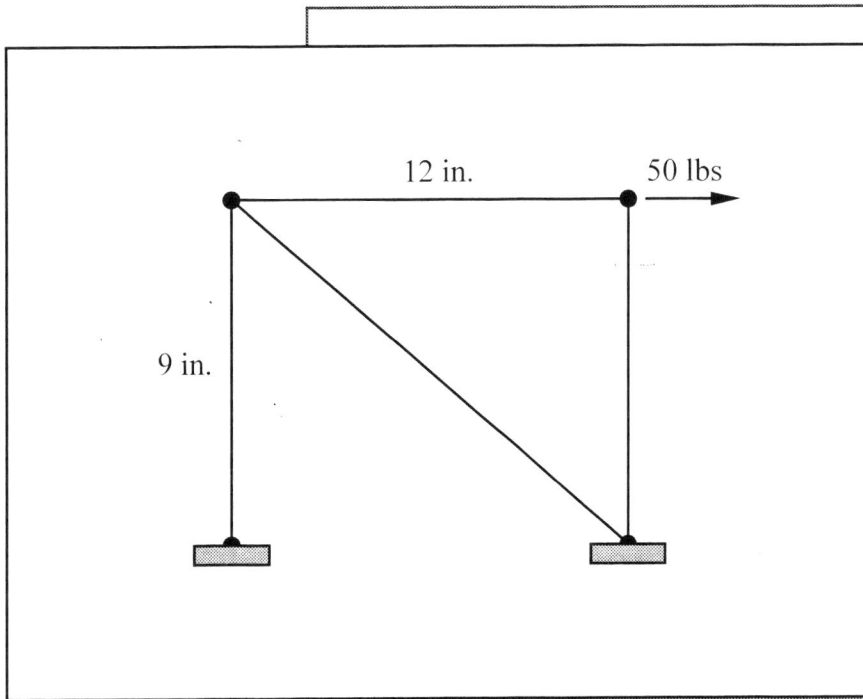

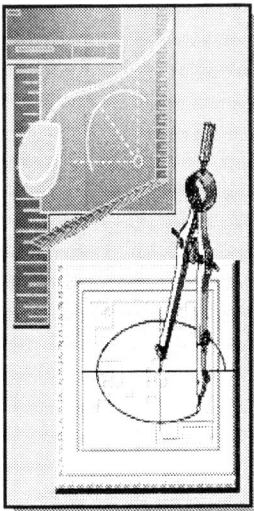

Learning Objectives

When you have completed this chapter, you will be able to:

◆ **Perform 2D Coordinates Transformation.**

◆ **Expand the Direct Stiffness Method to 2D Trusses.**

◆ **Derive the general 2D element Stiffness Matrix.**

◆ **Assemble the Global Stiffness Matrix for 2D Trusses.**

◆ **Solve 2D trusses using the Direct Stiffness Method.**

3.1 Introduction

This chapter presents the formulation of the direct stiffness method of truss elements in a two-dimensional space and the general procedure for solving two-dimensional truss structures using the direct stiffness method. The primary focus of this text is on the aspects of finite element analysis that are more important to the user than the programmer. However, for a user to utilize the software correctly and effectively, some understanding of the element formulation and computational aspects are also important. In this chapter, a two-dimensional truss structure consisting of two truss elements (as shown below) is used to illustrate the solution process of the direct stiffness method.

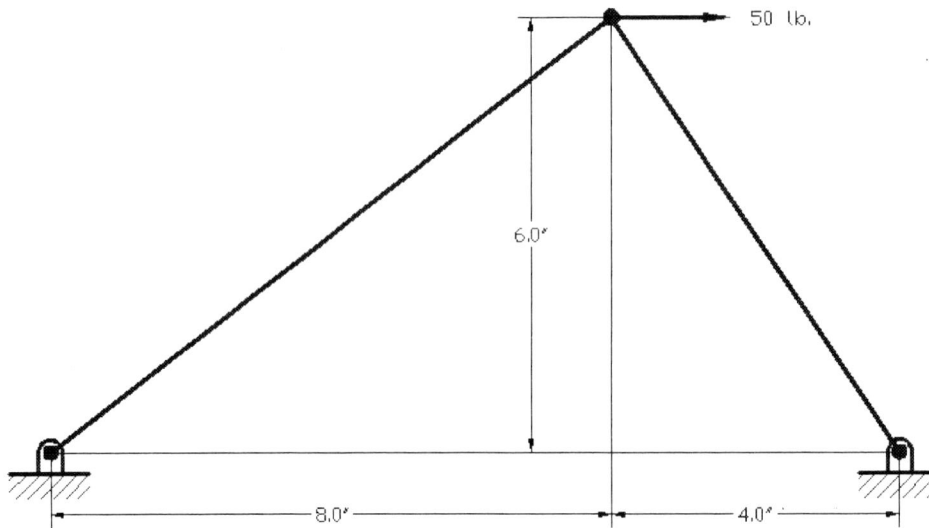

3.2 Truss Elements in Two-Dimensional Spaces

As introduced in Chapter 2, the system equations (stiffness matrix) of a truss element can be represented using the system equations of a linear spring in one-dimensional space.

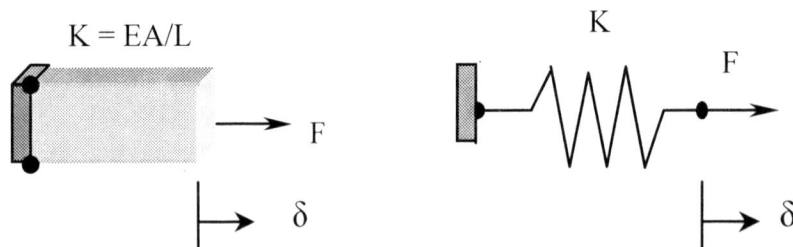

Free Body Diagram:

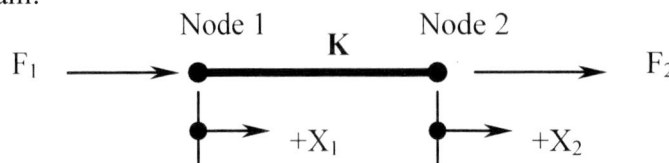

The general force-displacement equations in matrix form:

$$\left\{ \begin{array}{c} F_1 \\ F_2 \end{array} \right\} = \left[\begin{array}{cc} +K & -K \\ -K & +K \end{array} \right] \left\{ \begin{array}{c} X_1 \\ X_2 \end{array} \right\}$$

For a truss element, $\mathbf{K = EA/L}$

$$\left\{ \begin{array}{c} F_1 \\ F_2 \end{array} \right\} = \frac{\mathbf{EA}}{\mathbf{L}} \left[\begin{array}{cc} +1 & -1 \\ -1 & +1 \end{array} \right] \left\{ \begin{array}{c} X_1 \\ X_2 \end{array} \right\}$$

For truss members positioned in two-dimensional space, two coordinate systems are established:

1. The global coordinate system (**X** and **Y** axes) chosen to represent the entire structure.
2. The local coordinate system (**X** and **Y** axes) selected to align the **X**-axis along the length of the element.

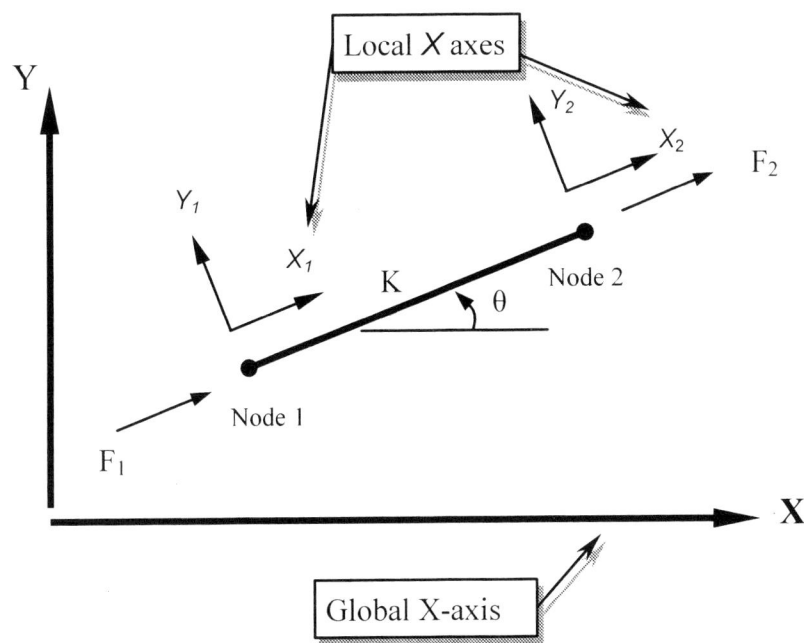

The force-displacement equations expressed in terms of components in the local *XY* coordinate system:

$$\left\{ \begin{array}{c} F_{1X} \\ F_{2X} \end{array} \right\} = \frac{\mathbf{EA}}{\mathbf{L}} \left[\begin{array}{cc} +1 & -1 \\ -1 & +1 \end{array} \right] \left\{ \begin{array}{c} X_1 \\ X_2 \end{array} \right\}$$

The above stiffness matrix (system equations in matrix form) can be expanded to incorporate the two force components at each node and the two displacement components at each node.

Force Components (Local Coordinate System)

$$\begin{Bmatrix} F_{1X} \\ F_{1Y} \\ F_{2X} \\ F_{2Y} \end{Bmatrix} = \frac{EA}{L} \begin{bmatrix} +1 & 0 & -1 & 0 \\ 0 & 0 & 0 & 0 \\ -1 & 0 & +1 & 0 \\ 0 & 0 & 0 & 0 \end{bmatrix} \begin{Bmatrix} X_1 \\ Y_1 \\ X_2 \\ Y_2 \end{Bmatrix}$$

Nodal Displacements (Local Coordinate System)

In regard to the expanded local stiffness matrix (system equations in matrix form):

1. It is always a square matrix.
2. It is always symmetrical for linear systems.
3. The diagonal elements are always positive or zero.

The above stiffness matrix, expressed in terms of the established 2D local coordinate system, represents a single truss element in a two-dimensional space. In a general structure, many elements are involved, and they would be oriented with different angles. The above stiffness matrix is a general form of a <u>SINGLE</u> element in a 2D local coordinate system. Imagine the number of coordinate systems involved for a 20-member structure. For the example that will be illustrated in the following sections, two local coordinate systems (one for each element) are needed for the truss structure shown below. The two local coordinate systems (X_1Y_1 & X_2Y_2) are aligned to the elements.

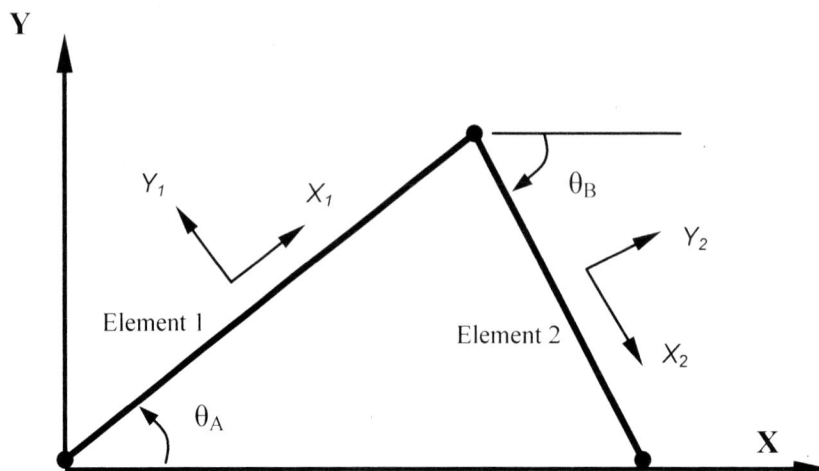

In order to solve the system equations of two-dimensional truss structures, it is necessary to assemble all elements' stiffness matrices into a **global stiffness matrix**, with all the equations of the individual elements referring to a common global coordinate system. This requires the use of *coordinate transformation equations* applied to system equations for all elements in the structure. For a one-dimensional truss structure (illustrated in chapter 2), the local coordinate system coincides with the global coordinate system; therefore, no coordinate transformation is needed to assemble the global stiffness matrix (the stiffness matrix in the global coordinate system). In the next section, the coordinate transformation equations are derived for truss elements in two-dimensional spaces.

3.3 Coordinate Transformation

A vector, in a two-dimensional space, can be expressed in terms of any coordinate system set of unit vectors.

For example,

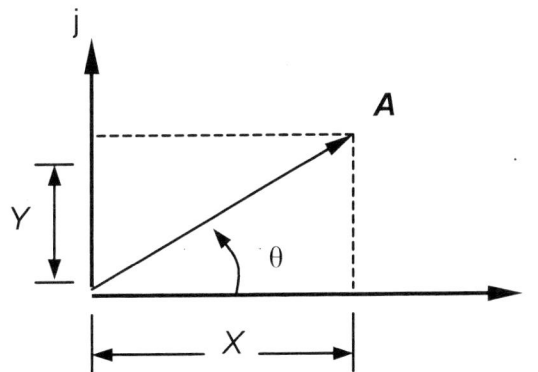

Vector **A** can be expressed as:

$$A = X\,i + Y\,j$$

Where *i* and *j* are unit vectors along the *X* and *Y* axes.

Magnitudes of *X and Y* can also be expressed as:

$$X = A\,cos\,(\theta)$$
$$Y = A\,sin\,(\theta)$$

Where *X, Y and A* are scalar quantities.

Therefore,

$$A = X\,i + Y\,j = A\,cos\,(\theta)\,i + A\,sin\,(\theta)\,j \quad \text{---------- (1)}$$

Next, establish a new unit vector (u) in the same direction as vector **A**.

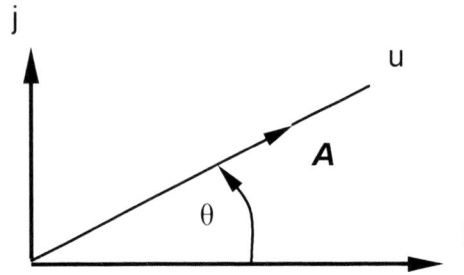

Vector *A* can now be expressed as: $A = A\ u$ ------------ (2)

Both equations (the above (1) and (2)) represent vector **A**:

$$A = A\ u = A \cos(\theta)\ i + A \sin(\theta)\ j$$

The unit vector *u* can now be expressed in terms of the original set of unit vectors *i* and *j*:

$$\boxed{u = \cos(\theta)\ i + \sin(\theta)\ j}$$

Now consider another vector **B**:

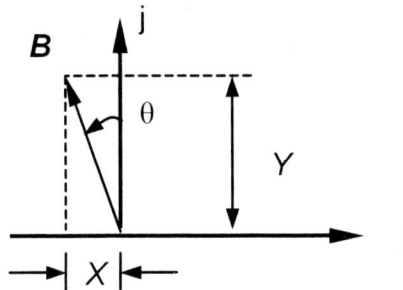

Vector **B** can be expressed as:

$$B = -X\ i + Y\ j$$

Where *i* and *j* are unit vectors along the *X* and *Y* axes.

Magnitudes of *X and Y* can also be expressed as components of the magnitude of the vector:

$$X = B \sin(\theta)$$
$$Y = B \cos(\theta)$$

Where *X, Y and B* are scalar quantities.

Therefore,

$$B = -X\ i + Y\ j = -B \sin(\theta)\ i + B \cos(\theta)\ j \quad \text{---------- (3)}$$

Next, establish a new unit vector (v) along vector **B**.

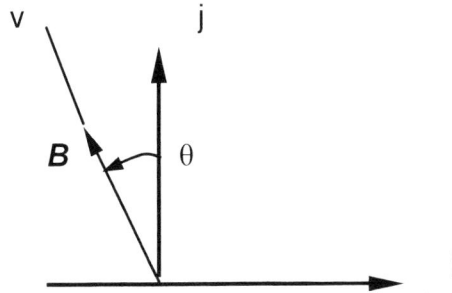

Vector **B** can now be expressed as: **B** = B v ------------ (4)

Equations (3) and (4) represent vector **B**:

$$\boldsymbol{B} = B\,v = -B\,sin\,(\theta)\,i + B\,cos\,(\theta)\,j$$

The unit vector v can now be expressed in terms of the original set of unit vectors i and j:

$$\boxed{v = -sin\,(\theta)\,i + cos\,(\theta)\,j}$$

We have established the coordinate transformation equations that can be used to transform vectors from ij coordinates to the rotated uv coordinates.

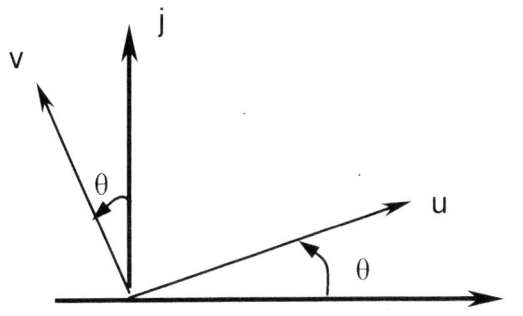

Coordinate Transformation Equations:

$$\boxed{\begin{array}{l} u = cos\,(\theta)\,i + sin\,(\theta)\,j \\ v = -sin\,(\theta)\,i + cos\,(\theta)\,j \end{array}}$$

Direction cosines

In matrix form,

$$\begin{Bmatrix} u \\ v \end{Bmatrix} = \begin{bmatrix} cos\,(\theta) & sin\,(\theta) \\ -sin\,(\theta) & cos\,(\theta) \end{bmatrix} \begin{Bmatrix} i \\ j \end{Bmatrix}$$

The above *direction cosines* allows us to transform vectors from the GLOBAL coordinates to the LOCAL coordinates. It is also necessary to be able to transform vectors from the LOCAL coordinates to the GLOBAL coordinates. Although it is possible to derive the LOCAL to GLOBAL transformation equations in a similar manner as demonstrated for the above equations, the *MATRIX operations* provide a slightly more elegant approach.

The above equations can be represent symbolically as:

$$\{a\} = [\,l\,]\;\{b\}$$

where $\{a\}$ and $\{b\}$ are direction vectors, [l] is the direction cosines.

Perform the matrix operations to derive the reverse transformation equations in terms of the above direction cosines:

$$\{b\} = [\,?\,]\;\{a\}.$$

First, multiply by $[\,l\,]^{-1}$ to remove the *direction cosines* from the right hand side of the original equation.

$$\{a\} = [\,l\,]\;\{b\}$$

$$[\,l\,]^{-1}\{a\} = [\,l\,]^{-1}\,[\,l\,]\;\{b\}$$

From matrix algebra, $[\,l\,]^{-1}\,[\,l\,] = [\,I\,]$ *and* $[\,I\,]\{b\} = \{b\}.$

The equation can now be simplified as

$$[\,l\,]^{-1}\{a\} = \{b\}$$

For *linear statics analyses*, the *direction cosines* is an *orthogonal matrix* and the *inverse of the matrix* is equal to the transpose of the matrix.

$$[\,l\,]^{-1} = [\,l\,]^{T}$$

Therefore, the transformation equation can be expressed as:

$$[\,l\,]^{T}\{a\} = \{b\}$$

The transformation equations that enable us to transform any vector from a *LOCAL coordinate system* to the *GLOBAL coordinate system* becomes:

LOCAL coordinates to the GLOBAL coordinates:

$$\begin{Bmatrix} i \\ j \end{Bmatrix} = \begin{bmatrix} \cos(\theta) & -\sin(\theta) \\ \sin(\theta) & \cos(\theta) \end{bmatrix} \begin{Bmatrix} u \\ v \end{Bmatrix}$$

The reverse transformation can also be established by applying the transformation equations that transform any vector from the *GLOBAL coordinate system* to the *LOCAL coordinate system*:

GLOBAL coordinates to the LOCAL coordinates:

$$\begin{Bmatrix} u \\ v \end{Bmatrix} = \begin{bmatrix} \cos(\theta) & \sin(\theta) \\ -\sin(\theta) & \cos(\theta) \end{bmatrix} \begin{Bmatrix} i \\ j \end{Bmatrix}$$

As it is the case with many mathematical equations, derivation of the equations usually appears to be much more complex than the actual application and utilization of the equations. The following example illustrates the application of the two-dimensional *coordinate transformation equations* on a point in between two coordinate systems.

EXAMPLE 3.1

Given:

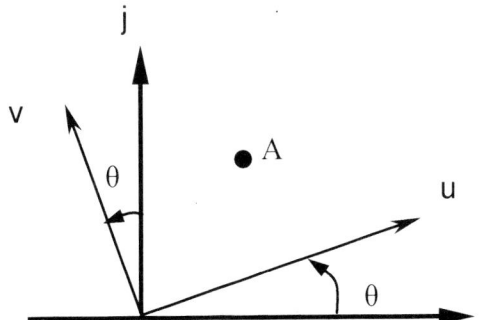

The coordinates of point A: (20 i, 40 j).

Find: The coordinates of point A if the local coordinate system is rotated 15 degrees relative to the global coordinate system.

Solution:

Using the coordinates transformation equations (GLOBAL coordinates to the LOCAL coordinates):

$$\begin{Bmatrix} u \\ v \end{Bmatrix} = \begin{bmatrix} \cos(\theta) & \sin(\theta) \\ -\sin(\theta) & \cos(\theta) \end{bmatrix} \begin{Bmatrix} i \\ j \end{Bmatrix}$$

$$= \begin{bmatrix} \cos(15°) & \sin(15°) \\ -\sin(15°) & \cos(15°) \end{bmatrix} \begin{Bmatrix} 20 \\ 40 \end{Bmatrix}$$

$$= \begin{Bmatrix} 29.7 \\ 33.5 \end{Bmatrix}$$

➤ On your own, perform a coordinate transformation to determine the global coordinates of point A using the *LOCAL* coordinates of (29.7,33.5) with the 15 degrees angle in between the two coordinate systems.

3.4 Global Stiffness Matrix

For a single truss element, using the coordinate transformation equations, we can proceed to transform the local stiffness matrix to the global stiffness matrix.

For a single truss element arbitrarily positioned in a two-dimensional space:

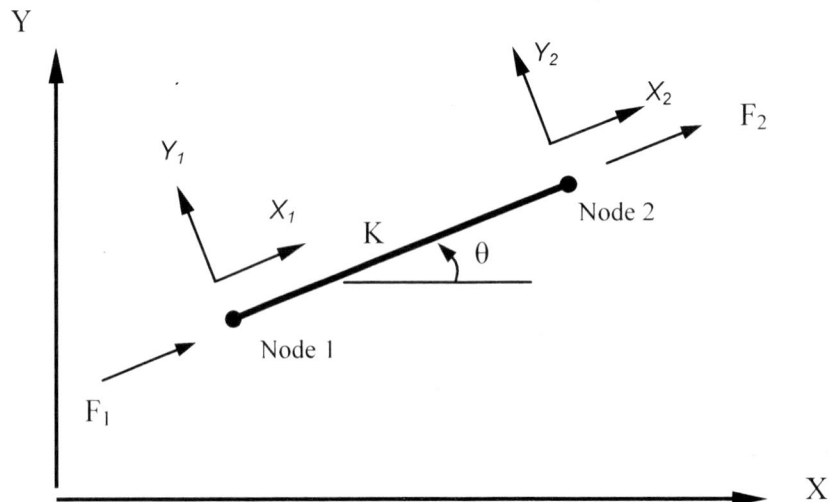

The force-displacement equations (in the local coordinate system) can be expressed as:

$$\begin{Bmatrix} F_{1X} \\ F_{1Y} \\ F_{2X} \\ F_{2Y} \end{Bmatrix} = \frac{EA}{L} \begin{bmatrix} +1 & 0 & -1 & 0 \\ 0 & 0 & 0 & 0 \\ -1 & 0 & +1 & 0 \\ 0 & 0 & 0 & 0 \end{bmatrix} \begin{Bmatrix} X_1 \\ Y_1 \\ X_2 \\ Y_2 \end{Bmatrix}$$

Local Stiffness Matrix

Next, apply the coordinate transformation equations to establish the general GLOBAL STIFFNESS MATRIX of a single truss element in a two-dimensional space.

First, the displacement transformation equations (GLOBAL to LOCAL):

$$\begin{Bmatrix} X_1 \\ Y_1 \\ X_2 \\ Y_2 \end{Bmatrix} = \begin{bmatrix} \cos(\theta) & \sin(\theta) & 0 & 0 \\ -\sin(\theta) & \cos(\theta) & 0 & 0 \\ 0 & 0 & \cos(\theta) & \sin(\theta) \\ 0 & 0 & -\sin(\theta) & \cos(\theta) \end{bmatrix} \begin{Bmatrix} X_1 \\ Y_1 \\ X_2 \\ Y_2 \end{Bmatrix}$$

Local   Global

The force transformation equations (GLOBAL to LOCAL):

$$\begin{Bmatrix} F_{1x} \\ F_{1y} \\ F_{2x} \\ F_{2y} \end{Bmatrix} = \begin{bmatrix} \cos(\theta) & \sin(\theta) & 0 & 0 \\ -\sin(\theta) & \cos(\theta) & 0 & 0 \\ 0 & 0 & \cos(\theta) & \sin(\theta) \\ 0 & 0 & -\sin(\theta) & \cos(\theta) \end{bmatrix} \begin{Bmatrix} F_{1x} \\ F_{1y} \\ F_{2x} \\ F_{2y} \end{Bmatrix}$$

Local   Global

The above three sets of equations can be represented as:

$\{F\} = [K]\{X\}$ ------- *Local force-displacement equation*

$\{X\} = [l]\{X\}$ ------- *Displacement transformation equation*

$\{F\} = [l]\{F\}$ ------- *Force transformation equation*

We will next perform *matrix operations* to obtain the GLOBAL stiffness matrix:
Starting with the local force-displacement equation

$$\{F\} = [K]\{X\}$$

Local   Local

Next, substituting the transformation equations for $\{F\}$ and $\{X\}$,

$$[l]\{F\} = [K][l]\{X\}$$

Multiply both sides of the equation with $[l]^{-1}$,

$$[l]^{-1}[l]\{F\} = [l]^{-1}[K][l]\{X\}$$

The equation can be simplified as:

$$\{F\} = [l]^{-1}[K][l]\{X\}$$

or   Global   Global

$$\{F\} = [l]^{T}[K][l]\{X\}$$

The GLOBAL force-displacement equation is then expressed as:

$$\{F\} = [l]^{T}[K][l]\{X\}$$

or

$$\{F\} = [K]\{X\}$$

The global stiffness matrix [**K**] can now be expressed in terms of the local stiffness matrix.

$$\boxed{[K] = [l]^{T}[K][l]}$$

For a single truss element in a two-dimensional space, the global stiffness matrix is

$$[K] = \frac{EA}{L}\begin{bmatrix} cos^2(\theta) & cos(\theta)sin(\theta) & -cos^2(\theta) & -cos(\theta)sin(\theta) \\ cos(\theta)sin(\theta) & sin^2(\theta) & -cos(\theta)sin(\theta) & -sin^2(\theta) \\ -cos^2(\theta) & -cos(\theta)sin(\theta) & cos^2(\theta) & cos(\theta)sin(\theta) \\ -cos(\theta)sin(\theta) & -sin^2(\theta) & sin(\theta)cos(\theta) & sin^2(\theta) \end{bmatrix}$$

➤ The above matrix can be applied to any truss element positioned in a two-dimensional space. We can now assemble the global stiffness matrix and analyze any two-dimensional multiple-elements truss structures. The following example illustrates, using the general global stiffness matrix derived above, the formulation and solution process of a 2D truss structure.

Example 3.2

Given: A two-dimensional truss structure as shown. (All joints are Pin Joints.)

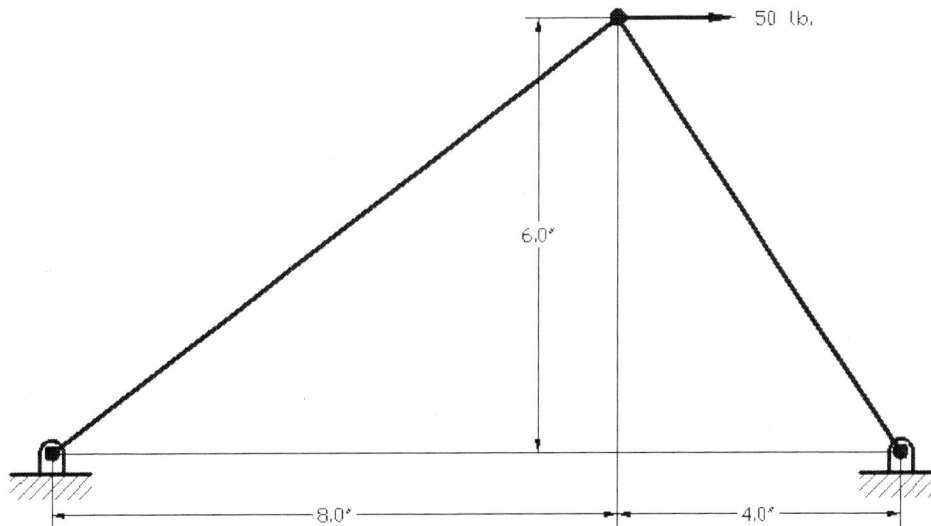

Material: Steel rod, diameter ¼ in.

Find: Displacements of each node and stresses in each member.

Solution:

The system contains two elements and three nodes. The nodes and elements are labeled as shown below.

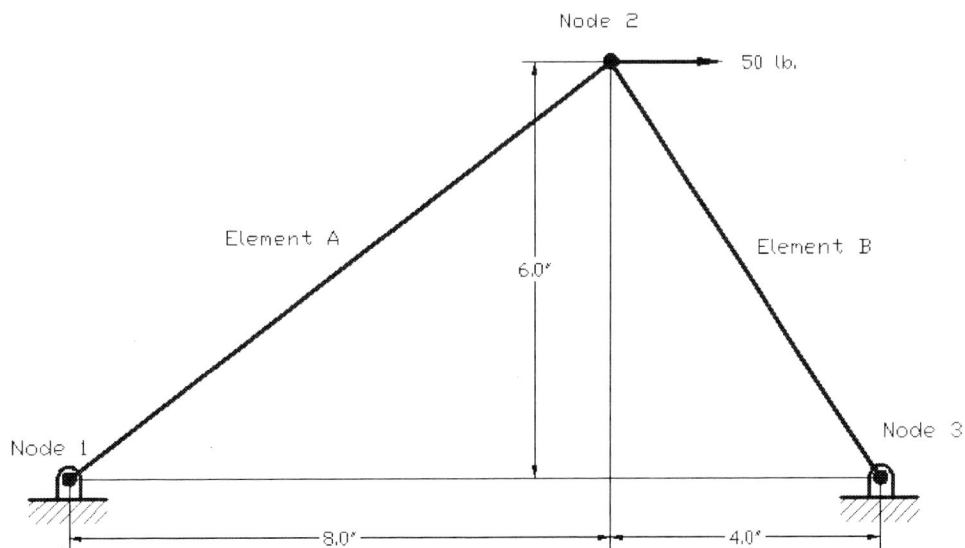

First, establish the GLOBAL stiffness matrix (system equations in matrix form) for each element.

Element A   (Node 1 to Node 2)

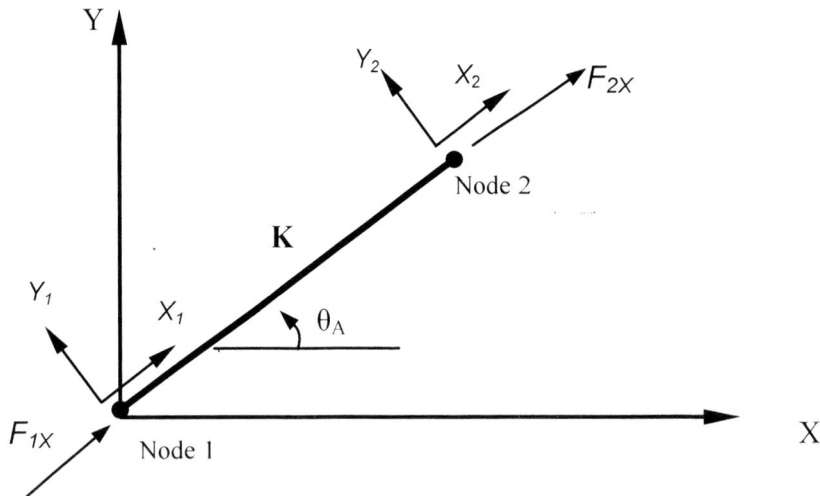

$\theta_A = \tan^{-1}(6/8) = 36.87°$,
E (Young's modulus) $= 30 \times 10^6$ psi
A (Cross sectional area) $= \pi r^2 = 0.049$ in^2.
L (Length of element) $= (6^2 + 8^2)^{1/2} = 10$ in.

Therefore, $\dfrac{EA}{L} = 147262$ lb/in.

The LOCAL force-displacement equations:

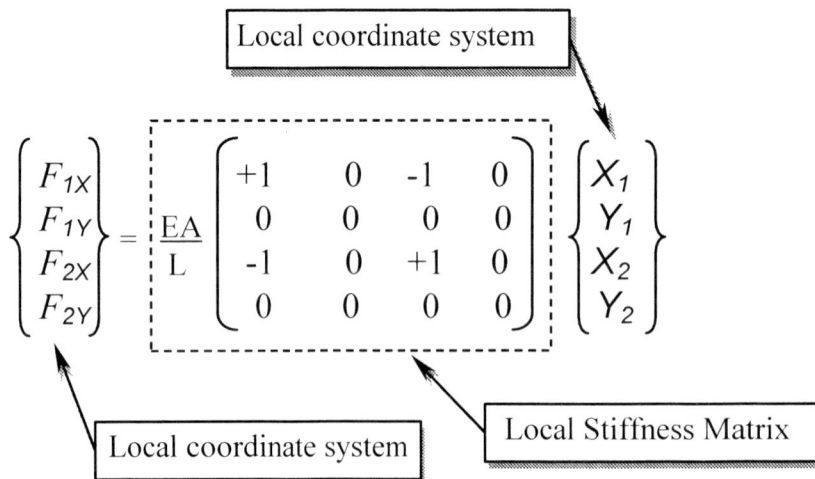

Local coordinate system

$$\begin{Bmatrix} F_{1X} \\ F_{1Y} \\ F_{2X} \\ F_{2Y} \end{Bmatrix} = \frac{EA}{L} \begin{bmatrix} +1 & 0 & -1 & 0 \\ 0 & 0 & 0 & 0 \\ -1 & 0 & +1 & 0 \\ 0 & 0 & 0 & 0 \end{bmatrix} \begin{Bmatrix} X_1 \\ Y_1 \\ X_2 \\ Y_2 \end{Bmatrix}$$

Local coordinate system

Local Stiffness Matrix

Using the equations we have derived, the GLOBAL system equations for *element A* can be expressed as:

$$\{ F \} = [K] \{ X \}$$

$$[K] = \frac{EA}{L} \begin{bmatrix} cos^2(\theta) & cos(\theta)sin(\theta) & -cos^2(\theta) & -cos(\theta)sin(\theta) \\ cos(\theta)sin(\theta) & sin^2(\theta) & -cos(\theta)sin(\theta) & -sin^2(\theta) \\ -cos^2(\theta) & -cos(\theta)sin(\theta) & cos^2(\theta) & cos(\theta)sin(\theta) \\ -cos(\theta)sin(\theta) & -sin^2(\theta) & sin(\theta)cos(\theta) & sin^2(\theta) \end{bmatrix}$$

Therefore,

Global

$$\begin{Bmatrix} F_{1X} \\ F_{1Y} \\ F_{2XA} \\ F_{2YA} \end{Bmatrix} = 147262 \begin{bmatrix} .64 & .48 & -.64 & -.48 \\ .48 & .36 & -.48 & -.36 \\ -.64 & -.48 & .64 & .48 \\ -.48 & -.36 & .48 & .36 \end{bmatrix} \begin{Bmatrix} X_1 \\ Y_1 \\ X_2 \\ Y_2 \end{Bmatrix}$$

Global

Global Stiffness Matrix

Element B   (Node 2 to Node 3)

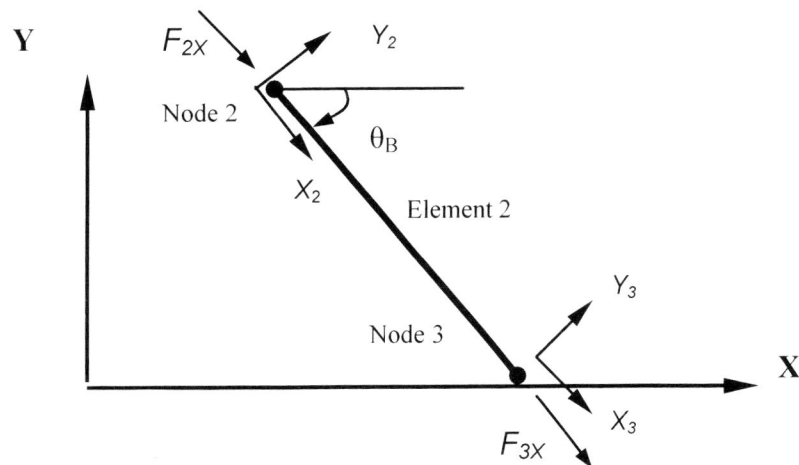

$\theta_B = tan^{-1}(6/4) = 56.31°$,

E (Young's modulus) = 30×10^6 psi

A (Cross sectional area) = $\pi r^2 = 0.049$ in^2

L (Length of element) = $(4^2 + 6^2)^{1/2} = 7.21$ in.

$$\frac{EA}{L} = 204216 \text{ lb/in.}$$

$$
\begin{Bmatrix} F_{2X} \\ F_{2Y} \\ F_{3X} \\ F_{3Y} \end{Bmatrix} = \frac{EA}{L} \begin{bmatrix} +1 & 0 & -1 & 0 \\ 0 & 0 & 0 & 0 \\ -1 & 0 & +1 & 0 \\ 0 & 0 & 0 & 0 \end{bmatrix} \begin{Bmatrix} X_2 \\ Y_2 \\ X_3 \\ Y_3 \end{Bmatrix}
$$

Local system — Local system — Local Stiffness Matrix — Local system

Using the equations we derived in the previous sections, the GLOBAL system equations for *element B* is:

$$
\{ F \} = [K] \{ X \}
$$

$$
[K] = \frac{EA}{L} \begin{bmatrix} cos^2(\theta) & cos(\theta)sin(\theta) & -cos^2(\theta) & -cos(\theta)sin(\theta) \\ cos(\theta)sin(\theta) & sin^2(\theta) & -cos(\theta)sin(\theta) & -sin^2(\theta) \\ -cos^2(\theta) & -cos(\theta)sin(\theta) & cos^2(\theta) & cos(\theta)sin(\theta) \\ -cos(\theta)sin(\theta) & -sin^2(\theta) & sin(\theta)cos(\theta) & sin^2(\theta) \end{bmatrix}
$$

Therefore,

Global

$$
\begin{Bmatrix} F_{2XB} \\ F_{2YB} \\ F_{3X} \\ F_{3Y} \end{Bmatrix} = 204216 \begin{bmatrix} 0.307 & -0.462 & -0.307 & 0.462 \\ -0.462 & 0.692 & 0.462 & -0.692 \\ -0.307 & 0.462 & 0.307 & -0.462 \\ 0.462 & -0.692 & -0.462 & 0.692 \end{bmatrix} \begin{Bmatrix} X_2 \\ Y_2 \\ X_3 \\ Y_3 \end{Bmatrix}
$$

Global — Global Stiffness Matrix

Now we are ready to assemble the overall global stiffness matrix of the structure.

Summing the two sets of global force-displacement equations:

$$
\begin{Bmatrix} F_{1X} \\ F_{1Y} \\ F_{2X} \\ F_{2Y} \\ F_{3X} \\ F_{3Y} \end{Bmatrix} = \begin{bmatrix} 94248 & 70686 & -94248 & -70686 & 0 & 0 \\ 70686 & 53014 & -70686 & -53014 & 0 & 0 \\ -94248 & -70686 & 157083 & -23568 & -62836 & 94253 \\ -70686 & -53014 & -23568 & 194395 & 94253 & -141380 \\ 0 & 0 & -62836 & 94253 & 62836 & -94253 \\ 0 & 0 & 94253 & -141380 & -94253 & 141380 \end{bmatrix} \begin{Bmatrix} X_1 \\ Y_1 \\ X_2 \\ Y_2 \\ X_3 \\ Y_3 \end{Bmatrix}
$$

Next, apply the following known boundary conditions into the system equations:

(a) Node 1 and Node 3 are fixed-points; therefore, any displacement components of these two node-points are zero (X_1, Y_1 and X_3, Y_3).

(b) The only external load is at Node 2: $F_{2X} = 50$ lbs.
Therefore,

$$\begin{Bmatrix} F_{1X} \\ F_{1Y} \\ 50 \\ 0 \\ F_{3X} \\ F_{3Y} \end{Bmatrix} = \begin{bmatrix} 94248 & 70686 & -94248 & -70686 & 0 & 0 \\ 70686 & 53014 & -70686 & -53014 & 0 & 0 \\ -94248 & -70686 & 157083 & -23568 & -628360 & 94253 \\ -70686 & -53014 & -23568 & 194395 & 94253 & -141380 \\ 0 & 0 & -62836 & 94253 & 62836 & -94253 \\ 0 & 0 & 94253 & -141380 & -94253 & 141380 \end{bmatrix} \begin{Bmatrix} 0 \\ 0 \\ X_2 \\ Y_2 \\ 0 \\ 0 \end{Bmatrix}$$

The two displacements we need to solve are X_2 and Y_2. Let's simplify the above matrix by removing the unaffected/unnecessary columns in the matrix.

$$\begin{Bmatrix} F_{1X} \\ F_{1Y} \\ 50 \\ 0 \\ F_{3X} \\ F_{3Y} \end{Bmatrix} = \begin{bmatrix} -94248 & -70686 \\ -70686 & -53014 \\ 157083 & -23568 \\ 23568 & 194395 \\ -62836 & 94253 \\ 94253 & -141380 \end{bmatrix} \begin{Bmatrix} X_2 \\ Y_2 \end{Bmatrix}$$

Solve for nodal displacements X_2 and Y_2:

$$\begin{Bmatrix} 50 \\ 0 \end{Bmatrix} = \begin{bmatrix} 157083 & -23568 \\ 23568 & 194395 \end{bmatrix} \begin{Bmatrix} X_2 \\ Y_2 \end{Bmatrix}$$

$$X_2 = 3.24 \text{ e}^{-4} \text{ in.}$$
$$Y_2 = 3.93 \text{ e}^{-5} \text{ in.}$$

Substitute the known X_2 and Y_2 values into the matrix and solve for the reaction forces:

$$\begin{Bmatrix} F_{1X} \\ F_{1Y} \\ F_{3X} \\ F_{3Y} \end{Bmatrix} = \begin{bmatrix} -94248 & -70686 \\ -70686 & -53014 \\ -62836 & 94253 \\ 94253 & -141380 \end{bmatrix} \begin{Bmatrix} 3.24 \text{ e}^{-4} \\ 3.93 \text{ e}^{-5} \end{Bmatrix}$$

Therefore,

$$F_{1X} = \text{-33.33 lbs.}, \quad F_{1Y} = \text{-25 lbs.}$$
$$F_{3X} = \text{-16.67 lbs.}, \quad F_{3Y} = \text{ 25 lbs.}$$

To determine the normal stress in each truss member, one option is to use the displacement transformation equations to transform the results from the global coordinate system back to the local coordinate system.

Element A

$\{X\} = [\,l\,]\,\{X\}$ ------- *Displacement transformation equation*

$$\begin{Bmatrix} X_1 \\ Y_1 \\ X_2 \\ Y_2 \end{Bmatrix} = \begin{bmatrix} .8 & .6 & 0 & 0 \\ -.6 & .8 & 0 & 0 \\ 0 & 0 & .8 & .6 \\ 0 & 0 & -.6 & .8 \end{bmatrix} \begin{Bmatrix} 0 \\ 0 \\ X_2 \\ Y_2 \end{Bmatrix}$$

Local   Global

$$X_2 = 2.83\ e^{-4}$$
$$Y_2 = \text{-1.63}\ e^{-4}$$

The LOCAL force-displacement equations:

Local system

$$\begin{Bmatrix} F_{1X} \\ F_{1Y} \\ F_{2X} \\ F_{2Y} \end{Bmatrix} = \frac{EA}{L} \begin{bmatrix} +1 & 0 & -1 & 0 \\ 0 & 0 & 0 & 0 \\ -1 & 0 & +1 & 0 \\ 0 & 0 & 0 & 0 \end{bmatrix} \begin{Bmatrix} X_1 \\ Y_1 \\ X_2 \\ Y_2 \end{Bmatrix}$$

Local system   Local Stiffness Matrix

$$F_{1X} = \text{-41.67 lbs.}, \ F_{1Y} = 0 \text{ lb.}$$
$$F_{2X} = \text{ 41.67 lbs.}, \ F_{2Y} = 0 \text{ lb.}$$

Therefore, the normal stress developed in *Element A* can be calculated as (41.67/0.049)=**850 psi**.

➤ On your own, calculate the normal stress developed in *Element B*.

Questions:

1. Determine the coordinates of point A if the local coordinate system is rotated 18 degrees relative to the global coordinate system. The global coordinates of point A: (30,50).

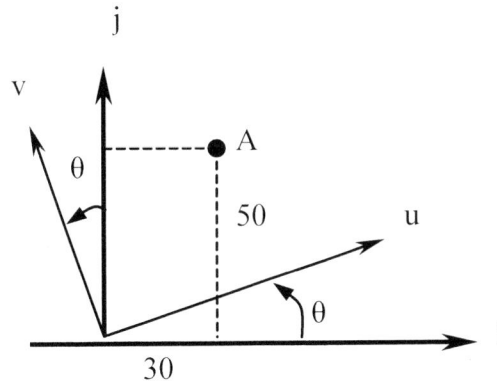

2. Determine the global coordinates of point B if the local coordinate system is rotated 25 degrees relative to the global coordinate system. The local coordinates of point B: (30,15).

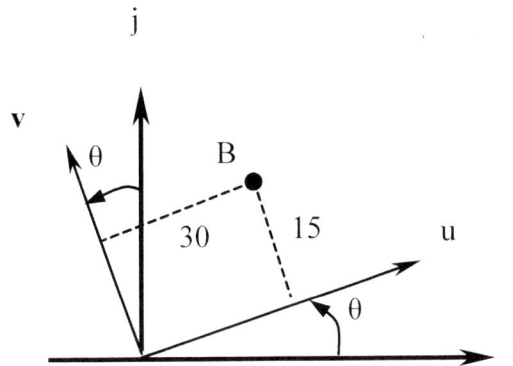

Exercises:

1. Given: two-dimensional truss structure as shown.

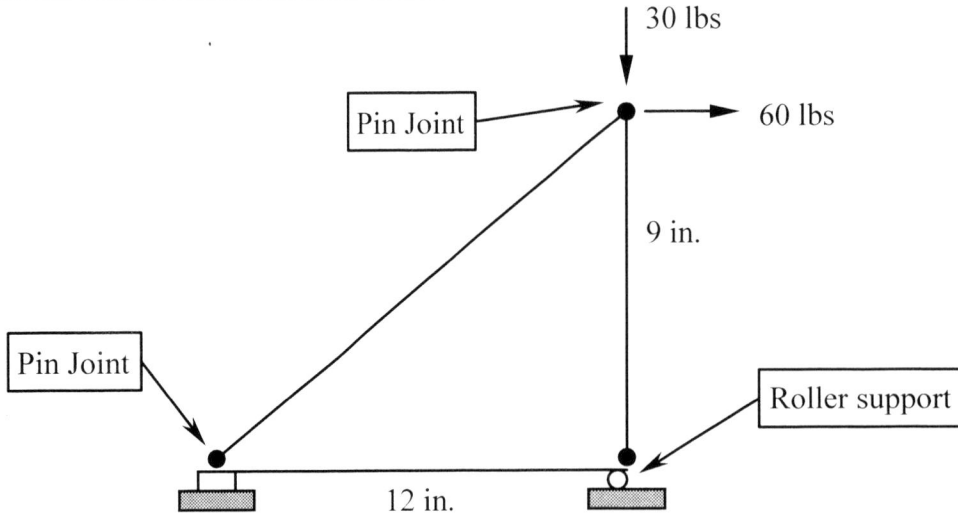

Material : Steel rod, diameter ¼ in.

Find: (a) Displacements of the nodes.
        (b) Normal stresses developed in the members.

2. Given: Two-dimensional truss structure as shown (All joints are pin joints).

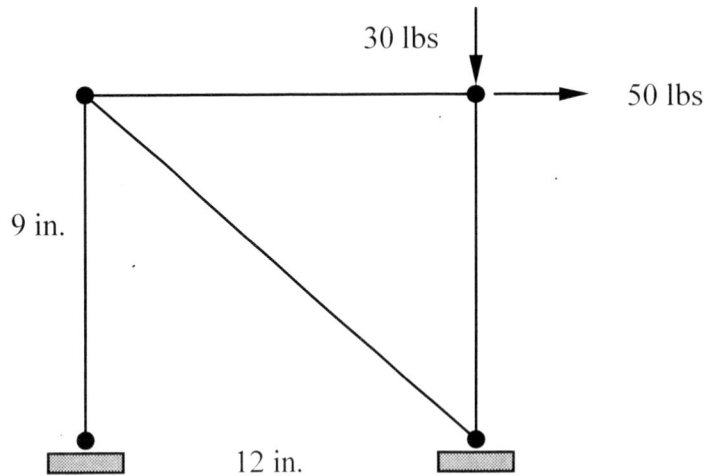

Material : Steel rod, diameter ¼ in.

Find: (a) Displacements of the nodes.
       (b) Normal stresses developed in the members.

Chapter 4
I-DEAS Two-Dimensional Truss Analysis

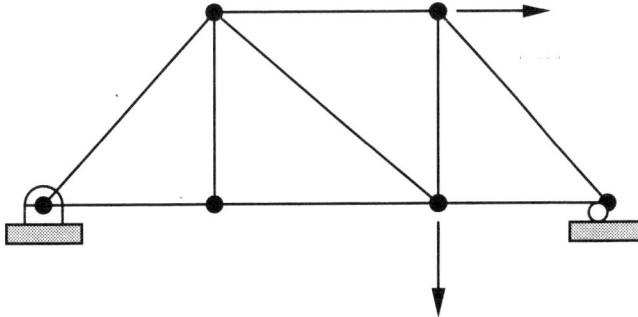

Learning Objectives

When you have completed this lesson, you will be able to:
- ◆ Create an I-DEAS FEModel.
- ◆ Create Nodes and Truss Elements in I-DEAS.
- ◆ Apply Loads and Boundary conditions at Nodes.
- ◆ Run the I-DEAS FEA Solver.
- ◆ View and Examine the I-DEAS FEA Results.
- ◆ Understand the General Computer FEA Procedure.

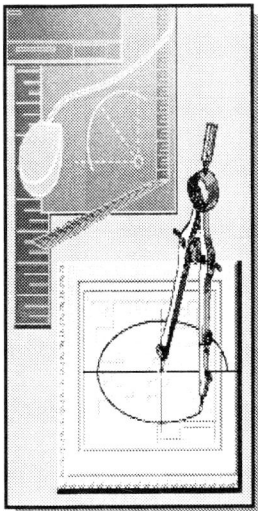

4.1 Finite Element Analysis Procedure

While all real-life structures are three-dimensional in nature, solutions of many stress analyses are done on two-dimensional spaces and sometimes one-dimensional spaces. Quite often, the approximations represent the three-dimensional members very well and there is no need to do a three-dimensional analysis.

This chapter demonstrates the entire process of creating, solving, and viewing the results of a finite element analysis on a two-dimensional truss structure using the FEA application software available in *I-DEAS*. The following illustration follows the typical procedure of performing finite element analysis:

    a. Preliminary analysis of the system:
        Perform an approximate calculation to gain some insights about the system.

    b. Preparation of the finite element model:
        1. Prescribe the geometric and material information of the system.
        2. Prescribe how the system is supported.
        3. Determine how the loads are applied to the system.

    c. Perform the calculations:
        Solve the system equations and compute displacements, strains and stresses.

    d. Post-processing of the results:
        Viewing the stresses and displacements.

➤ Before going through the tutorial, perform a preliminary analysis of the two-dimensional truss structure as shown below. First create a free body diagram of the entire structure to find the reactions at the supports. Then use either the classical *joint method* or the *section method* to find the forces in each of the members. Which member would you expect to have the highest stress? On your own, calculate the stress for the member you identified as the highest-stress member and compare to the computer solution at the end of this chapter.

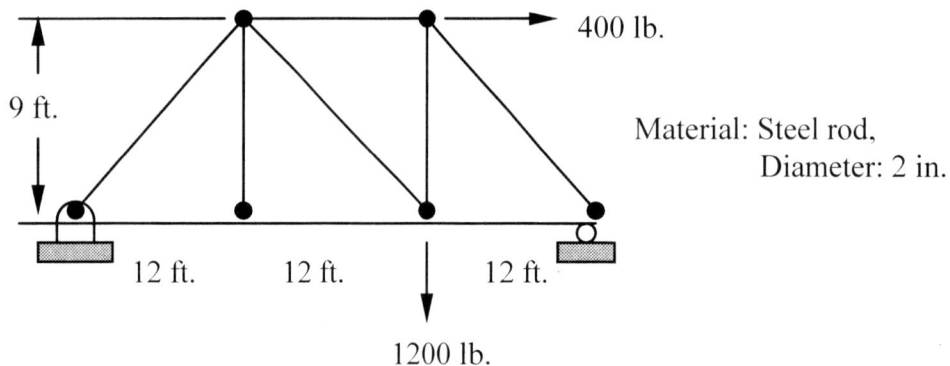

9 ft.   400 lb.

Material: Steel rod,
Diameter: 2 in.

12 ft.   12 ft.   12 ft.

1200 lb.

4.2 Preliminary Analysis

Determine the normal stress in each member of the truss structure shown.

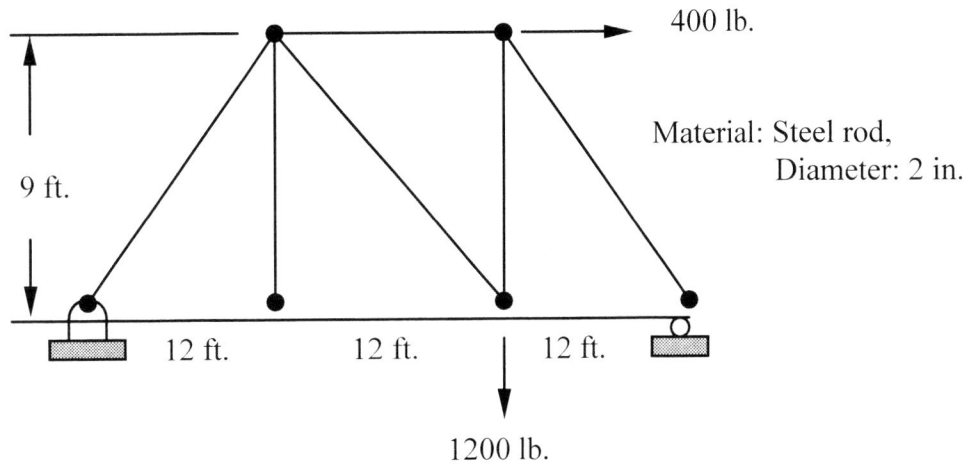

Prior to carrying out the finite element analysis, it is important to do an approximate preliminary analysis to gain some insights into the problem and as a means to check the finite element analysis results.

Free Body Diagram of the structure:

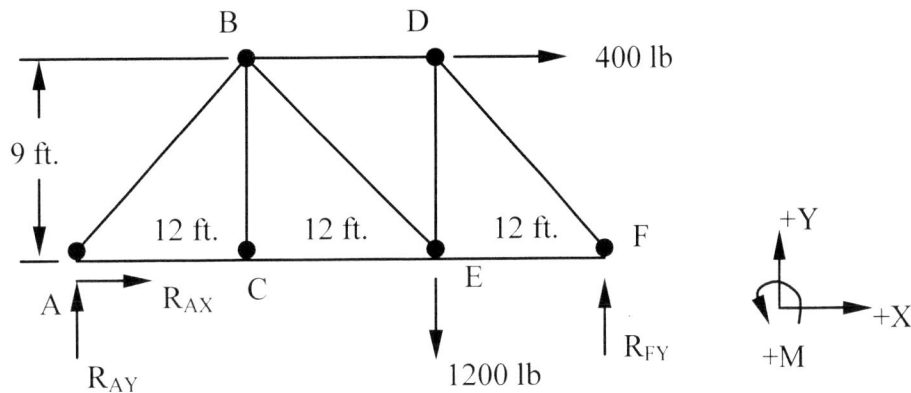

By inspection, member BC can be identified as a ZERO-FORCE member. Therefore, the stress in BC will be zero.

$$\Sigma M_A = 36 \times R_{FY} - 24 \times 1200 - 9 \times 400 = 0$$

Solving for R_{FY}:

$$R_{FY} = 900 \text{ lb.}$$

Next, using the *Joint Method* (Conventional Statics analysis technique), solve for internal forces in members DF and EF.

FBD of point F:

$$\Sigma F_x = -F_{EF} - F_{DF} \times \frac{4}{5} = 0$$

$$\Sigma F_Y = 900 + F_{DF} \times \frac{3}{5} = 0$$

F

900 lb

F_{DF}

F_{EF}

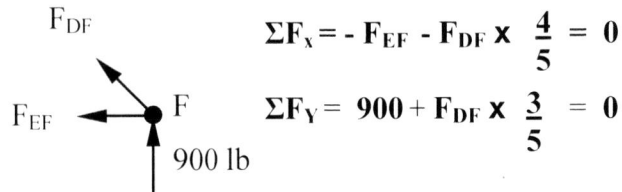

Solving the two simultaneous equations with two unknowns:

$F_{DF} = $ **- 1500 lb.**; therefore, normal stress $\sigma_{DF} = -1500/\pi = $ **477.5 psi**

$F_{EF} = $ **- 1200 lb.**; therefore, normal stress $\sigma_{EF} = -1200/\pi = $ **382 psi**

➢ It is not necessary to solve the entire problem by hand. We will compare the three results we have calculated so far to the computer solution in the following sections.

4.3 Starting *I-DEAS*

1. Login to the computer and bring up *I-DEAS 9*. In the *I-DEAS Start* window, start a new model file by filling in and selecting the items as shown below:

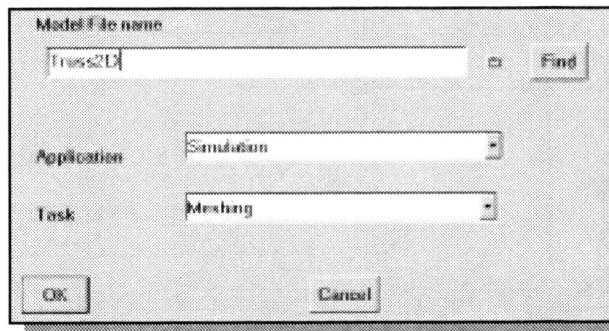

Model File name

Truss2D

Find

Application   Simulation

Task   Meshing

OK   Cancel

2. After you click **OK**, a *warning window* will appear indicating a new model file will be created. Click **OK** to exit the window and proceed to create the new file.

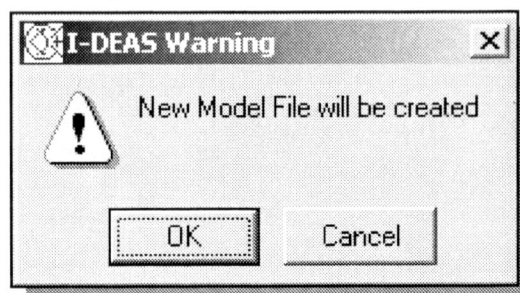

I-DEAS Warning   ✕

! New Model File will be created

OK   Cancel

4.4 Units Setup

When starting a new model, the first thing we should do is to determine the set of units we would like to use. *I-DEAS* displays the default set of units in the list window.

1. Use the left-mouse-button and select the **Options** menu in the icon panel.

2. Select the **Units** option.

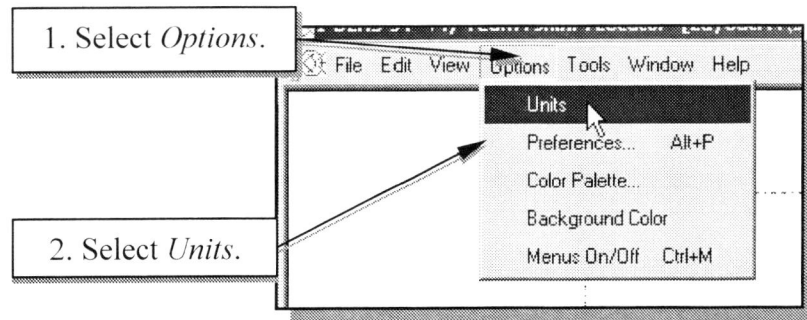

3. Inside the graphics window, pick **Inch (pound f)** from the pop-up menu. The set of units is stored with the model file when you save.

❖ This should be the first thing you set up when doing finite element analysis; the units you use <u>MUST</u> be consistent throughout the analysis.

4.5 Create an FE Model

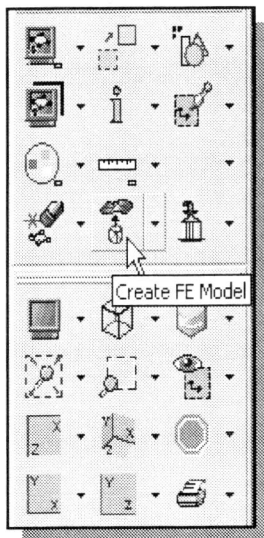

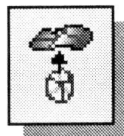

1. Choose **Create FE Model** in the icon panel. (The icon is located in the last row of the application specific icon panel.) The *FE Model Create* window appears.

2. In the *FE Model Create* window, enter **Truss2D** in the *FE Model Name* box.

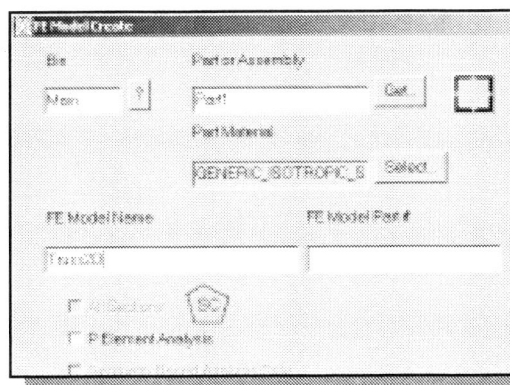

3. Click on the **OK** button to accept the settings.

4. In the *I-DEAS* warning window, click on the **OK** button to create a new part.

4.6 Workplane Appearance

The workplane is a construction tool; it is a coordinate system that can be moved in space. The size of the workplane display is only for our visual reference, since we can sketch on the entire plane, which extends to infinity.

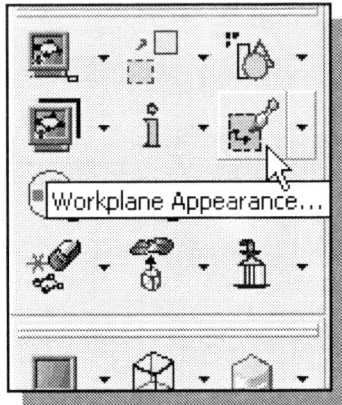

1. Choose **Workplane Appearance** in the icon panel. (The icon is located in the second row of the application icon panel.) The *Workplane Attributes* window appears.

2. Toggle **on** the *Display Border* switch, if it is not turned on already, as shown.

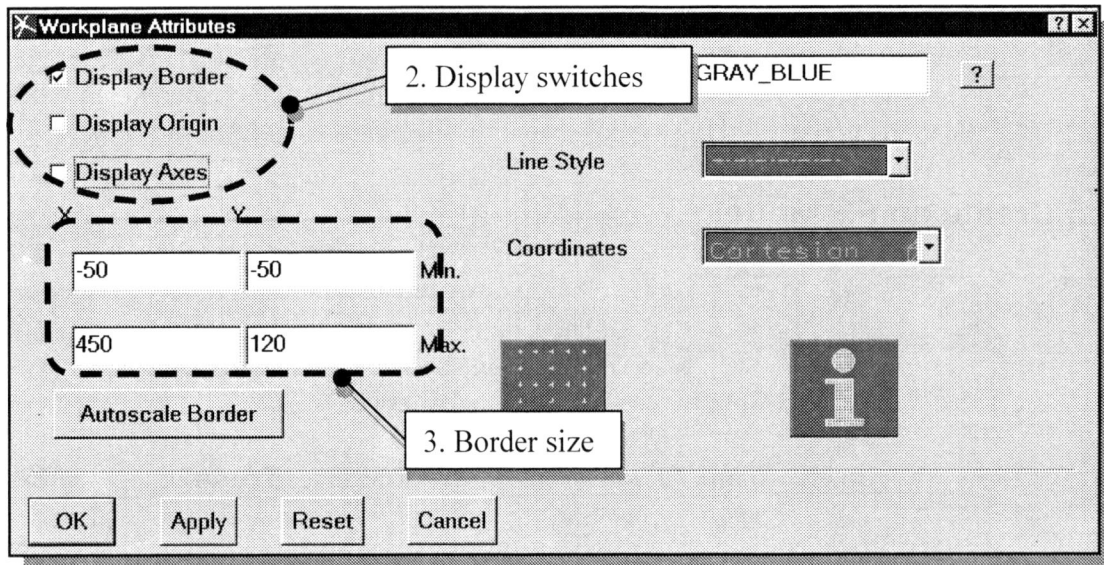

3. Adjust the **workplane border size** by entering the *Min. & Max.* values as shown in the figure above.

4. Click on the **OK** button to exit the *Workplane Attributes* window.

❖   The structure is 36 feet long and 9 feet tall. The values we entered for the border size are in inches since we set the system units to inches. Again, the units we use <u>MUST</u> be consistent throughout the analysis.

4.7 Creating Node Points by entering coordinates

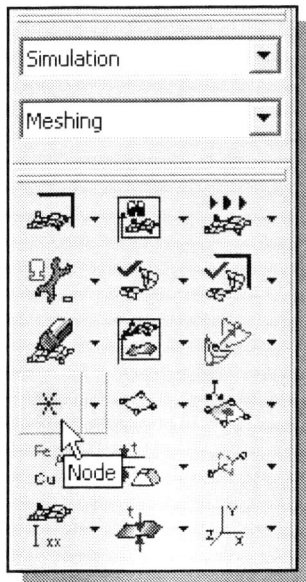

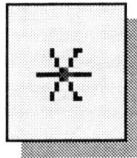

1. Choose **Node** in the icon panel. (The icon is located in the fourth row of the task specific icon panel. If the icon is not on top of the stack, press and hold down the left-mouse-button on the displayed icon to display all the choices. Slide the cursor up and down to switch to different options. Select the icon by releasing the left-mouse-button when the icon is highlighted.) The *Node* window appears.

2. Click on the **OK** button to accept the default settings, which will begin with *"Node Number 1"* and automatically increment the node numbers.

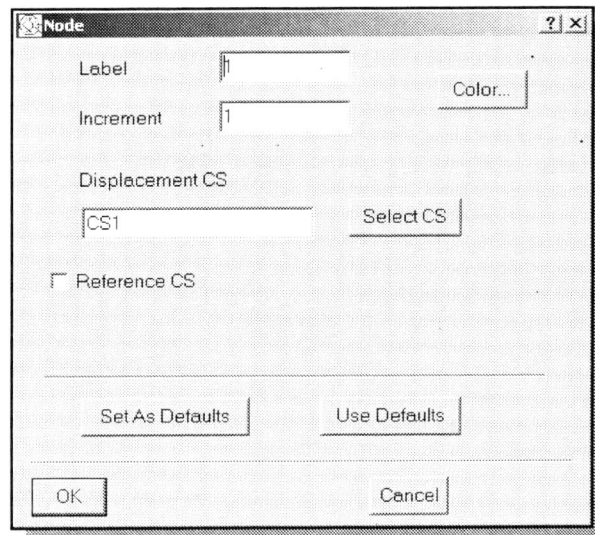

3. At the prompt window, the messages *"Enter Node 1 location"* and *"Enter X, Y, Z (0.0,0.0,0.0)"* are displayed. Click inside the prompt window with the left-mouse-button and press the **ENTER** key once to place *Node 1* at the origin.

4. For *Node 2*, enter **12*12,12*9,0**. *I-DEAS* will automatically do the calculations and position the node at the entered coordinates.

5. For *Node 3*, enter **12*24,12*9,0**.

6. Pick **Done** in the popup menu to end the *Node* command.

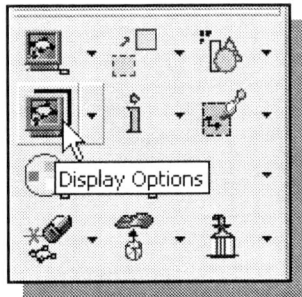

7. Choose ***Display Options*** in the icon panel. (The icon is located in the second row of the application specific icon panel.).

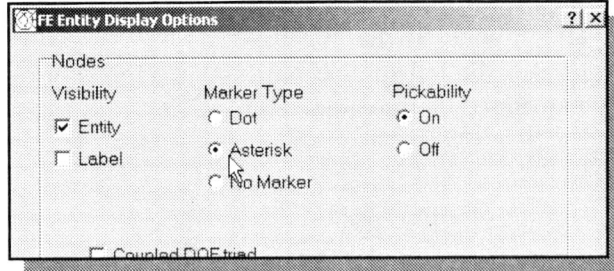

8. In the *FE Entity Display Options* window, set the *Marker Type* for *Nodes* to ***Asterisk*** as shown.

9. Click on the **OK** button to accept the settings and exit the *FE Entity Display Options* window.

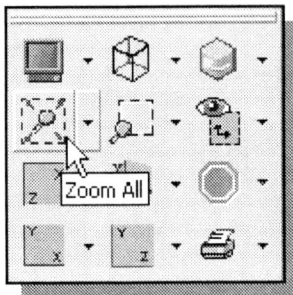

10. Choose ***Zoom-All*** in the display icon panel.

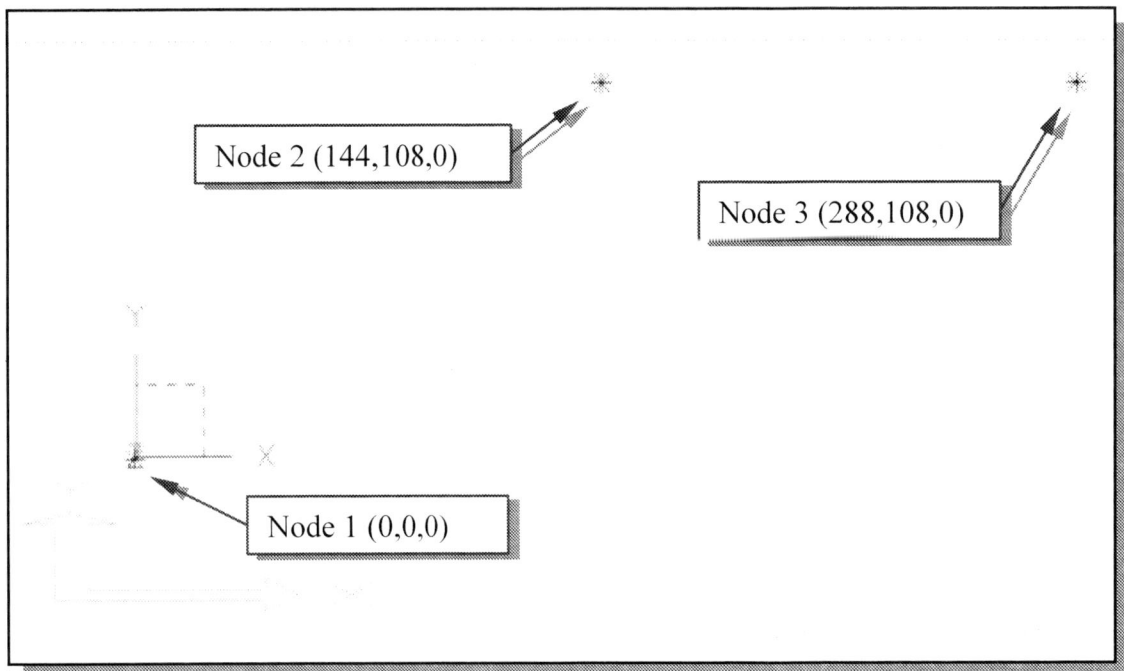

4.8 Making Copies of Node Points

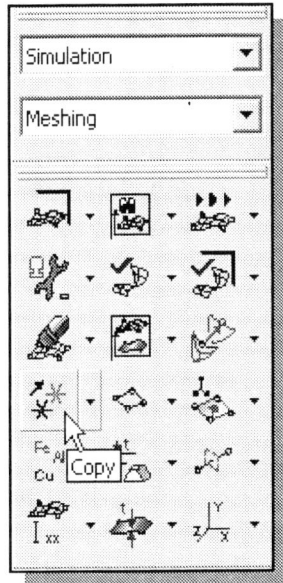

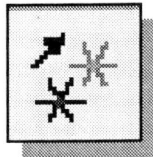

1. Choose **Copy Node** in the icon panel. (The icon is located in the same location as the *Node* icon.) The message "*Pick Nodes*" is displayed in the prompt window.

2. Pick **Node 1** as shown in the above figure. Move the cursor near the origin of the coordinate system and select *Node Point 1* when the **N1** symbol is displayed.

3. The message "*Pick Nodes (Done)*" is displayed in the prompt window. Press the **ENTER** key to accept the selection.

4. The message "*Enter number of copies (1)*" is displayed in the prompt window. Enter **3** to make three copies of *Node Point 1*.

5. The message "*Enter node start label, inc (4,1)*" is displayed in the prompt window. Press the **ENTER** key to accept the default setting and continue with the *Copy* option.

6. The message "*Enter delta X, Y, Z (0.0,0.0,0.0)*" is displayed in the prompt window. Enter **12*12,0,0** to make the copies along the X-axis.

7. The message "*OK to keep these additions (Yes)*" is displayed in the prompt window. Press the **ENTER** key to continue.

8. Choose **Zoom-All** in the display viewing icon panel. Your screen should appear as shown below.

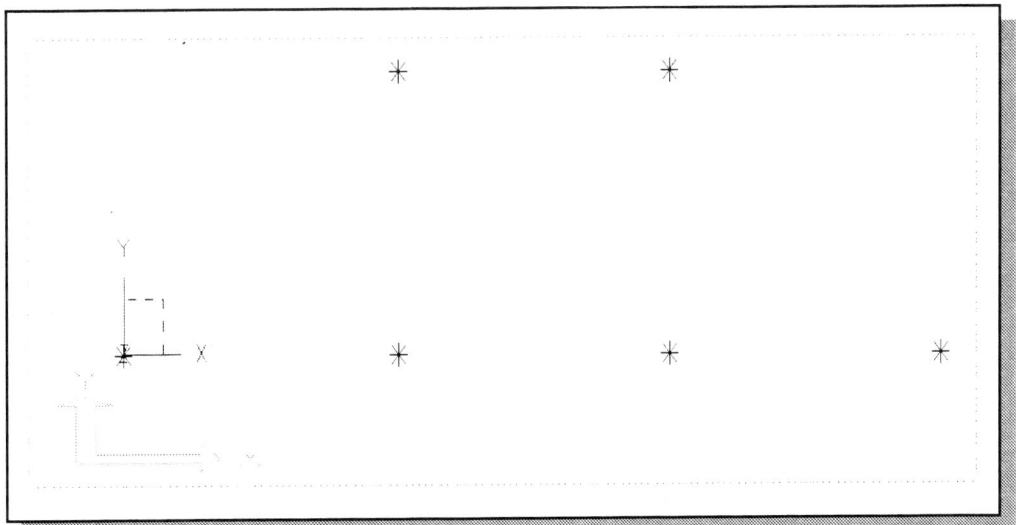

4.9 Material Property Table

Before creating elements, we will first set up a *Material Property* table. The *Material Property* table contains general material information, such as *Modulus of Elasticity, Poisson's Ratio,* etc.

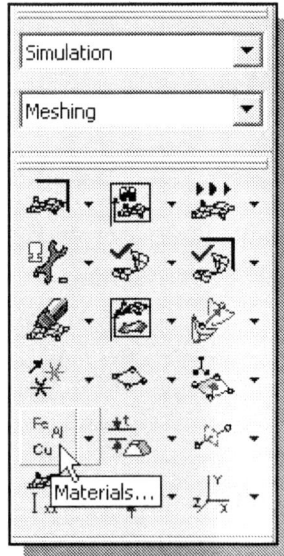

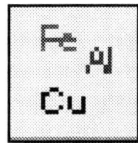

1. Choose **Materials** in the icon panel. (The icon is located in the fifth row of the task icon panel.) The *Materials* window appears.

2. Choose **Create** in the *Materials* window. The *Create Material* window appears.

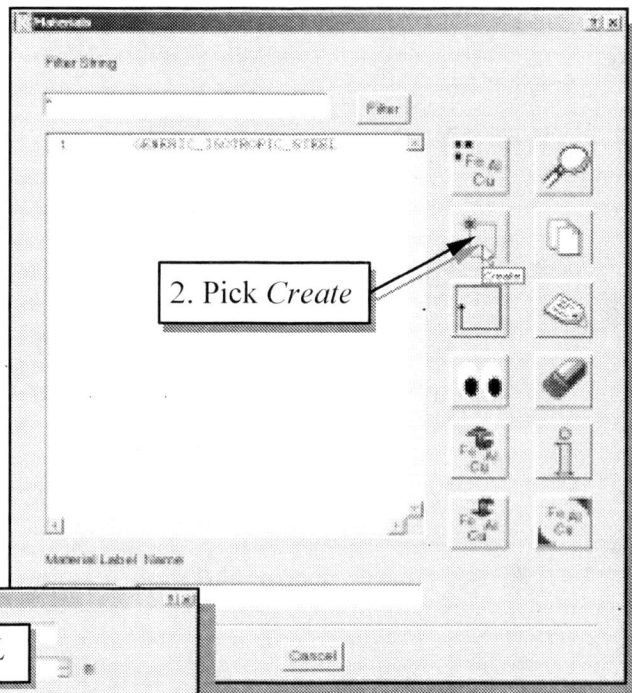

2. Pick *Create*

3. Type **STEEL** in the *Material Name* box.

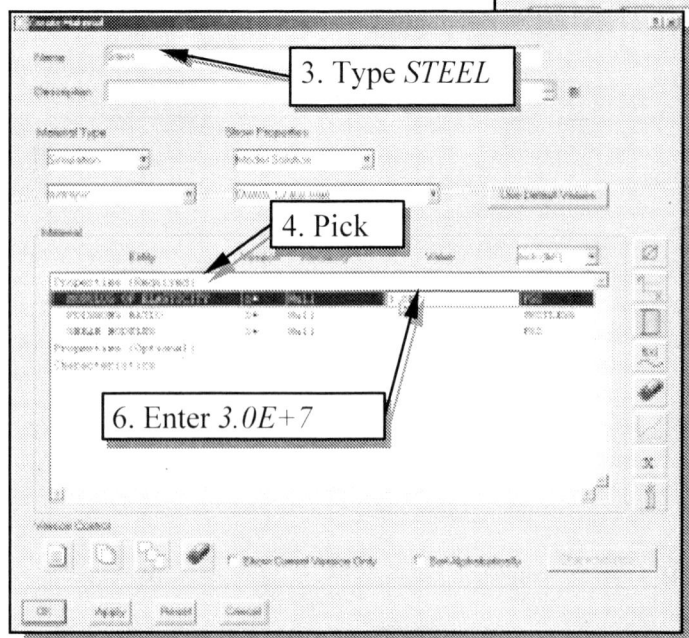

3. Type *STEEL*

4. Pick

6. Enter *3.0E+7*

4. Pick **Modulus of Elasticity** in the *Properties* list.

5. Click the **Value** box of the selected property.

6. Type **3.0E+7** in the property *Value* box.

7. Repeat the above steps and pick **Poisson's Ratio** in the *Properties* list.

8. Enter **0.3** in the property *Value* box.

Material					
Entity	Version	Variability		Value	Inch (lbF) ▼
Properties (Required)					
MODULUS OF ELASTICITY	1*	Constant		3.0e7	PSI
POISSONS RATIO	1*	Constant		0.3	UNITLESS
SHEAR MODULUS	1*	Null			PSI
Properties (Optional)...					
Characteristics...					

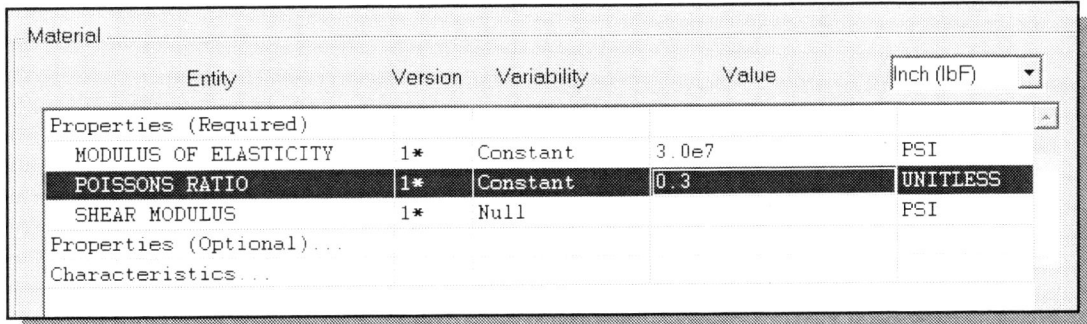

9. Click on the **OK** button to exit the *Create Material* window.

10. Click on the **OK** button to exit the *Materials* window.

4.10 Setting Up an Element Cross Section

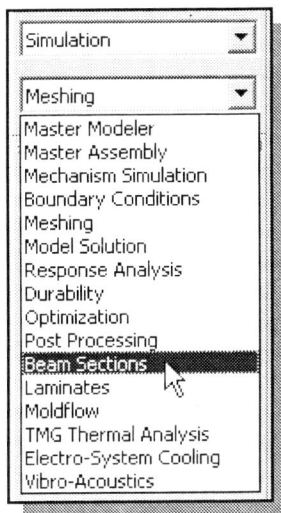

1. Switch to the **Beam Sections** task by selecting the task menu in the icon panel as shown.

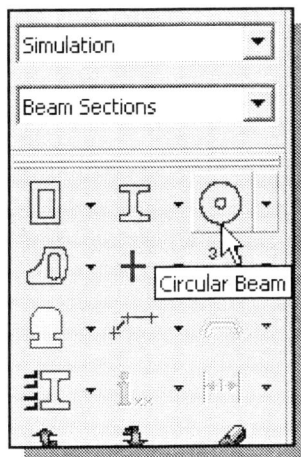

2. Choose **Circular Beam** in the icon panel. (The icon is located in the first row of the task icon panel. If the icon is not on top of the stack, press and hold down the left-mouse-button on the displayed icon to display all the choices.) The message "*ENTER outside diameter (1.0)*" is displayed.

3. Enter **2.0** for the outside diameter.

4. Enter **0.0** for the inside diameter.

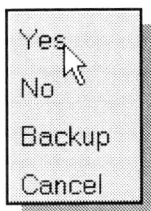

5. Pick **Yes** in the popup menu to complete creating the cross section.

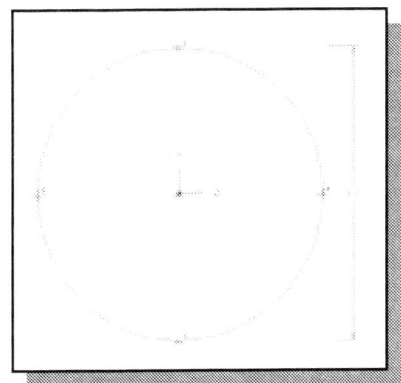

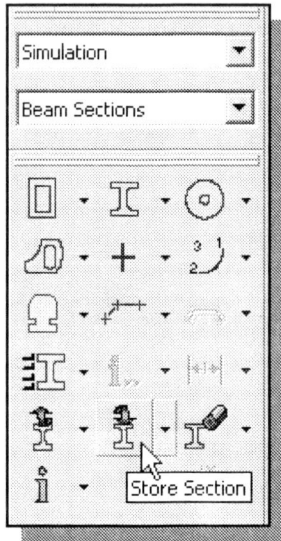

6. Choose **Store Section** in the icon panel. (The icon is located in the fifth row of the task icon panel.) The message *"Enter beam cross sect prop name or no (1-CIRCULAR 2.0 x 0.0)."* is displayed in the prompt window.

7. Press the **ENTER** key to accept the default name *(1-CIRCULAR 2.0 x 0.0)*.

4.11 Creating Elements

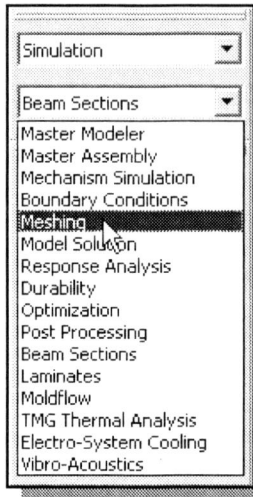

1. Switch back to the **Meshing** task by selecting the Meshing task in the task menu as shown.

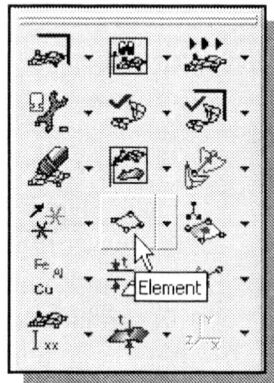

2. Choose **Element** in the icon panel. (The icon is located in the fourth row of the task specific icon panel.) The *Element* window appears.

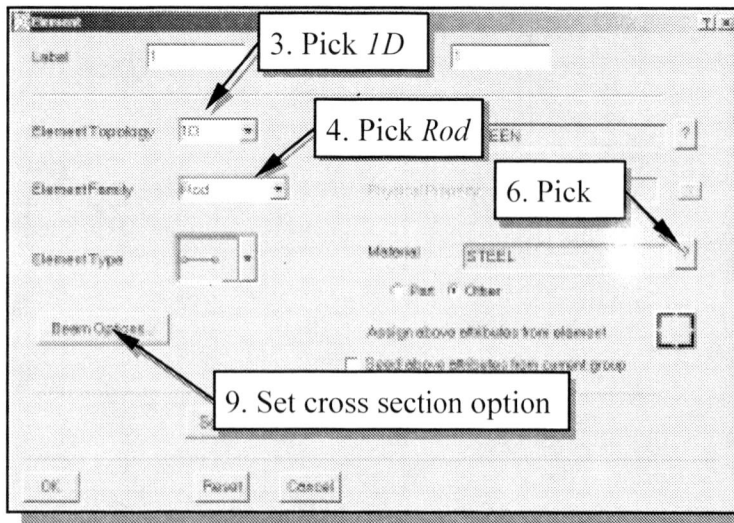

3. Pick *1D*

4. Pick *Rod*

6. Pick

9. Set cross section option

3. In the *Element* window, select **1D** element.

4. Pick **Rod** from the *Element Family* list.

5. Set the *Material* option to *Other* as shown.

6. Click on the [**?**] button to set the material property.

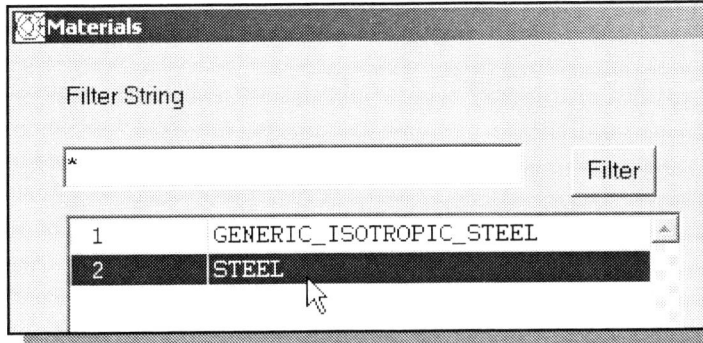

7. Choose **STEEL** in the *Materials* window.

8. Click on the **OK** button to accept the selection.

9. Click on the **Beam Options** button. The *Beam Options* window appears.

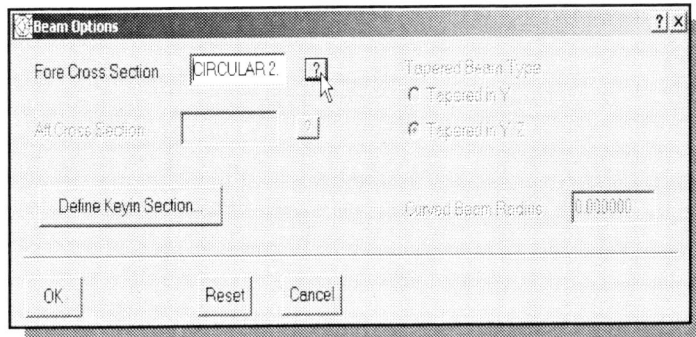

10. Click on the [**?**] button and confirm the circular section we just created is selected.

11. Click on the **OK** button to exit the *Fore Cross Section* window.

12. Click on the **OK** button to exit the *Beam Options* window.

13. Click on the **OK** button to exit the *Element* window. The message "*Pick Nodes*" is displayed in the prompt window.

14. *I-DEAS* expects us to select two node points to create an element between. Create the 9 elements as shown below. Note the **Nx** symbol as the cursor is moved next to the node points.

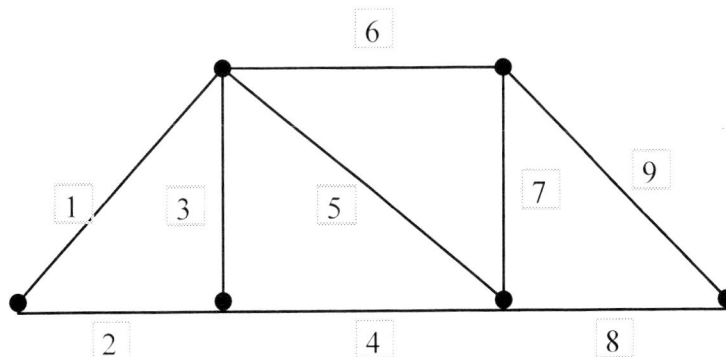

15. Hit the **ENTER** key to end the *Element* command.

➤ Note that *truss members* are two-force elements. The loads can only be applied at the nodes. The base of the structure contains **THREE** elements, **NOT** one element.

4.12 Applying Boundary Conditions

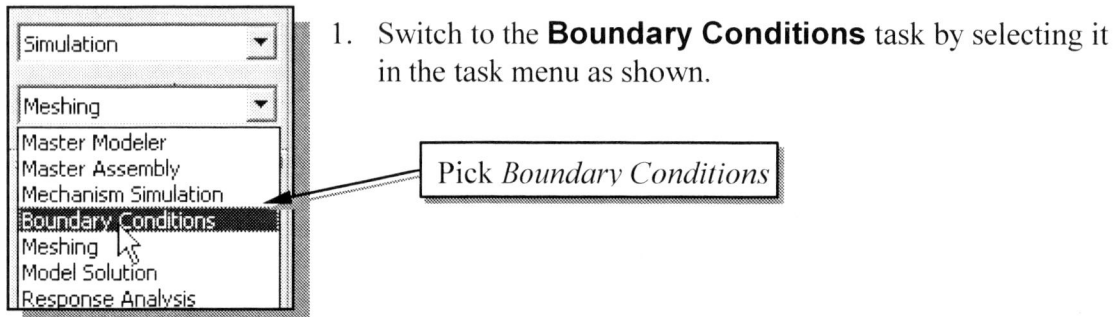

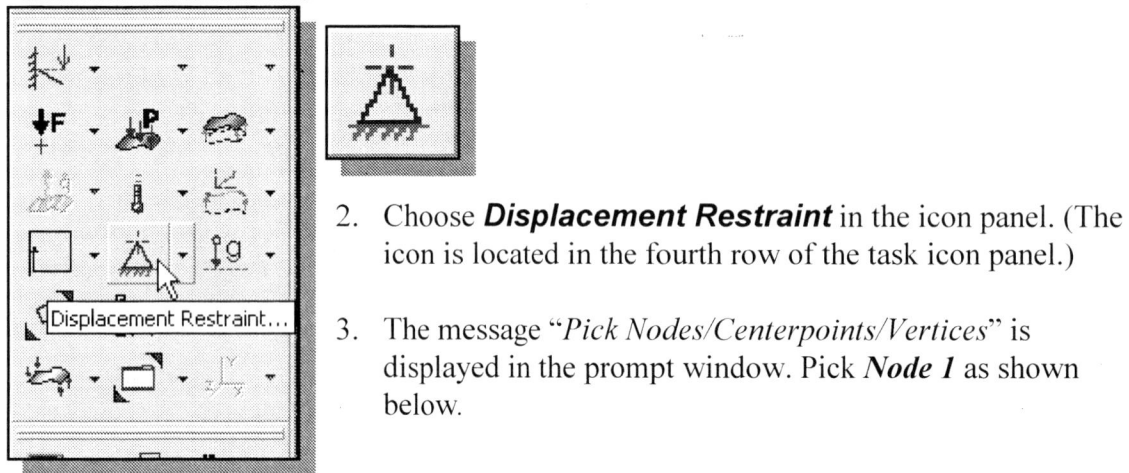

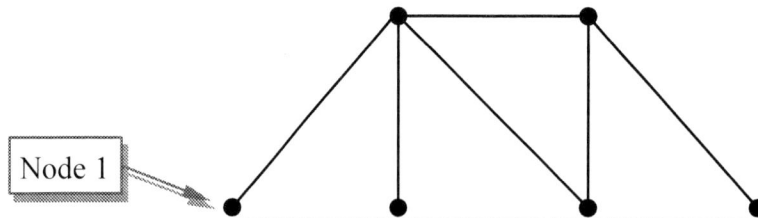

1. Switch to the **Boundary Conditions** task by selecting it in the task menu as shown.

Pick *Boundary Conditions*

2. Choose **Displacement Restraint** in the icon panel. (The icon is located in the fourth row of the task icon panel.)

3. The message "*Pick Nodes/Centerpoints/Vertices*" is displayed in the prompt window. Pick *Node 1* as shown below.

Node 1

4. Press the **ENTER** key to accept the selection. The *Displacement Restraint on Node* window appears.

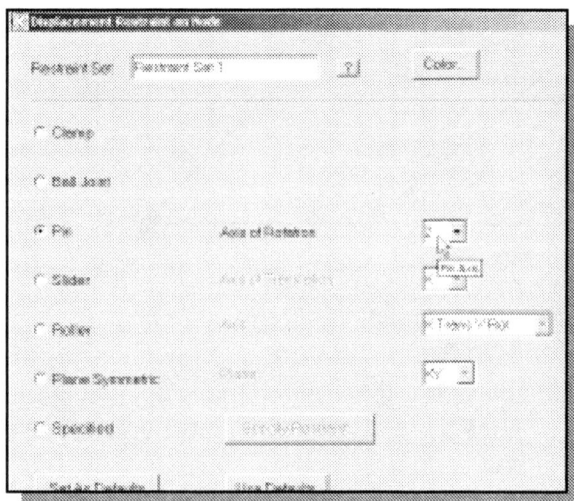

5. Pick *Pin Joint* as the type of restraint to be applied at the selected node.

6. Set *Axis of Rotation* to **Z-axis**.

7. Click on the **OK** button to exit the *Displacement Restraint on Node* window.

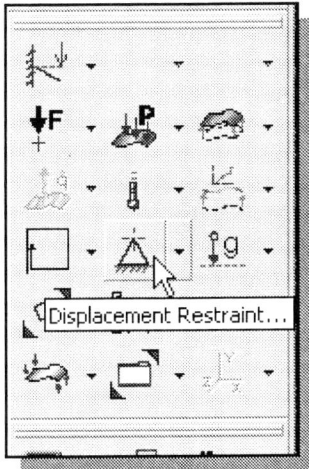

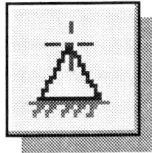

8. Choose **Displacement Restraint** in the icon panel. (The icon is located in the fourth row of the task icon panel.)

9. The message "*Pick Nodes/Centerpoints/Vertices*" is displayed in the prompt window. Pick **Node 6** as shown below.

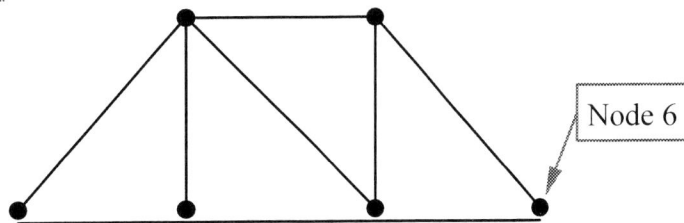

10. Press the **ENTER** key to accept the selection. The *Displacement Restraint on Node* window appears.

11. Pick **Roller** as the type of restraint to be applied at the selected node.

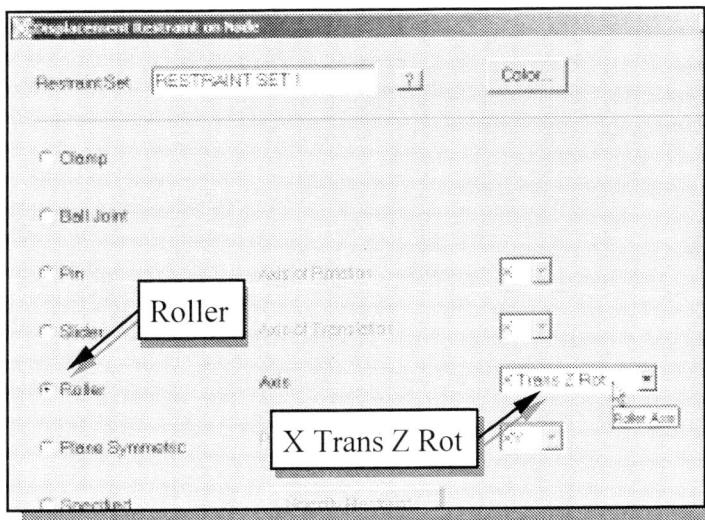

12. Set *Roller Axis* to **X Trans Z Rot** as shown.

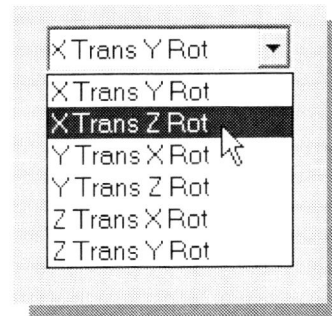

13. Click on the **OK** button to exit the *Displacement Restraint on Node* window.

➢   On your own, confirm that proper boundary conditions are applied to the *I-DEAS* FE model by examining the original structure on page 4-3.

4.13 Applying External Loads

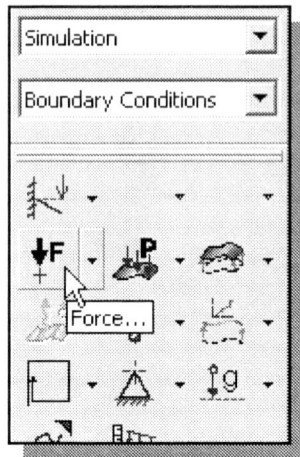

1. Choose **Force on Node** in the icon panel. (The icon is located in the second row of the task icon panel.)

2. The message "*Pick entities*" is displayed in the prompt window. Pick *Node 5* as shown below.

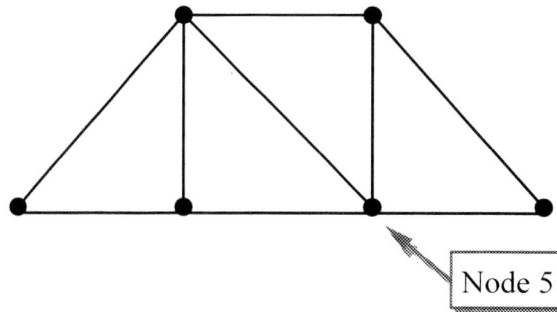

Node 5

3. Press the **ENTER** key to accept the selection. The *Force on Node* window appears.

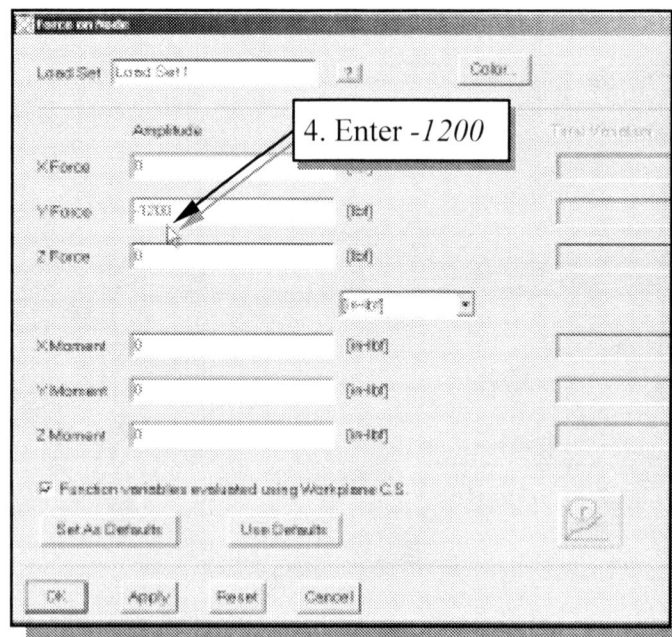

4. Enter -*1200*

4. The load at *Node 5* is 1200lb. in the negative Y direction. Enter **-1200** in the *Y Force* box.

5. Click on the **OK** button to exit the *Force on Node* window.

Simulation

Boundary Conditions

Force...

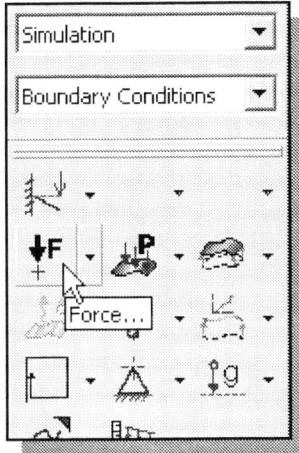

6. Choose **Force On Node** in the icon panel. (The icon is located in the second row of the task icon panel.)

7. The message "*Pick entities*" is displayed in the prompt window. Pick *Node 3* as shown below.

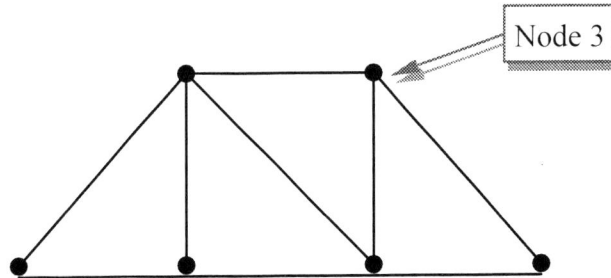

Node 3

8. Press the **ENTER** key to continue. The *Force on Node* window appears.

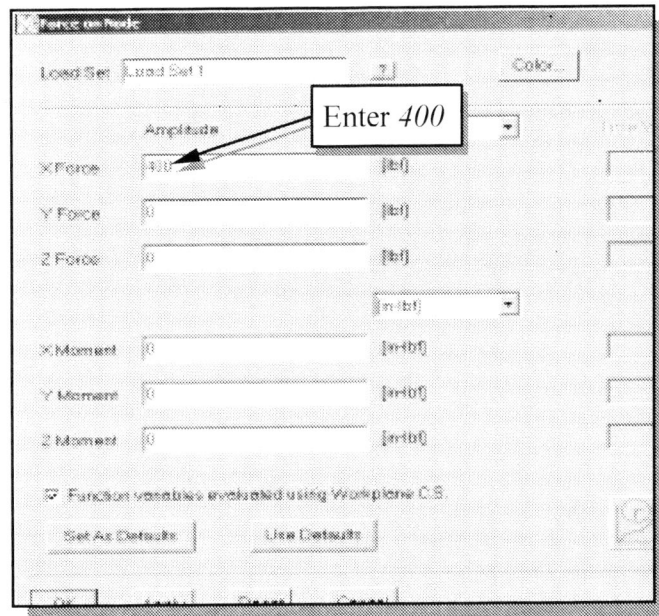

Enter *400*

9. The load at *Node 3* is 400 lb. in the X direction. Enter **400** in the *X Force* box.

10. Click on the **OK** button to exit the *Force on Node* window.

➤ On your own, confirm that proper loads are applied to the *I-DEAS* FE model by examining the original structure on page 4-3.

4.14 Running the Solver

1. Switch to the **Model Solution** task by selecting the task in the task menu as shown.

   Pick *Model Solution*

2. Choose **Solution Set** in the icon panel. (The icon is located in the first row of the task icon panel.) The *Manage Solution Sets* window appears.

3. Click on the **Create** button. The *Solution Set* window appears.

4. Click on the **Output Selection** button. The *Output Selection* window appears.

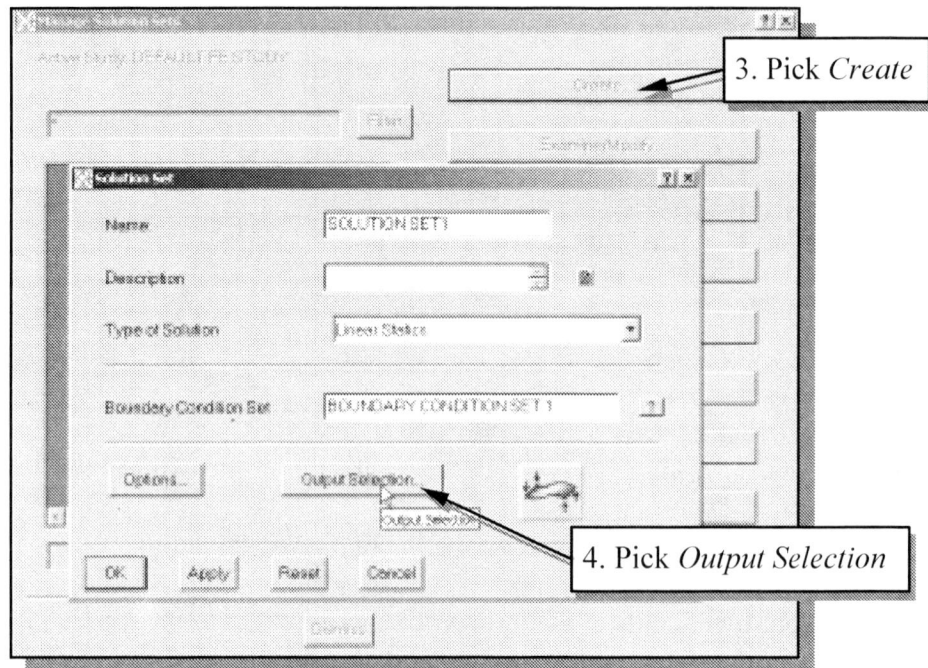

3. Pick *Create*

4. Pick *Output Selection*

5. Pick **Reaction Forces** in the *Output* list.

6. Press the left-mouse-button on the *Store/List* button to show the different output options that are available. We will accept the default output option, **Store**.

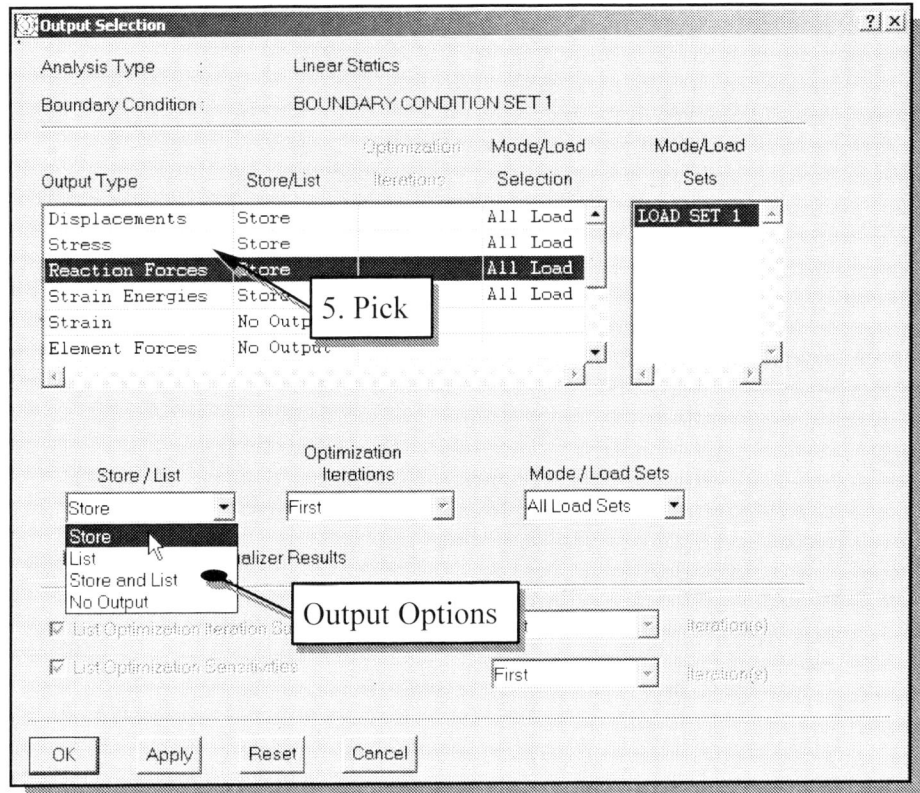

7. Click on the **OK** button to exit the *Output Selection* window.

8. Click on the **OK** button to exit the *Solution Set* window.

9. Pick **Dismiss** to exit the *Manage Solution Sets* window.

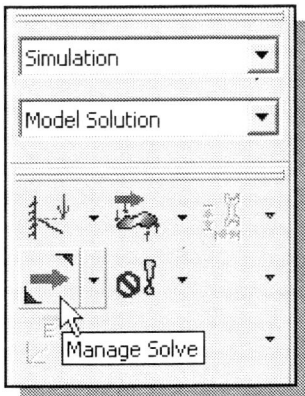

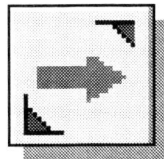

10. Choose **Manage Solve** in the icon panel. (The icon is located in the second row of the task icon panel.) The *Solve* window appears.

11. In the *Solve* window, pick the **Solve** icon to find the solutions.

➢ *I-DEAS* will begin the solving process. <u>DO NOT</u> close any windows. Any errors or warnings are displayed in the list window.

4.15 Viewing the results

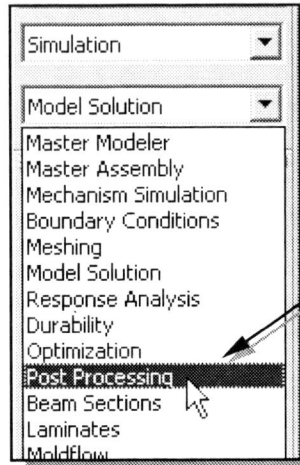

1. Switch to the **Post Processing** task by selecting it from the task menu as shown.

Pick *Post Processing*

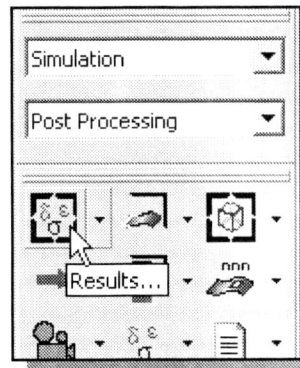

2. Choose **Results Selection** in the icon panel. (The icon is located in the first row of the task icon panel.)

3. The *Results Selection* window appears. Pick **Reaction Force** from the list.

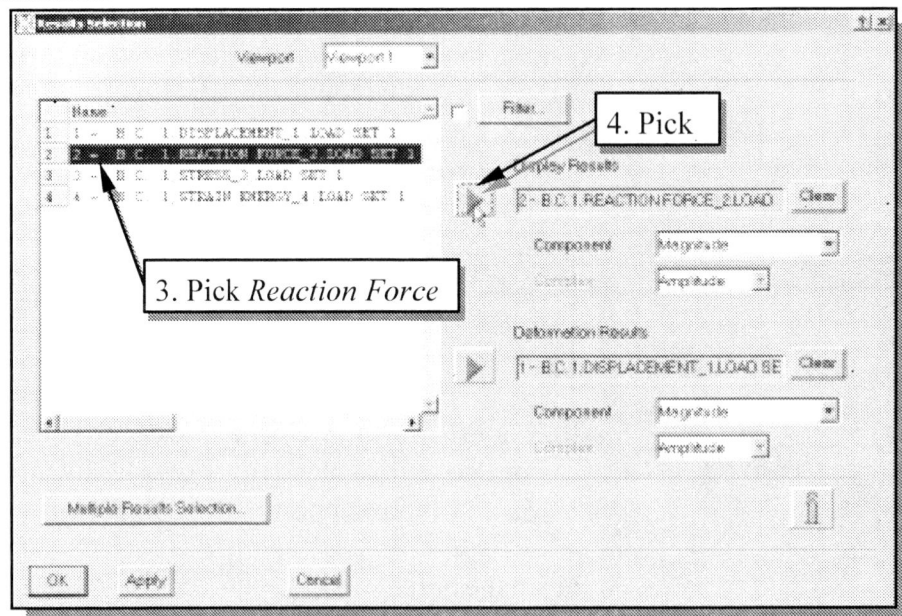

4. Pick

3. Pick *Reaction Force*

4. Click on the **Triangle** button to set the *Reaction Force* as the *Display Results*.

5. Click on the **OK** button to exit the *Results Selection* window.

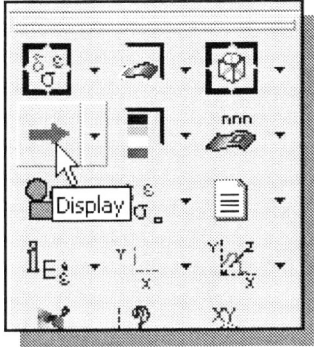

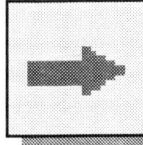

6. Choose **Results Display** in the icon panel. (The icon is located in the second row of the task icon panel.)

7. The message "*Pick elements*" is displayed in the prompt window. Pick **Element 8** as shown below; note the E8 symbol displayed near the center of the element.

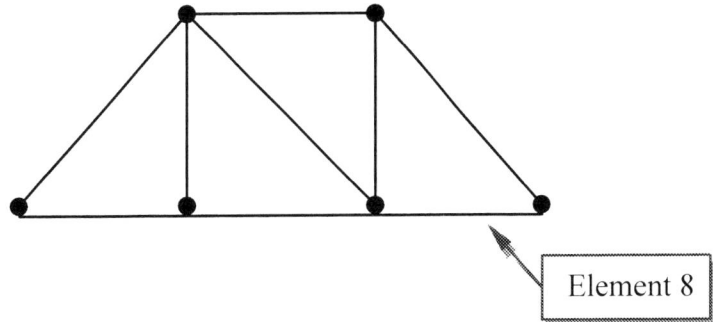

Element 8

8. Press the **ENTER** key to accept the selection and display any reaction force associated with this element.

❖ The *I-DEAS* solution for the reaction at *Node 6* is 900 lb., which matches with our hand calculation (see section **4.2 Preliminary Analysis**).

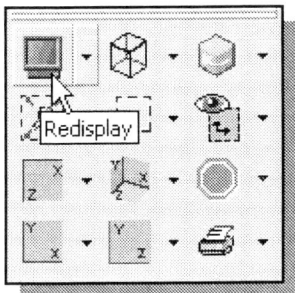

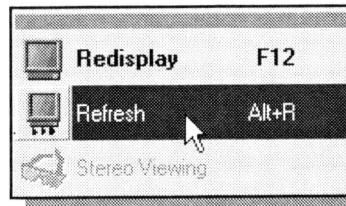

9. In the icon panel, press and hold the left-mouse-button on the **Redisplay** icon and select **Refresh**.

❖ Use **Refresh** to redisplay the FEA model.

➤ Next, we will examine the stresses in the elements.

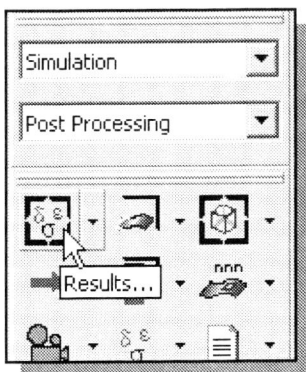

10. Choose **Results Selection** in the icon panel. (The icon is located in the first row of the task icon panel.)

11. The *Results Selection* window appears. Pick **Stress** from the list.

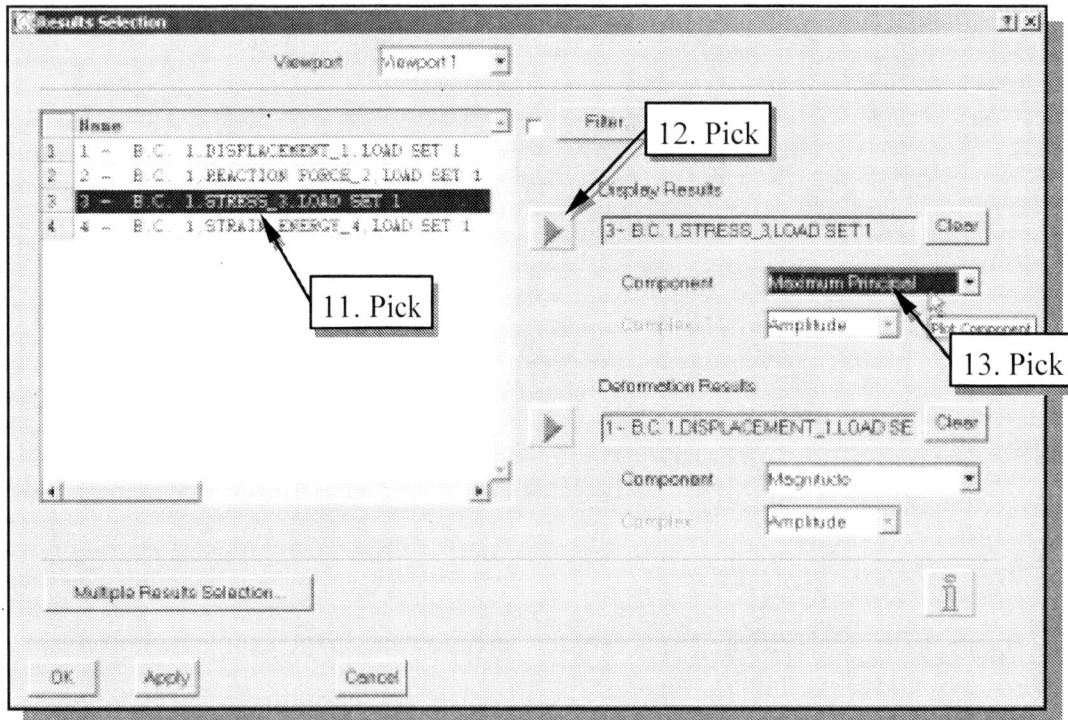

12. Click on the **Triangle** button to set the selection.

13. Select **Maximum Principal** in the component list.

14. Click on the **OK** button to exit the *Results Selection* window.

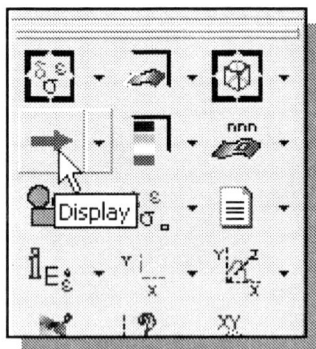

15. Choose ***Results Display*** in the icon panel. (The icon is located in the second row of the task icon panel.)

16. The message "*Pick elements*" is displayed in the prompt window. Pick ***Element 9*** as shown below.

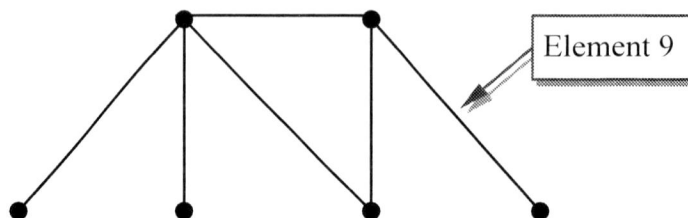

17. Press the **ENTER** key to accept the selection to display the maximum principal stress force associated with this element.

18. A warning window appears, which is with regard to the fast contour option. Click **OK** to proceed.

➢ The *I-DEAS* solution is **477 psi.**, which again matches our hand calculation (see section **4.2 Preliminary Analysis**).

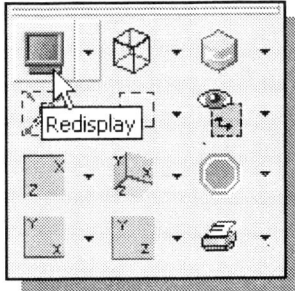

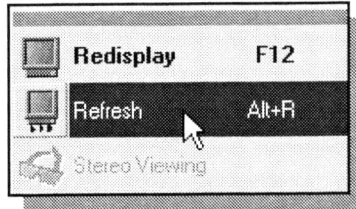

19. Choose **Refresh** in the icon panel.

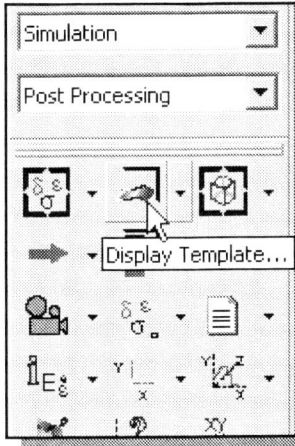

20. Choose **Display Template** in the icon panel. (The icon is located in the first row of the task icon panel.)

21. Click on the **Contour** button.

22. Click on *Fast Display* to switch *off* the *Fast Display* option as shown.

23. Click the **OK** button to exit the *Contour Options* window.

24. Click the **OK** button to exit the *Display Template* window.

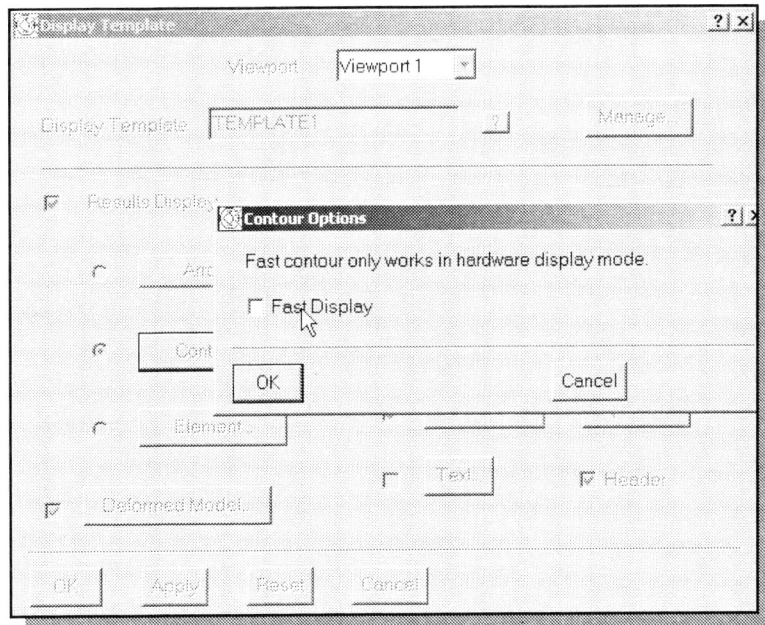

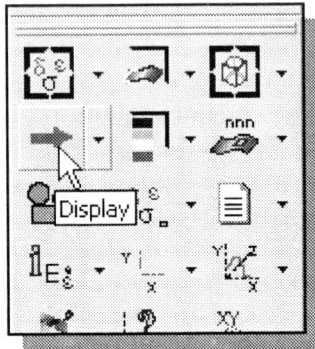

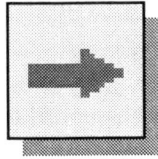

25. Choose **Results Display** in the icon panel. (The icon is located in the second row of the task icon panel.)

26. The message "*Pick elements*" is displayed in the prompt window. Press the **ENTER** key without selecting any elements. *I-DEAS* will automatically select all elements and display the results. Does the maximum normal stress occur in the element as you have expected?

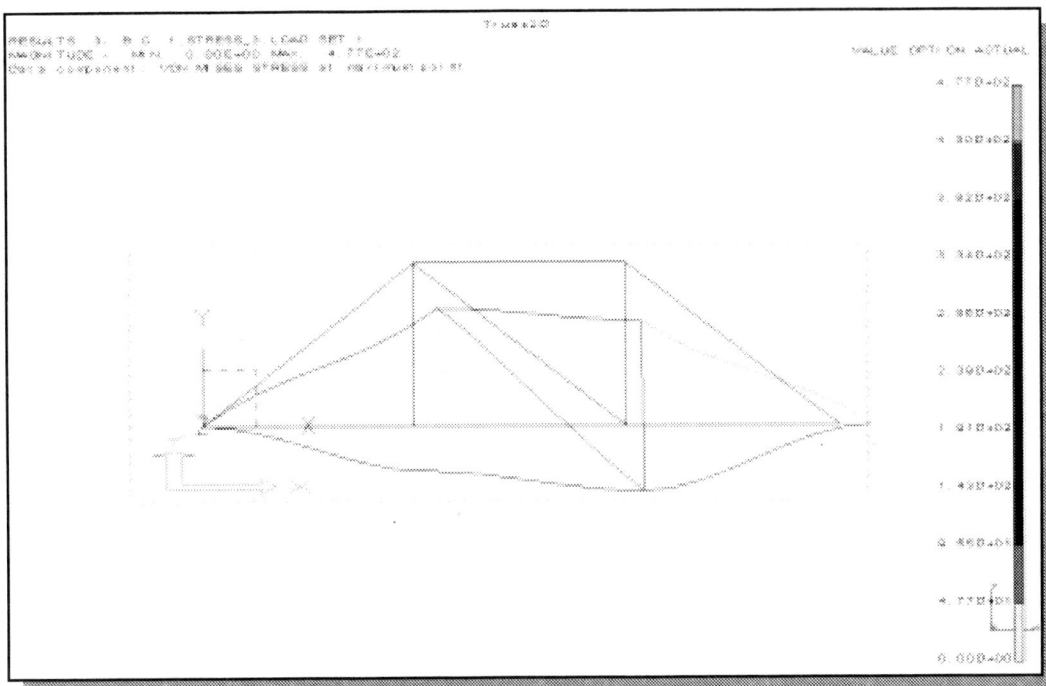

➢ On your own, examine the stresses developed in the other elements and compare them to the preliminary analysis we did at the beginning of this chapter. What is the stress in member BC? Does *I-DEAS* display the stress value for the zero-force member correctly? Which member has the highest stress and what is the value of the stress?

Questions:

1. Describe the typical finite element analysis steps.

2. Nodes and elements are created in which *I-DEAS* task software?

3. Can loads be applied to the midpoint of a truss element?

4. How do you identify the ZERO-FORCE members in a truss structure?

5. What are the basic assumptions, as defined in Statics, for truss members?

6. Identify the following commands:

   (a)

   (b)

   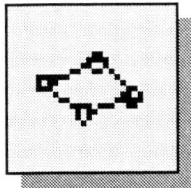

   (c)

Exercises:

Determine the normal stress in each member of the truss structures shown.

1.   Material: Steel rod,
          Diameter: 2.5 in.

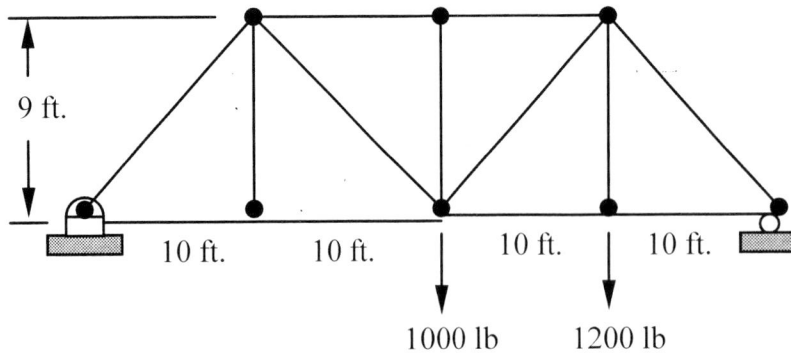

2.   Material: Steel rod,
          Diameter: 2 cm.

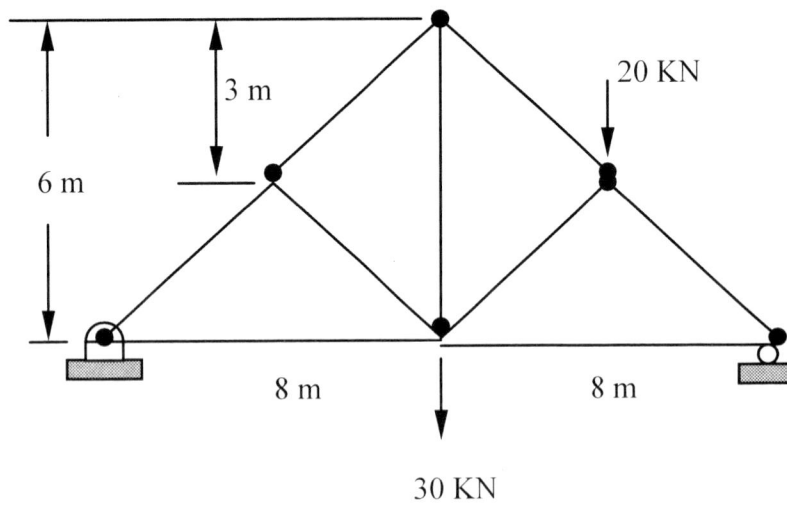

Chapter 5
Three-Dimensional Truss Analysis

Learning Objectives

When you have completed this lesson, you will be able to:
- ◆ Determine the Number of Degrees of Freedom in Elements.
- ◆ Create 3D FE Truss Models.
- ◆ Apply proper boundary conditions to FE Models
- ◆ Use I-DEAS Solver for 3D Trusses.
- ◆ Setup I-DEAS to determine Axial Loads.
- ◆ Use the Cascading Menus System.

5.1 Three-Dimensional Coordinate Transformation Matrix

For truss members positioned in a three-dimensional space, the coordinate transformation equations are more complex than the transformation equations for truss members positioned in two-dimensional space. The coordinate transformation matrix is necessary to obtain the global stiffness matrix of a truss element.

1. The global coordinate system (X, Y and Z axes), chosen for representation of the entire structure.
2. The local coordinate system (*X, Y* and *Z* axes), with the *X*-axis aligned along the length of the element.

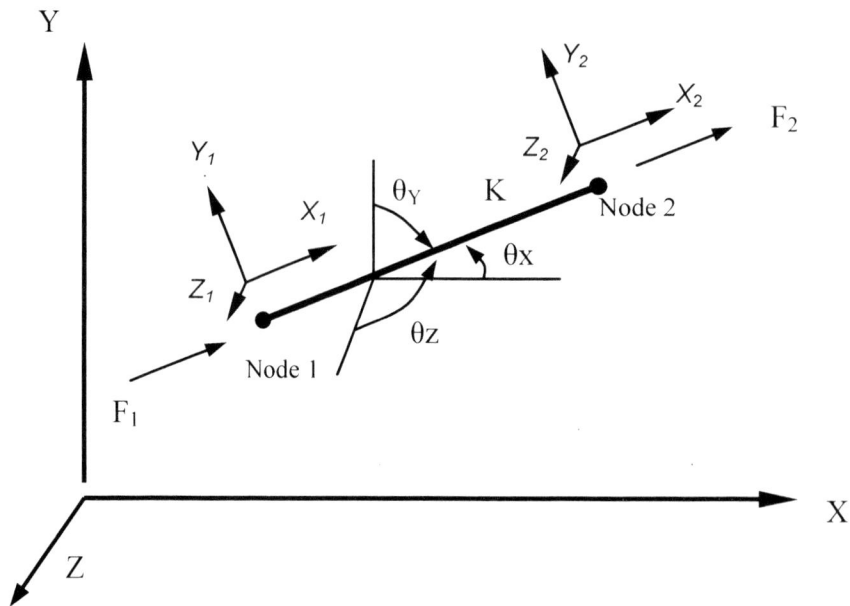

Since truss elements are two-force members, the displacements occur only along the local *X*-axis. The GLOBAL to LOCAL transformation matrix can be written as:

$$
\left\{ \begin{array}{c} X_1 \\ X_2 \end{array} \right\} = \left[\begin{array}{cccccc} \cos(\theta x) & \cos(\theta_Y) & \cos(\theta_Z) & 0 & 0 & 0 \\ 0 & 0 & 0 & \cos(\theta x) & \cos(\theta_Y) & \cos(\theta_Z) \end{array} \right] \left\{ \begin{array}{c} X_1 \\ Y_1 \\ Z_1 \\ X_2 \\ Y_2 \\ Z_2 \end{array} \right\}
$$

Local

Global

5.2 Stiffness Matrix

The displacement and force transformations can be expressed as:

$$\{X\}' = [\,l\,]\,\{X\} \quad \text{------- } \textit{Displacement transformation equation}$$

$$\{F\} = [\,l\,]\,\{F\} \quad \text{------- } \textit{Force transformation equation}$$

Combined with the local stiffness matrix, $\{F\} = [K]\{X\}$, we can then derive the global stiffness matrix for an element:

$$[K] = \frac{EA}{L}\begin{pmatrix} C_X^2 & C_X C_Y & C_X C_Z & -C_X^2 & C_X C_Y & C_X C_Z \\ & C_Y^2 & C_Y C_Z & -C_X C_Y & -C_Y^2 & -C_Y C_Z \\ & & C_Z^2 & -C_X C_Z & -C_Y C_Z & -C_Z^2 \\ & & & C_X^2 & C_X C_Y & C_X C_Z \\ & \text{Symmetry} & & & C_Y^2 & C_Y C_Z \\ & & & & & C_Z^2 \end{pmatrix}$$

where $C_X = cos(\theta_X)$, $C_Y = cos(\theta_Y)$, $C_Z = cos(\theta_Z)$.

The resulting matrix is a 6 x 6 matrix. The size of the stiffness matrix is related to the number of nodal displacements. The nodal displacements are also used to determine the number of **degrees of freedom** at each node.

5.3 Degrees of Freedom

♦ For a truss element in a one-dimensional space:

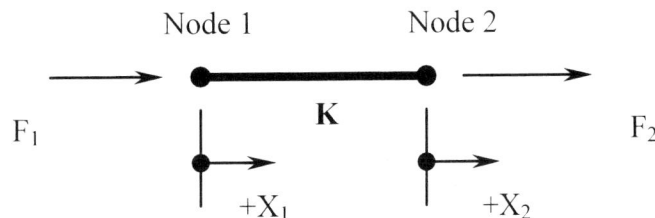

One nodal displacement at each node: one degree of freedom at each node. Each element possesses two degrees of freedom, which forms a 2 x 2 stiffness matrix for the element. The global coordinate system coincides with the local coordinate system.

♦ For a truss element in a two-dimensional space:

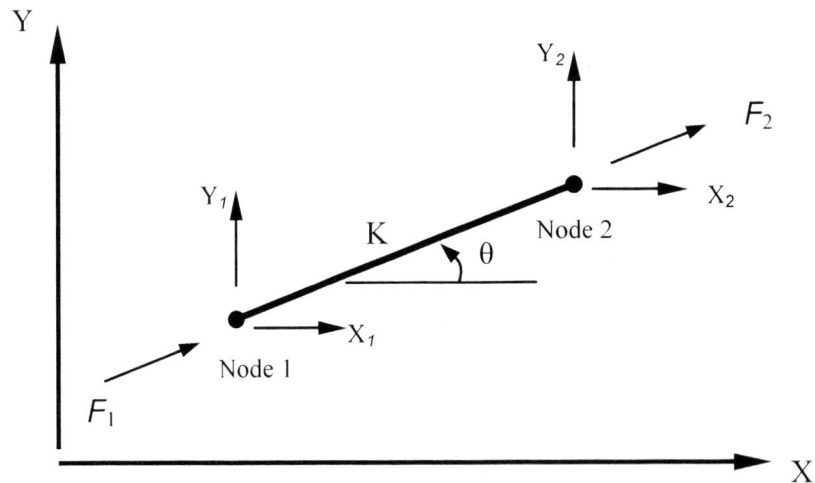

Two nodal displacements at each node: two degrees of freedom at each node. Each element possesses four degrees of freedom, which forms a 4 x 4 stiffness matrix for the element.

♦ For a truss element in a three-dimensional space:

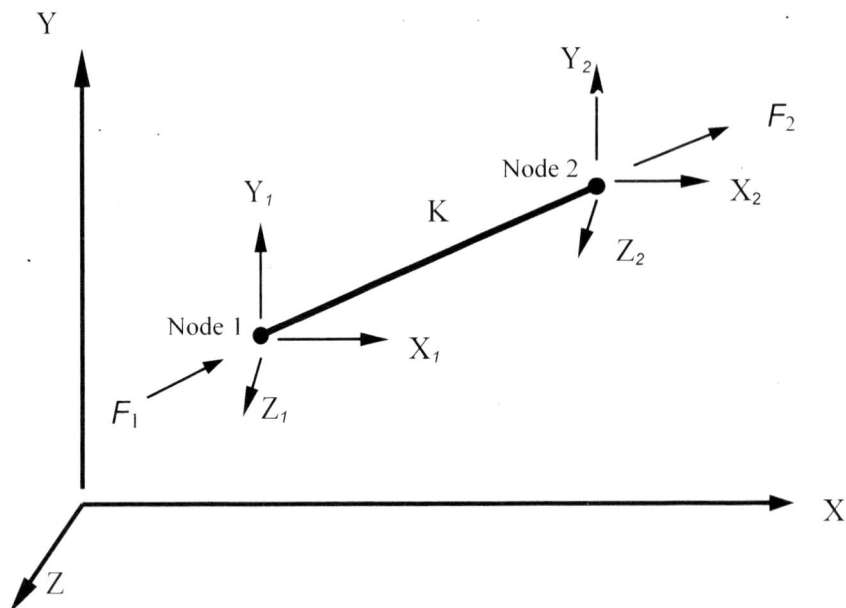

Three nodal displacements at each node: three degrees of freedom at each node. Each element possesses six degrees of freedom, which forms a 6 x 6 stiffness matrix for the element.

5.4 Problem Statement

➤ Determine the normal stress in each member of the truss structure shown. All joints are ball-joints.

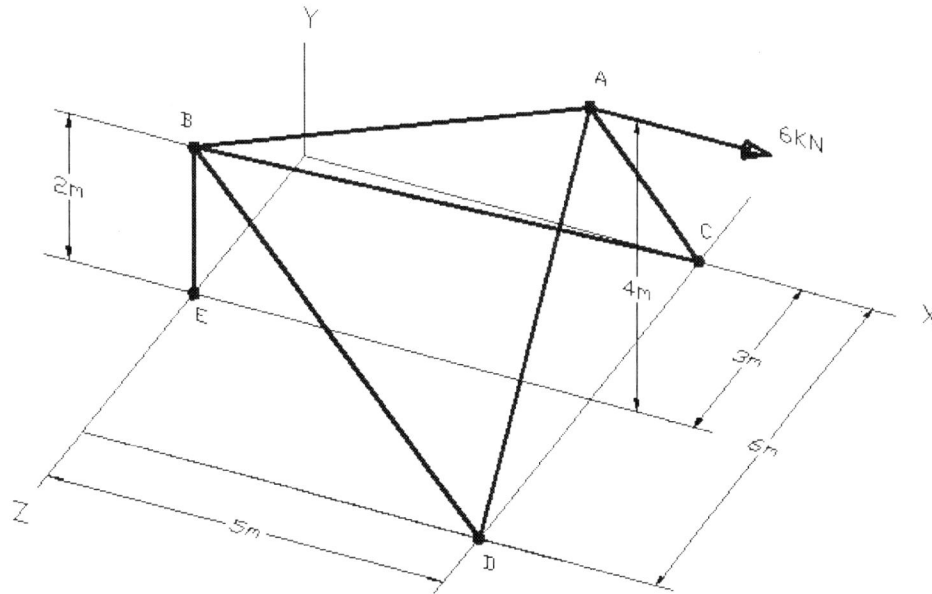

Material: Steel   Cross Section: [ ] a   a = b = 7.5mm

b

5.5 Preliminary Analysis

❖ This section demonstrates using conventional *vector algebra* to solve the three-dimensional truss problem.

The coordinates of the nodes:
     A(5,4,3), B(0,2,3), C(5,0,0), D(5,0,6), E(0,0,3)

Free Body Diagram of Node A:

Position vectors AB, AC and AD:

$$AB = (0-5) i + (2-4) j + (3-3) k = (-5) i + (-2) j$$
$$AC = (5-5) i + (0-4) j + (0-3) k = (-4) j + (-3) k$$
$$AD = (5-5) i + (0-4) j + (6-3) k = (-4) j + (3) k$$

Unit vectors along AB, AC and AD:

$$u_{AB} = \frac{(-5)\,i + (-2)\,j}{\sqrt{(-5)^2 + (-2)^2}} = -0.9285\,i - 0.371\,j$$

$$u_{AC} = \frac{(-4)\,j + (-3)\,k}{\sqrt{(-4)^2 + (-3)^2}} = -0.8\,j - 0.6\,k$$

$$u_{AD} = \frac{(-4)\,j + (3)\,k}{\sqrt{(-4)^2 + (3)^2}} = -0.8\,j + 0.6\,k$$

Forces in each member:

$$F_{AB} = F_{AB}\,u_{AB} = F_{AB}\,(-0.9285\,i - 0.371\,j)$$
$$F_{AC} = F_{AC}\,u_{AC} = F_{AC}\,(-0.8\,j - 0.6\,k)$$
$$F_{AD} = F_{AD}\,u_{AD} = F_{AD}\,(-0.8\,j + 0.6\,k)$$

(F_{AB}, F_{AC} and F_{AD} are magnitudes of vectors F_{AB}, F_{AC} and F_{AD})

Applying the equation of equilibrium at node A:

$$\sum F_{@A} = 0 = 6000\,i + F_{AB} + F_{AC} + F_{AD}$$
$$= 6000\,i - 0.9285\,F_{AB}\,i - 0.371\,F_{AB}\,j - 0.8\,F_{AC}\,j - 0.6\,F_{AC}\,k$$
$$- 0.8\,F_{AD}\,j + 0.6\,F_{AD}\,k$$
$$= (6000 - 0.9285\,F_{AB})i + (-0.371\,F_{AB} - 0.8\,F_{AC} - 0.8\,F_{AD})j$$
$$+ (-0.6\,F_{AC} + 0.6\,F_{AD})k$$

Also, since the structure is symmetrical, $F_{AC} = F_{AD}$

Therefore,

$6000 - 0.9285\,F_{AB} = 0$, $\boxed{F_{AB} = \textbf{6462 N}}$

$-0.371\,F_{AB} - 0.8\,F_{AC} - 0.8\,F_{AD} = 0$,

$\boxed{F_{AC} = F_{AD} = \textbf{-1500 N}}$

The stresses:

$$\boxed{\begin{array}{l} \sigma_{AB} = 6460 / (5.63 \times 10^{-5}) = \textbf{115 MPa} \\ \sigma_{AC} = \sigma_{AD} = -1500 / (5.63 \times 10^{-5}) = \textbf{-26.7 MPa} \end{array}}$$

5.6 Starting *I-DEAS*

1. Login to the computer and bring up *I-DEAS*. Start a new model file by filling in the items as shown below in the *I-DEAS Start* window:

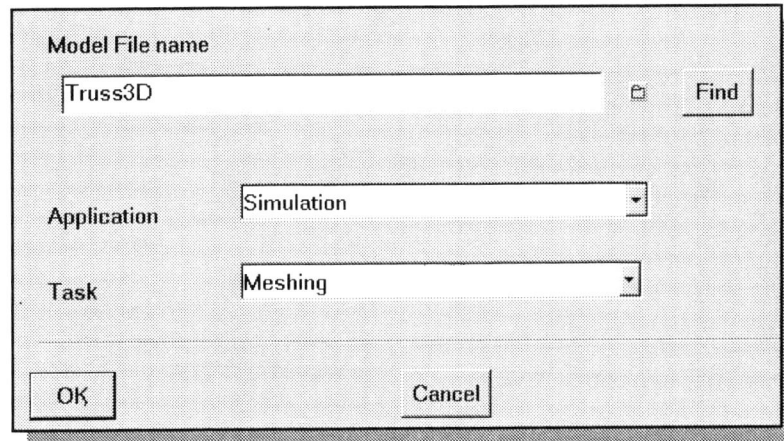

Model File name		
Truss3D		Find
Application	Simulation	
Task	Meshing	
OK	Cancel	

2. After you click **OK**, a *warning window* will appear indicating a new model file will be created. Click **OK** to proceed.

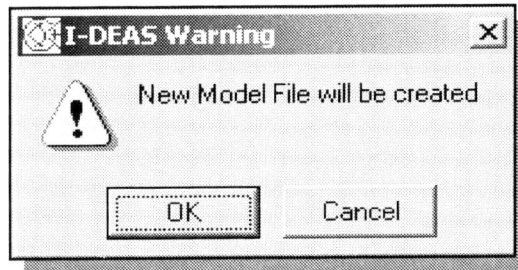

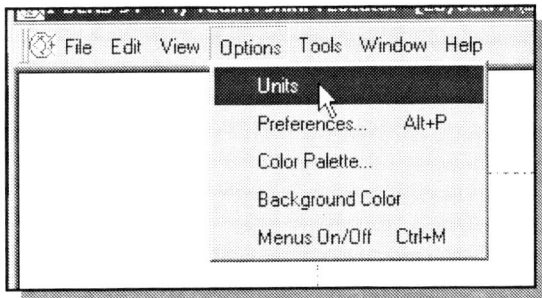

I-DEAS Warning

New Model File will be created

OK   Cancel

File   Edit   View   Options   Tools   Window   Help

Units
Preferences...   Alt+P
Color Palette...
Background Color
Menus On/Off   Ctrl+M

3. Use the left-mouse-button and select the **Options** menu in the icon panel.

4. Select the **Units** option.

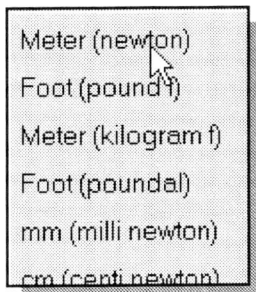

Meter (newton)
Foot (pound f)
Meter (kilogram f)
Foot (poundal)
mm (milli newton)
cm (centi newton)

5. Set the units to **Meter (newton)** by selecting from the menu of choices. This should be the first thing you set up when doing finite element analysis. The units you use MUST be consistent throughout the analysis.

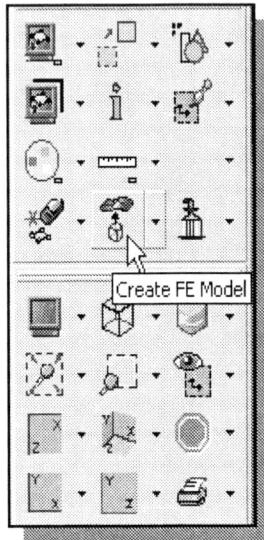

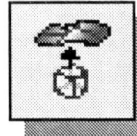

6. Choose **Create FE Model** in the icon panel. (The icon is located in the last row of the application specific icon panel.) The *FE Model Create* window appears.

7. In the *FE Model Create* window, enter **Truss3D** in the *FE Model Name* box.

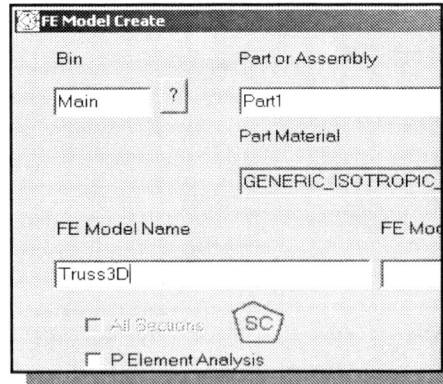

8. Click on the **OK** button to accept the settings.

9. In the *I-DEAS* warning window, click on the **OK** button to create a new part.

5.7 Workplane Appearance

❖ The workplane is a construction tool; it is a coordinate system that can be moved in space. The size of the workplane display is only for our visual reference, since we can sketch on the entire plane, which extends to infinity.

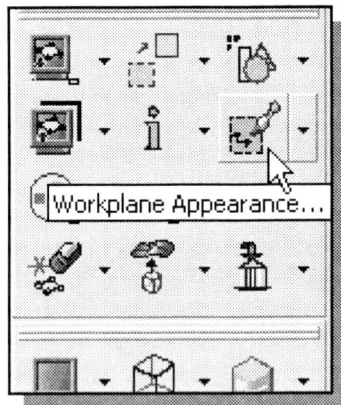

1. Choose **Workplane Appearance** in the icon panel. (The icon is located in the second row of the application icon panel.) The *Workplane Attributes* window appears.

2. Toggle *on* the *Display Border* switch as shown.

3. Adjust the **workplane border size** by entering the *Min. & Max.* values as shown in the above figure.

4. Click on the **OK** button to exit the *Workplane Attributes* window.

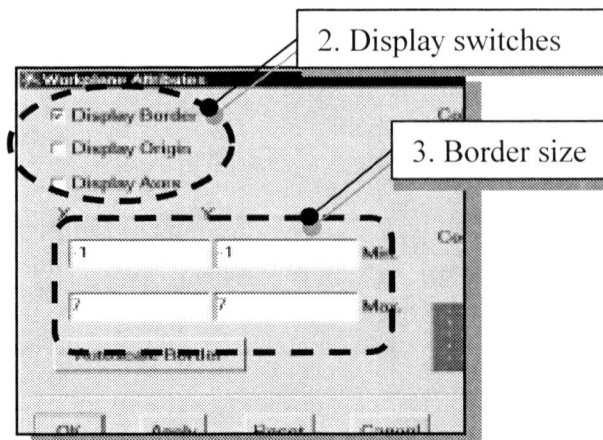

2. Display switches

3. Border size

5.8 Creating Node points

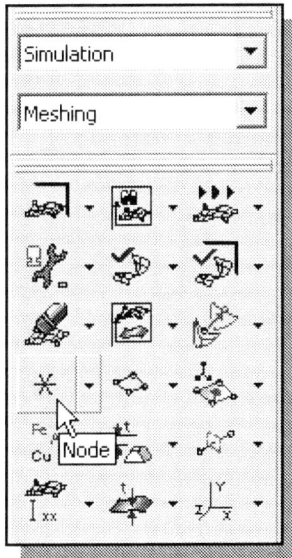

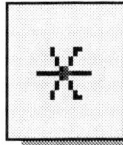

1. Choose **Node** in the icon panel. (The icon is located in the fourth row of the application icon panel.) The *Node* window appears.

2. Click on the **OK** button to accept the default settings, which will begin with *Node Number 1* and automatically increment the node numbers.

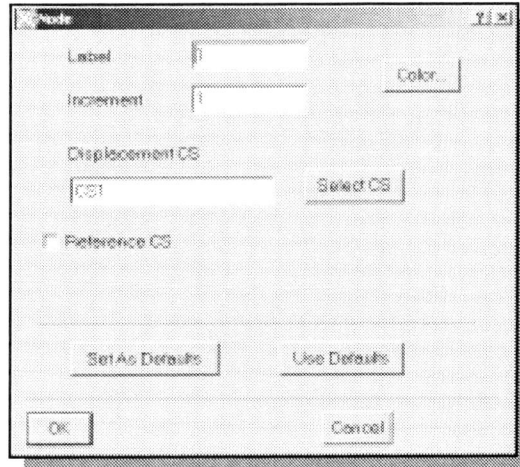

3. At the prompt window, the messages "*Enter Node 1 location*" and "*Enter X, Y, Z (0.0,0.0,0.0)*" are displayed. Enter **5,4,3** for *Node 1*.

4. For *Node 2*, enter **0,2,3**.

5. For *Node 3*, enter **5,0,0**.

6. For *Node 4*, enter **5,0,6**.

7. For *Node 5*, enter **0,0,3**.

8. Pick **Done** in the popup menu.

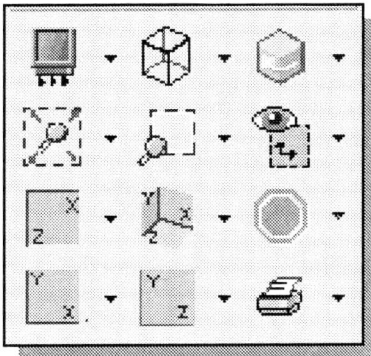

9. Choose *Isometric View* in the *Display Viewing* icon panel.

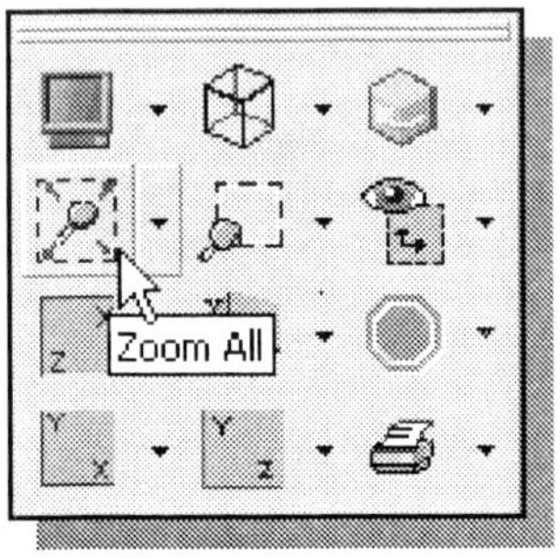

10. Choose **Zoom-All** in the display viewing icon panel.

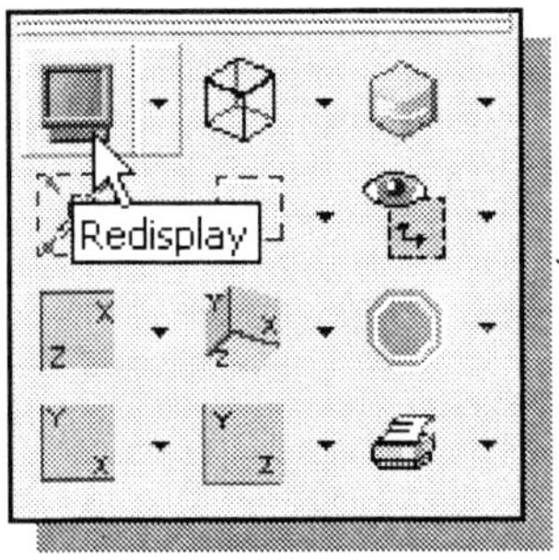

11. Choose **Redisplay** in the display viewing icon panel. Your screen should appear as shown below

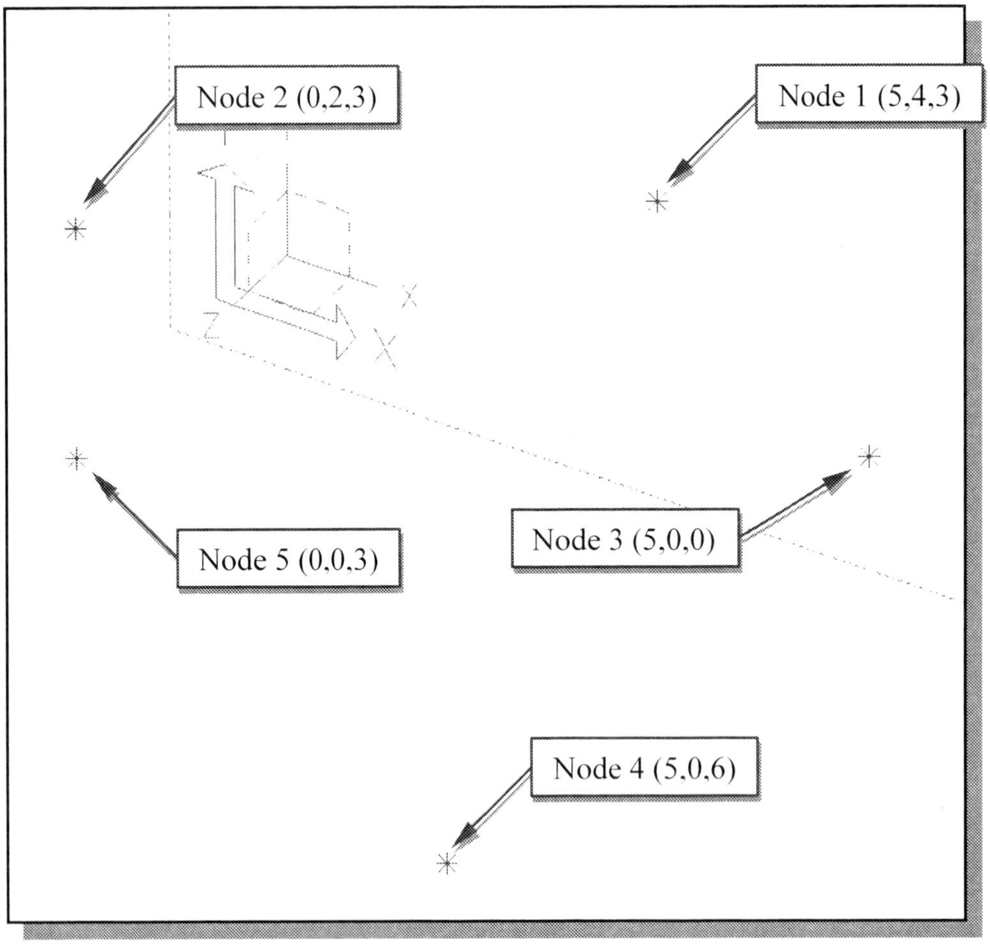

Node 2 (0,2,3)

Node 1 (5,4,3)

Node 5 (0,0,3)

Node 3 (5,0,0)

Node 4 (5,0,6)

5.9 Material Property Table

➢ Before creating elements, we will set up a *Material Property* table. The *Material Property* table contains general material information, such as *Modulus of Elasticity*, *Poisson's Ratio*, etc.

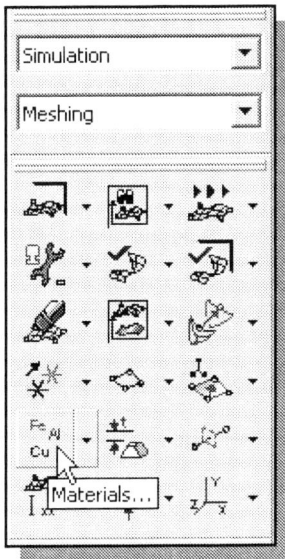

1. Choose **Materials** in the icon panel. (The icon is located in the fifth row of the task icon panel.) The *Materials* window appears.

2. Choose **Create** in the *Materials* window. The *Create Material* window appears.

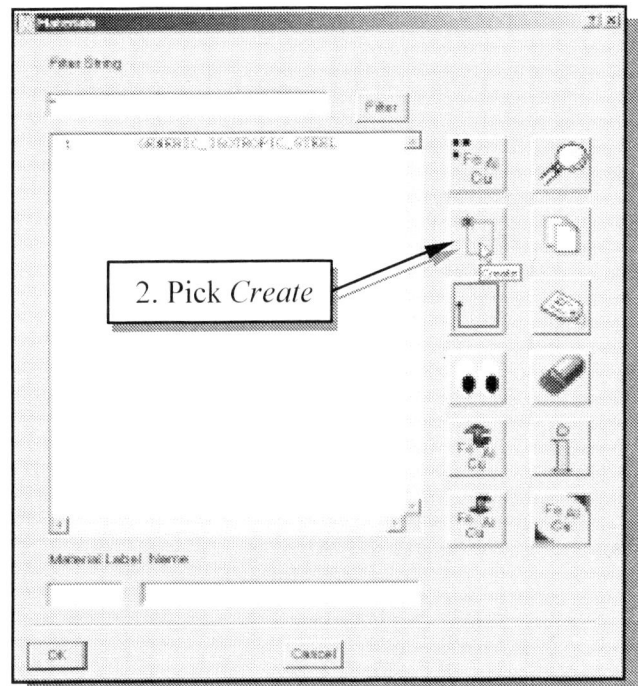

2. Pick *Create*

3. Type **STEEL_SI** in the material *Name* box.

4. Pick **Modulus of Elasticity** in the *Properties* list.

3. *STEEL_SI*

4. Pick

5. Click on the *Value* box to enter new value for the **Modulus of Elasticity** in the *Properties* list.

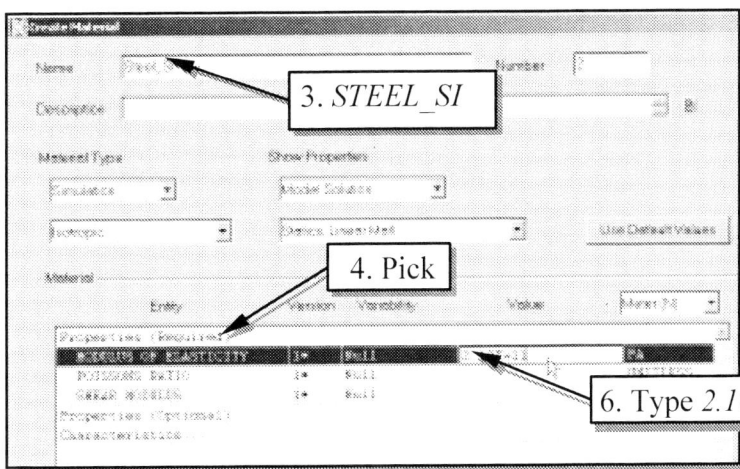

6. Type **2.10E+11** in the *Properties Value* box.

6. Type *2.10E+11*

7. Repeat the above steps and pick **Poisson's Ratio** in the *Properties* list.

8. Type **0.3** in the *Properties Value* box.

```
Material
         Entity          Version  Variability       Value        Meter (N)    ▼

Properties (Required)
  MODULUS OF ELASTICITY     1*     Constant      2.10e+11       PA
  POISSONS RATIO            1*     Null          0.3            UNITLESS
  SHEAR MODULUS             1*     Null                         PA
Properties (Optional)...
Characteristics...
```

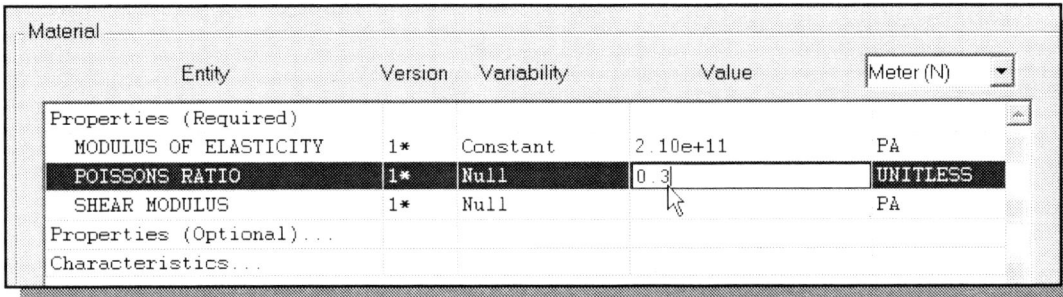

9. Click on the **OK** button to exit the *Create Material* window.

10. Click on the **OK** button to exit the *Materials* window.

5.10 Setting up an Element Cross Section

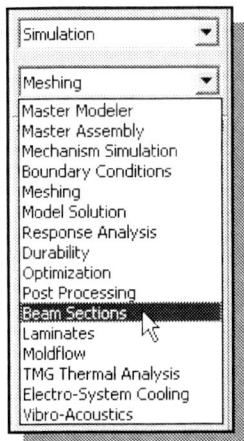

1. Switch to the **Beam Sections** task by selecting the task menu in the icon panel as shown.

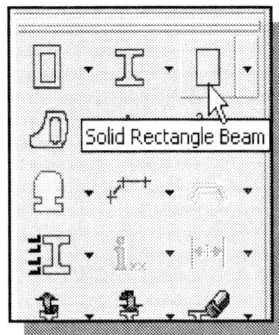

2. Choose **Solid Rectangular Beam** in the icon panel. (The icon is located in the first row of the task icon panel.) The message "*ENTER base*" is displayed.

3. Enter **7.5E-3** for the *base dimension*.

4. Press the **ENTER** key to make the height dimension the same as the base.

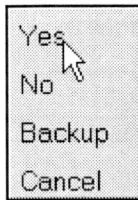

5. Pick **Yes** to complete the creation of the cross section.

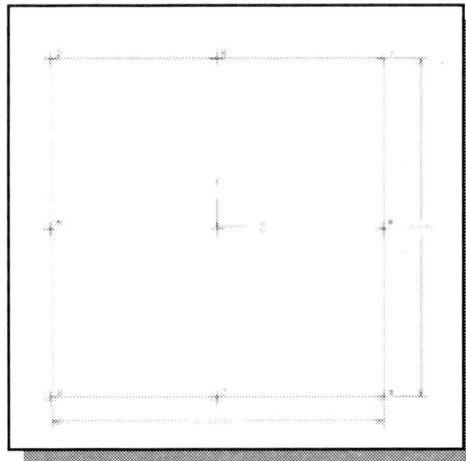

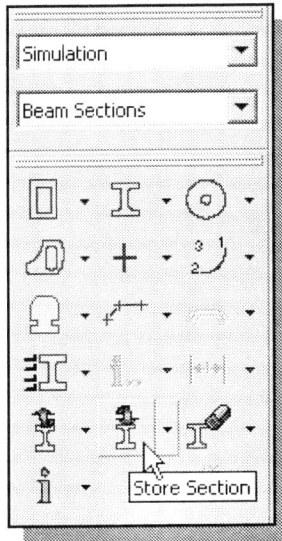

6. Choose **Store Section** in the icon panel. (The icon is located in the fifth row of the task icon panel.) The message *"Enter beam cross sect prop name or no (1-Rectangle 0.0075 x 0.0075)."* is displayed in the prompt window.

7. Press the **ENTER** key to accept the default name *(1-Rectangle 0.0075 x 0.0075).*

5.11 Creating Elements

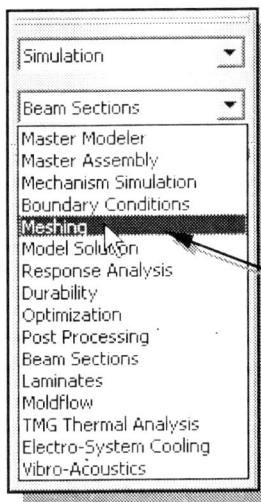

1. Switch back to the **Meshing** task by selecting the Meshing task in the task menu as shown.

Pick Meshing

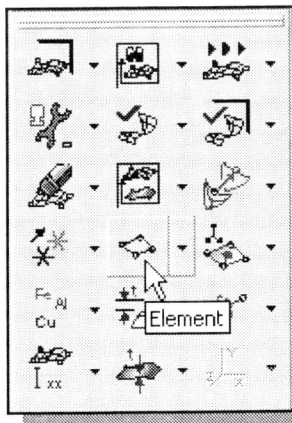

2. Choose **Element** in the icon panel. (The icon is located in the fourth row of the task specific icon panel. If the icon is not on top of the stack, press and hold down the left-mouse-button on the displayed icon to display all the choices.) The *Element* window appears.

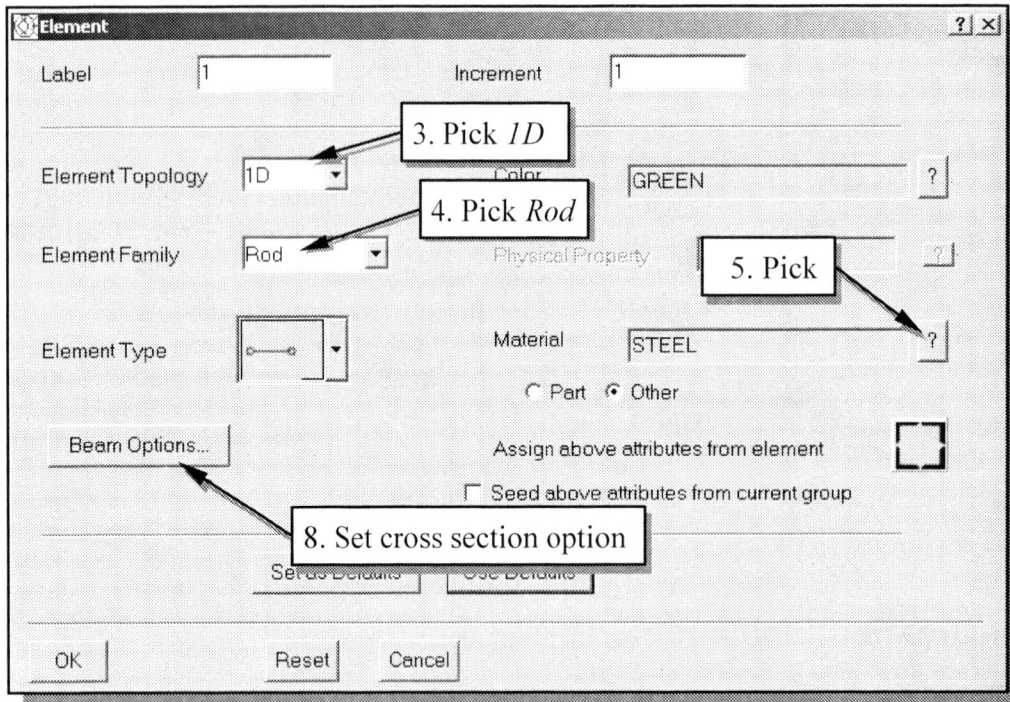

3. In the element window, select **1D** element.

4. Pick **Rod** from the *Element Family* list.

5. Click on the **Material Selection** button.

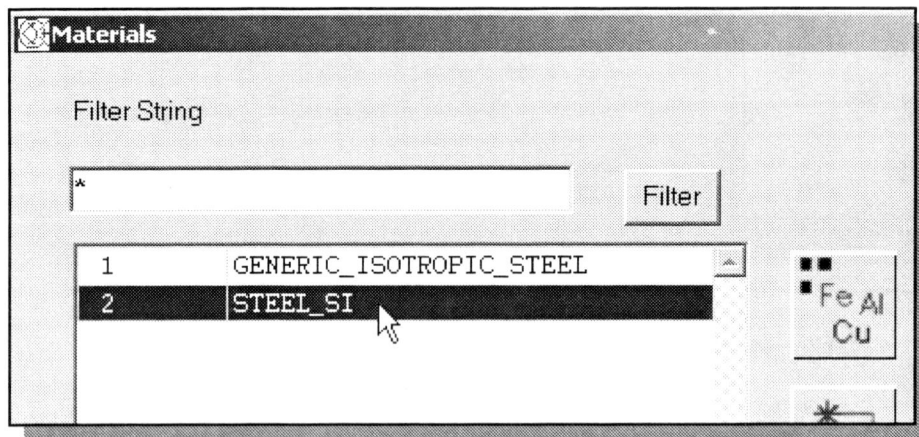

6. Choose **STEEL_SI** in the *Materials* window.

7. Click on the **OK** button to accept the selection.

8. Click on the **Beam Options** button; the *Beam Options* window appears.

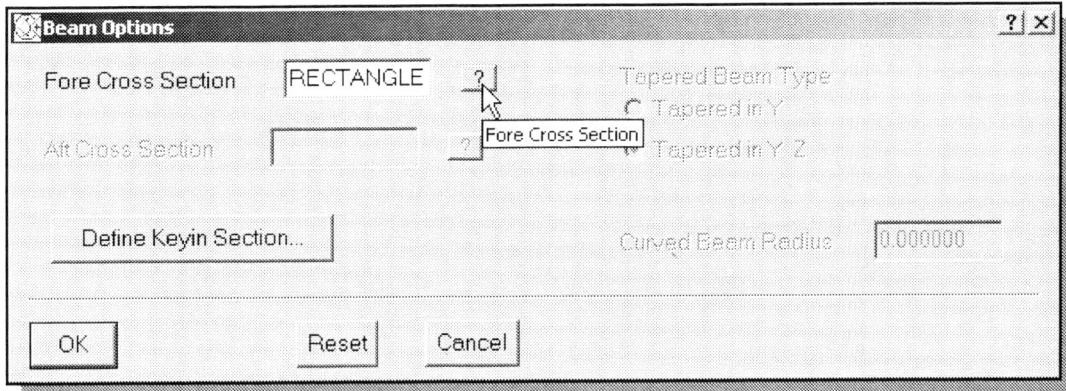

Beam Options

Fore Cross Section   RECTANGLE   ?   Tapered Beam Type
                                              ○ Tapered in Y
Aft Cross Section | ? [Fore Cross Section] ○ Tapered in Y Z

Define Keyin Section...   Curved Beam Radius   0.000000

OK   Reset   Cancel

9. Click on the [**?**] button and confirm the rectangular section we just created is selected.

10. Click on the **OK** button to exit the *Fore Cross Section* window.

11. Click on the **OK** button to exit the *Beam Options* window.

12. Click on the **OK** button to exit the *Element* window. The message "*Pick Nodes*" is displayed in the prompt window.

13. *I-DEAS* expects us to select two node points to create an element in between. Create the 6 elements as shown below. Note the **Nx** symbol as the cursor is moved next to the node points.

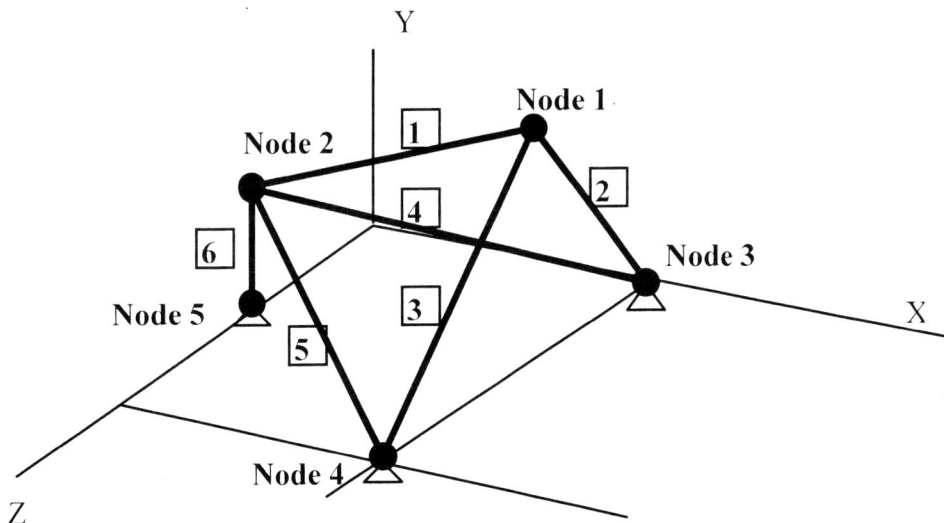

5.12 Applying Loads and Boundary Conditions

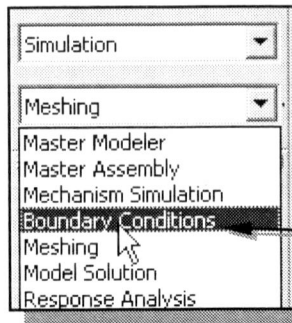

1. Switch to the **Boundary Conditions** task by selecting it in the task menu as shown.

Pick *Boundary Conditions*

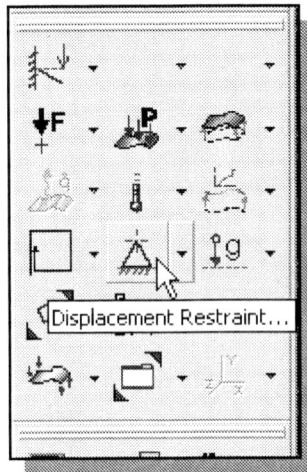

2. Choose **Displacement Restraint** in the icon panel. (The icon is located in the fourth row of the task icon panel.) The message "*Pick Nodes/Centerpoints/Vertices*" is displayed in the prompt window.

3. Pick *Node 3*, *Node 4*, and *Node 5* by pressing down the **SHIFT** key and left-clicking the three nodes.

SHIFT   +   Left-mouse-button

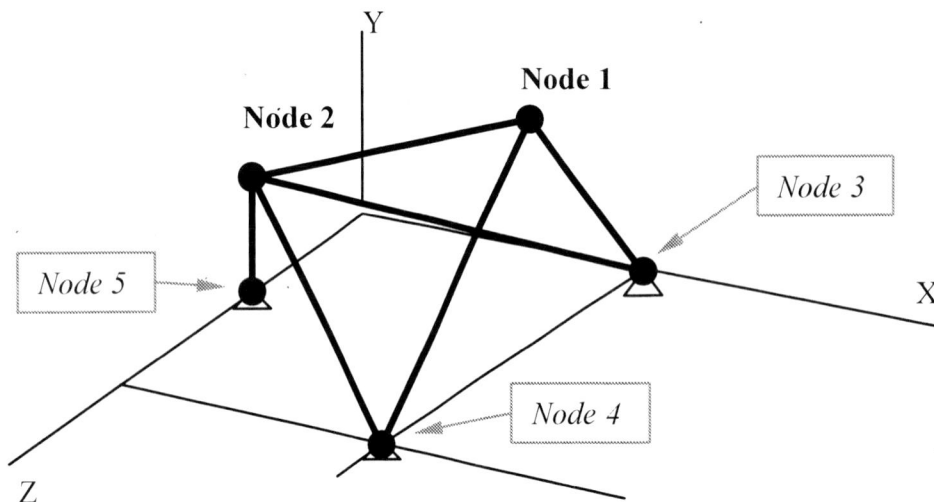

Y

Node 1

Node 2

Node 3

Node 5

X

Node 4

Z

4. Press the **ENTER** key to continue. The *Displacement Restraint on Node* window appears.

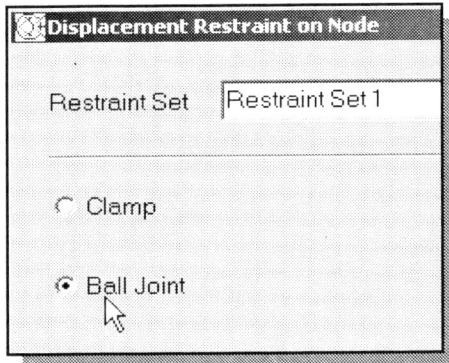

5. Pick **Ball Joint** in the *Displacement Restraint on Node* window.

6. Click on the **OK** button to exit the *Displacement Restraint on Node* window.

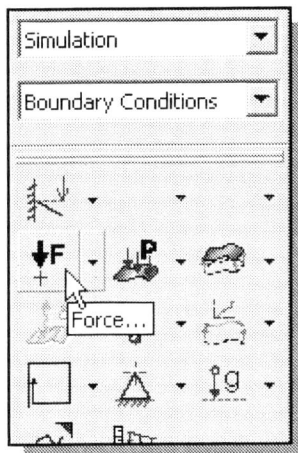

7. Choose **Force** in the icon panel. (The icon is located in the second row of the task icon panel.)

8. The message "*Pick entities*" is displayed in the prompt window. Pick *Node 1*.

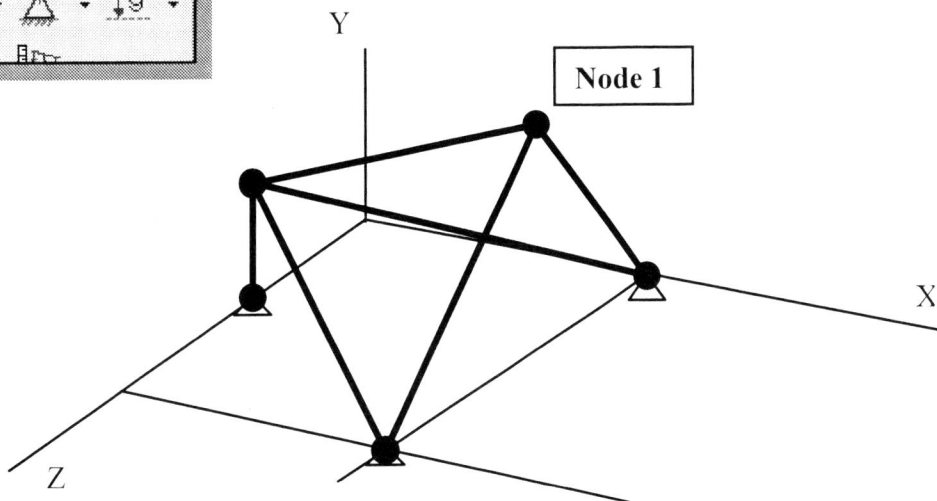

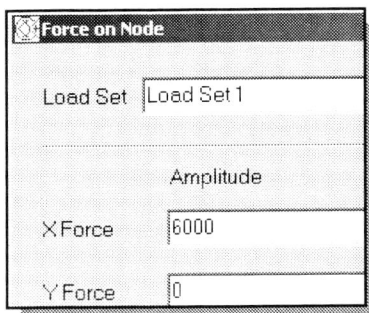

9. Press the **ENTER** key to accept the selection. The *Force on Node* window appears.

10. The load at *Node 1* is 6000 N in the positive X direction. Enter **6000** in the *X Force* box.

11. Click on the **OK** button to proceed.

5.13 Running the Solver

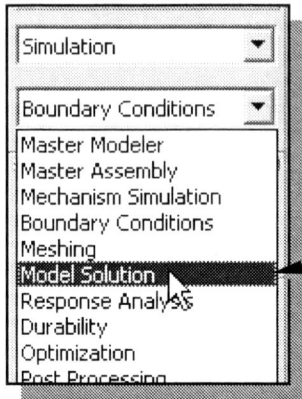

1. Switch to the **Model Solution** task by selecting the task in the task menu as shown.

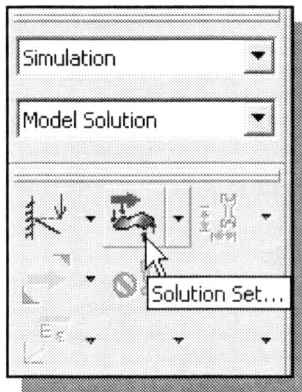

Pick *Model Solution*

2. Choose ***Solution Set*** in the icon panel. (The icon is located in the first row of the task icon panel.) The *Manage Solution Sets* window appears.

3. Click on the **Create** button. The *Solution Set* window appears.

4. Click on the **Output Selection** button. The *Output Selection* window appears.

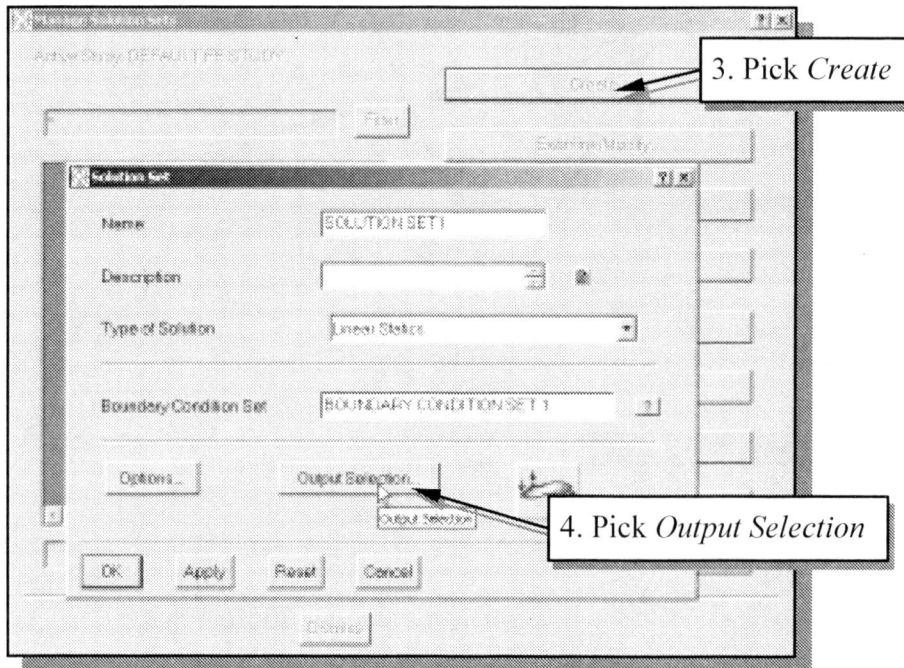

3. Pick *Create*

4. Pick *Output Selection*

5. Pick **Element Forces** in the *Output* list.

6. Press the left-mouse-button on the ***Store/List*** button to show the different output options that are available. We will accept the default output option, **Store**.

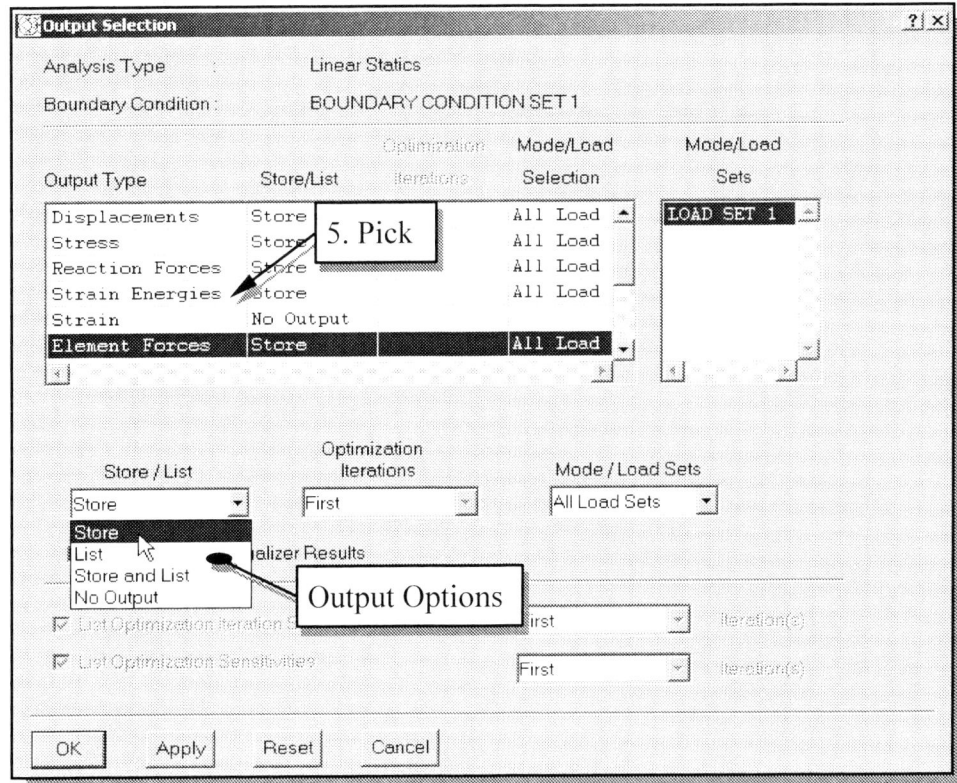

7. Click on the **OK** button to exit the *Output Selection* window.

8. Click on the **OK** button to exit the *Solution Set* window.

9. Pick **Dismiss** to exit the *Manage Solution Sets* window.

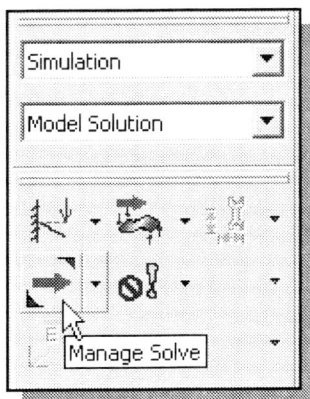

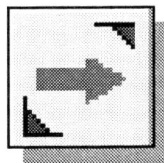

10. Choose ***Manage Solve*** in the icon panel. (The icon is located in the second row of the task icon panel.) The *Solve* window appears.

11. In the *Solve* window, pick the ***Solve*** icon to find the solutions.

➤ *I-DEAS* will begin the solving process. <u>DO NOT</u> close any windows. Any errors or warnings are displayed in the list window.

5.14 Viewing the Results – Axial Forces

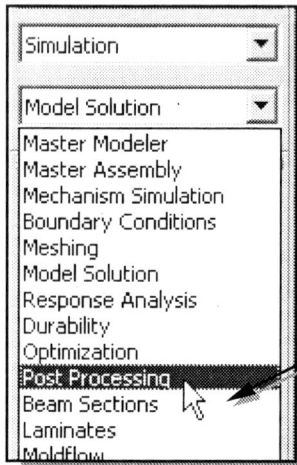

1. Switch to the **Post Processing** task by selecting it in the task menu as shown.

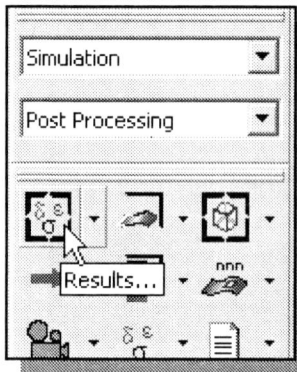

Pick *Post Processing*

2. Choose **Results Selection** in the icon panel. (The icon is located in the first row of the task icon panel.)

3. The *Results Selection* window appears. Pick **Element Force** from the list.

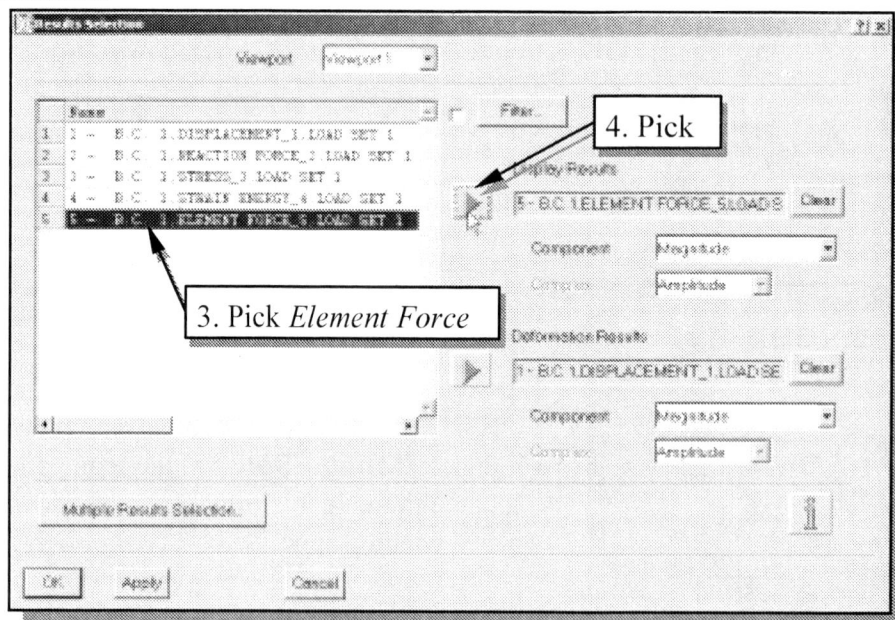

4. Pick

3. Pick *Element Force*

4. Click on the **Triangle** button to set the *Element Force* as *the Display Results*.

5. Click on the **OK** button to exit the *Results Selection* window.

6. Select **Menus On/Off** under the **Options** toolbar in the pull-down menu to switch *on* the cascading menu system. (If the cascading menu is not displayed on the screen, select Preferences under the Options toolbar in the icon panel, and set the *Menu Type* to Long under the *Menus Preference* option.)

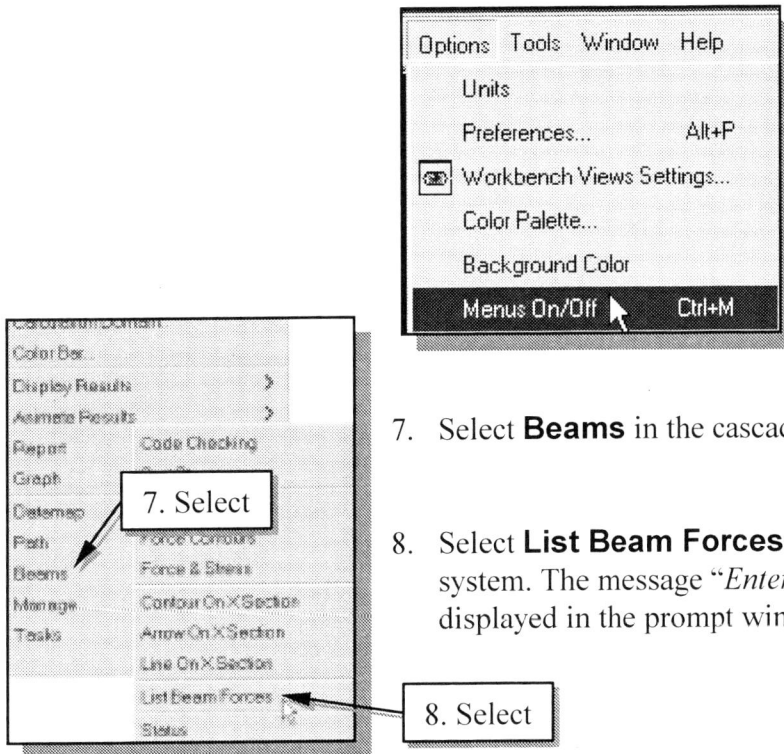

7. Select **Beams** in the cascading menu system.

8. Select **List Beam Forces** in the cascading menu system. The message "*Enter Result name or no.*" is displayed in the prompt window.

9. Select **B.C.1, Element Force 5, Load Set 1** in the popup menu.

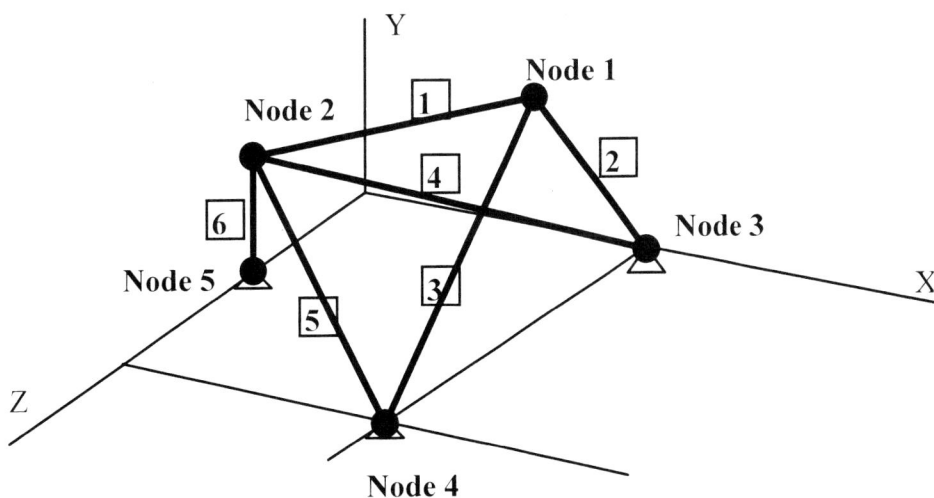

10. The message "*Pick beam for forces*" is displayed in the prompt window. Pick **Element 1**. (Note the E1 symbol associated with the element.)

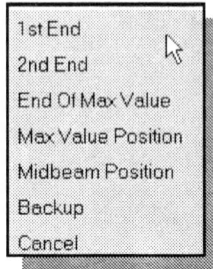

```
1st End
2nd End
End Of Max Value
Max Value Position
Midbeam Position
Backup
Cancel
```

11. Select **1ST End** in the popup menu to display the element force associated with this element.

12. Use the scroll bar of the *List* window to examine the result.

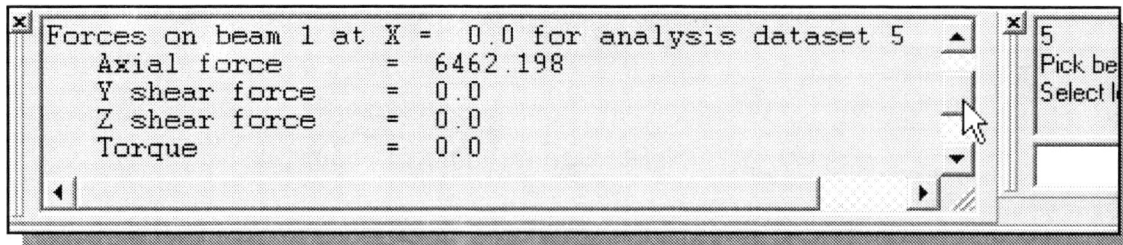

```
Forces on beam 1 at X =   0.0 for analysis dataset 5
    Axial force      =  6462.198
    Y shear force    =  0.0
    Z shear force    =  0.0
    Torque           =  0.0
```

➤ The *I-DEAS* solution is **6462.198 N.**, which matches with our hand calculation (see section **5.4 Preliminary Analysis**).

13. Pick **Element 3**.

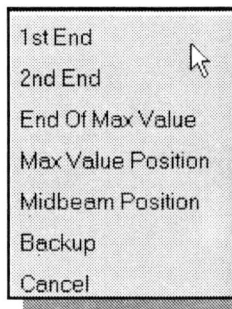

```
1st End
2nd End
End Of Max Value
Max Value Position
Midbeam Position
Backup
Cancel
```

14. Select **1ST End** in the popup menu to display the element force associated with this element.

15. Enlarge the *List* window to examine the result. Does the result match the hand calculation?

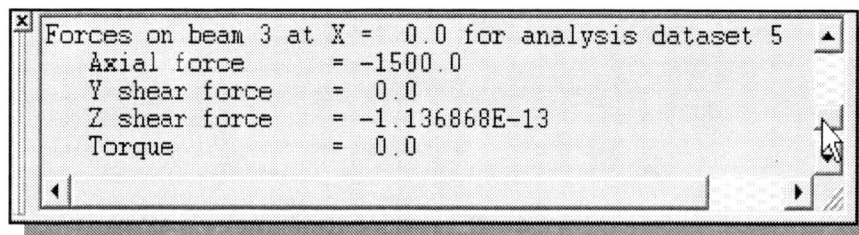

```
Forces on beam 3 at X =   0.0 for analysis dataset 5
    Axial force      = -1500.0
    Y shear force    =  0.0
    Z shear force    = -1.136868E-13
    Torque           =  0.0
```

16. Switch back to the prompt window by left-clicking inside the window.

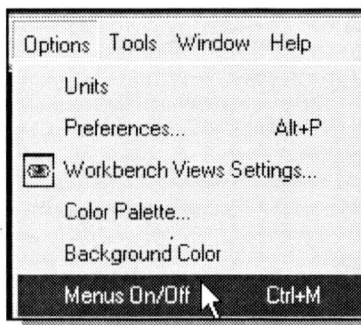

```
Options  Tools  Window  Help
    Units
    Preferences...          Alt+P
👁 Workbench Views Settings...
    Color Palette...
    Background Color
    Menus On/Off            Ctrl+M
```

17. Press the **ENTER** key to exit the *List Beam Force* command.

18. Select **Menus On/Off** under the **Options** toolbar in the icon panel to switch *off* the cascading menu system.

5.15 Viewing the Results – Stresses

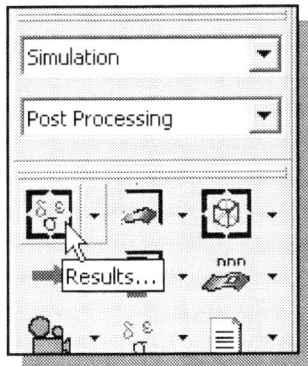

1. Choose **Results Selection** in the icon panel. (The icon is located in the first row of the task icon panel.) The *Results Selection* window appears.

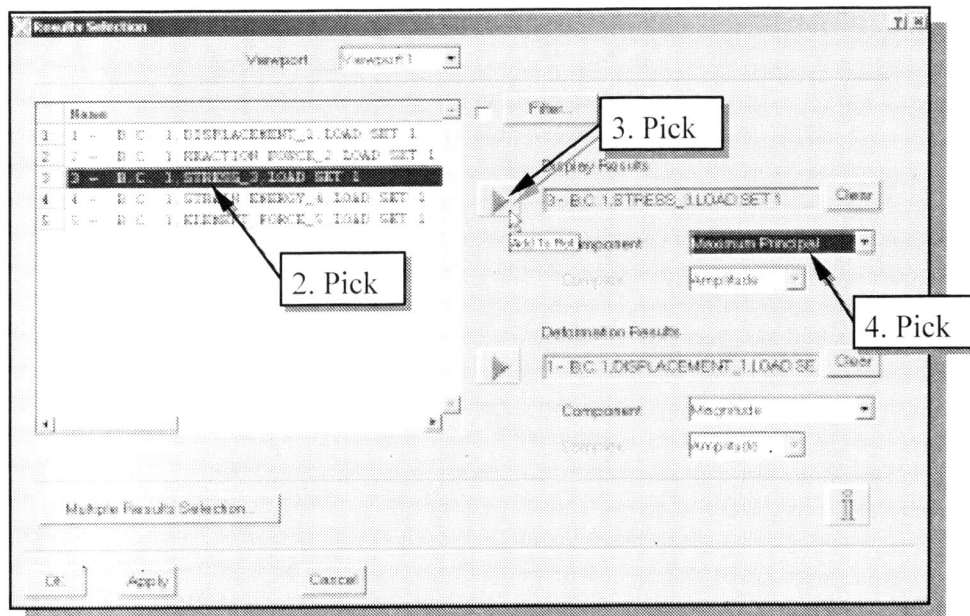

2. Pick **Stress** from the list.

3. Click on the **Triangle** button to set the display option.

4. Select the **Maximum Principal** option.

5. Click on the **OK** button to exit the *Results Selection* window.

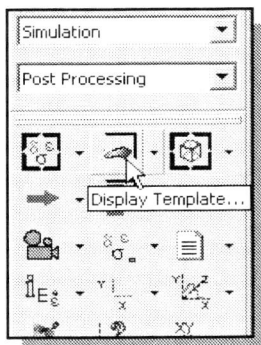

6. Choose **Display Template** in the icon panel. (The icon is located in the first row of the task icon panel.)

7. Click on the **Contour** button.

8. Click on the **Fast Display** to switch off the *Fast Display* option as shown.

9. Click the **OK** button to exit the *Contour Options* window.

10. Click the **OK** button to exit the *Display Template* window.

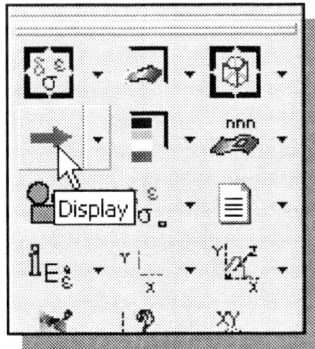

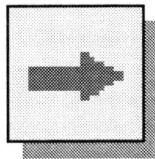

11. Choose **Results Display** in the icon panel. (The icon is located in the second row of the task icon panel.) The message "*Pick elements*" is displayed in the prompt window.

12. Pick *Element 1* and *Element 3* by pressing down the **SHIFT** key and left-clicking on the two elements.

13. Press the **ENTER** key to display the stresses associated with the selected elements.

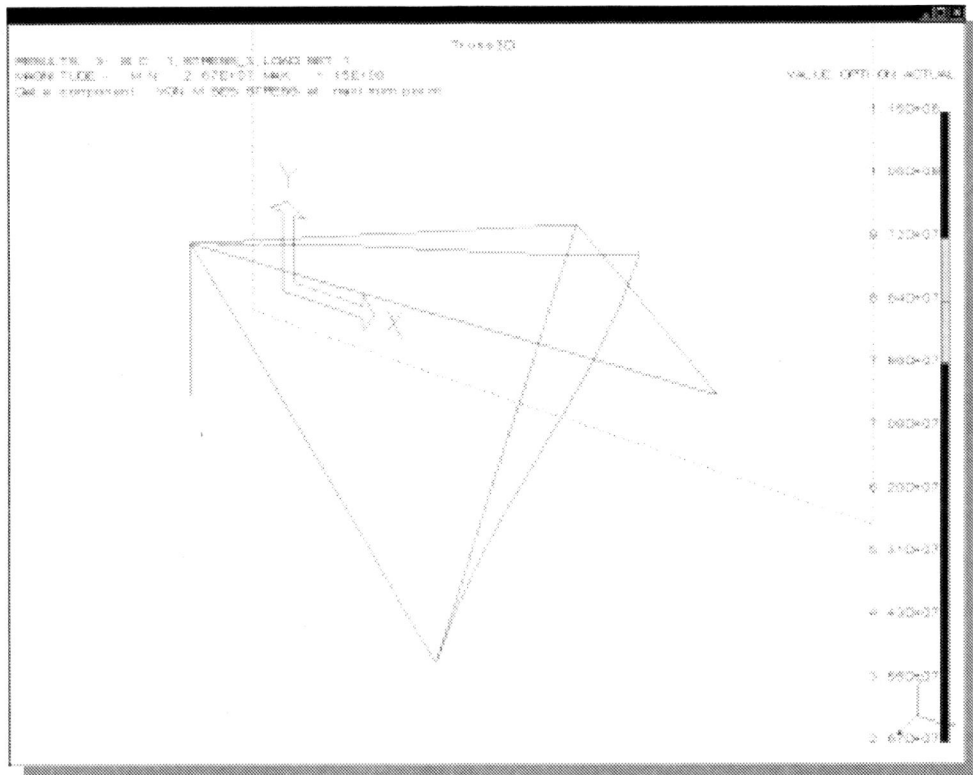

Questions:

1. For a truss element in three-dimensional space, what is the number of degrees of freedom for the element?

2. How does the number of degrees of freedom affect the size of the stiffness matrix?

3. Will the *I-DEAS* FEA software calculate the reaction forces at the supports?

4. Describe the procedure in setting up additional solution options.

5. Which *I-DEAS* task software is used to define the cross section of truss elements?

6. Identify and describe the following commands:

    (a)

    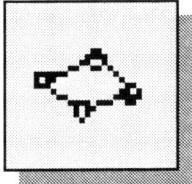

    (b)

    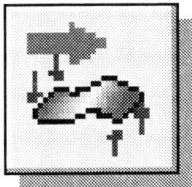

    (c)

    (d)

    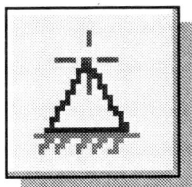

Exercises: Determine the normal stress in each member of the truss structures.

1. All joints are ball-joint.
    Material: Steel   b   a = b = 2 cm

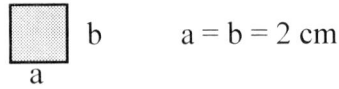

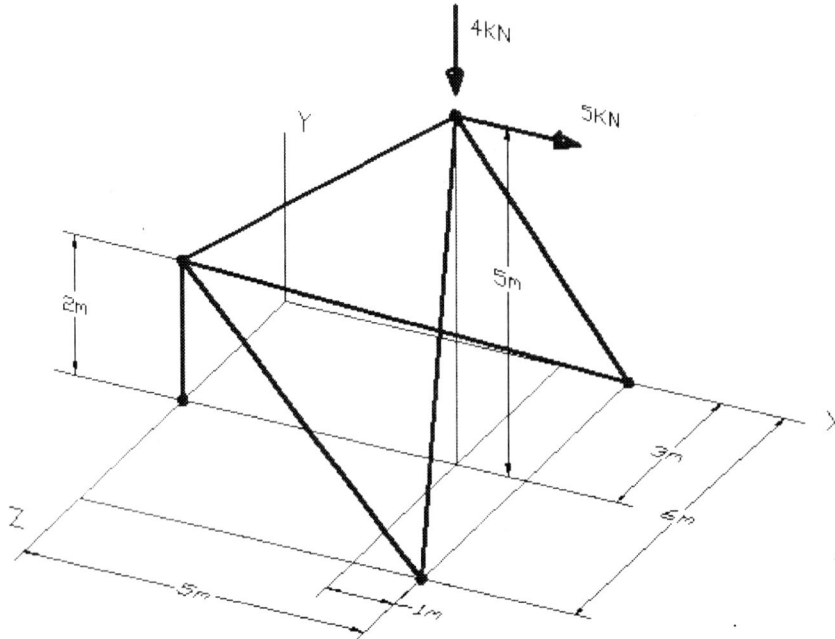

4KN

5KN

Y

5m

2m

X

3m

6m

Z

5m

1m

2.   All joints are ball-joint.
     Material: Steel rod.
            Diameter: 1 in.

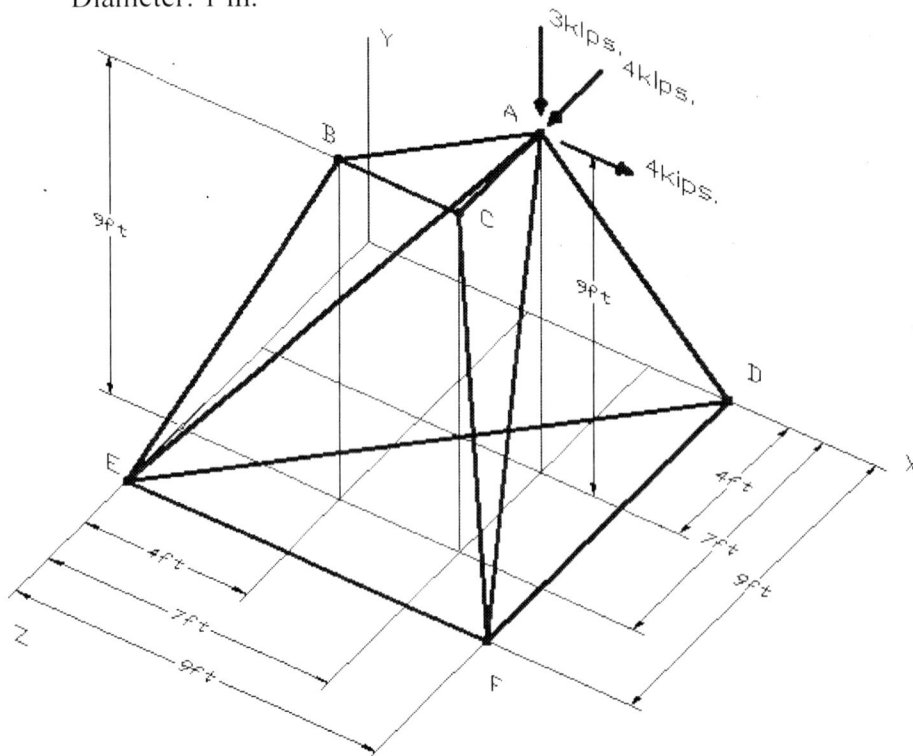

3klps.
4klps.

Y

B   A

9ft   C

9ft   4kips.

D

E   X

4ft

4ft

7ft

7ft

Z

9ft

9ft

F

Chapter 6
Basic Beam Analysis

Shear Diagram

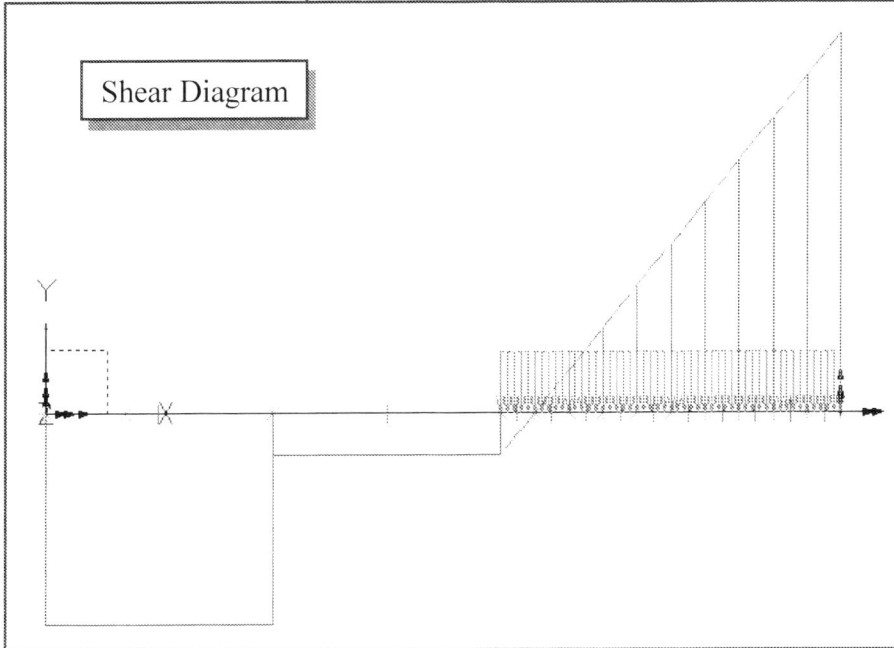

Learning Objectives

When you have completed this lesson, you will be able to:
◆ **Understand the basic assumptions for Beam elements.**
◆ **Apply Constraints and Forces on FE Models.**
◆ **Apply Distributed Loads on FE Models.**
◆ **Create and Display a Shear Diagram.**
◆ **Create and Display a Moment Diagram.**
◆ **Perform Basic Beam Analysis using I-DEAS.**

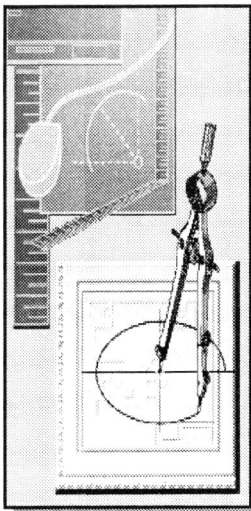

6.1 Introduction

The truss element discussed in the previous chapters does have a practical application in structural analysis, but it is a very limiting element since it can only transmit axial loads. The second type of finite element to study in this text is the *beam element*. Beams are used extensively in engineering. As the members are rigidly connected, the members at a joint transmit not only axial loads but also bending and shear. This contrasts with truss elements where all loads are transmitted by axial force only. Furthermore, beams are often designed to carry loads both at the joints and along the lengths of the members, whereas truss elements can only carry loads that are applied at the joints. A beam element is a long, slender member generally subjected to transverse loading that produces significant bending effects as opposed to axial or twisting effects.

6.2 Modeling Considerations

Consider a simple beam element positioned in a two-dimensional space:

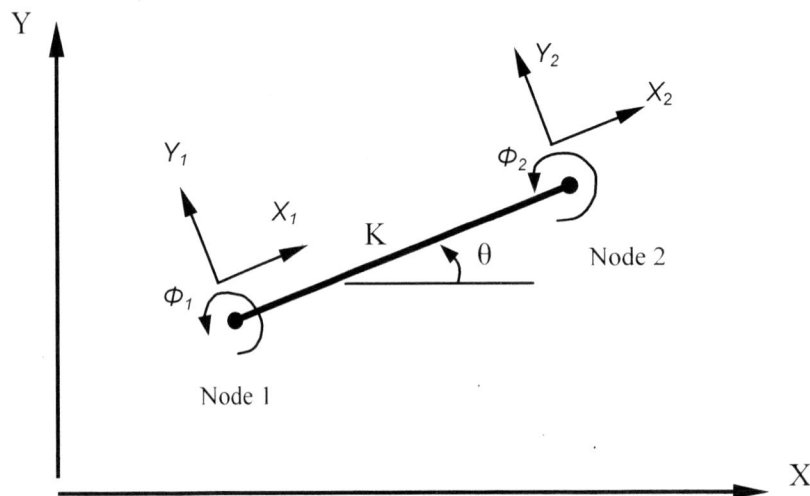

A beam element positioned in two-dimensional space is typically modeled to possess three nodal displacements (two translational and one rotational) at each node: three degrees of freedom at each node. Each element therefore possesses six degrees of freedom. A finite element analysis using beam elements typically provides a solution to the displacements, reaction forces, and moments at each node. The formulation of the beam element is based on the elastic beam theory, which implies the beam element is initially straight, linearly elastic, and loads (forces and moments) are applied at the ends. Therefore, in modeling considerations, place nodes at all locations that concentrated forces and moments are applied. For a distributed load, most finite element procedures replace the distributed load with an equivalent load set, which is applied to the nodes available along the beam span. Accordingly, in modeling considerations, place more nodes along the beam spans with distributed loads to lessen the errors.

6.3 Problem Statement

Determine the maximum normal stress that loading produces in the steel member.

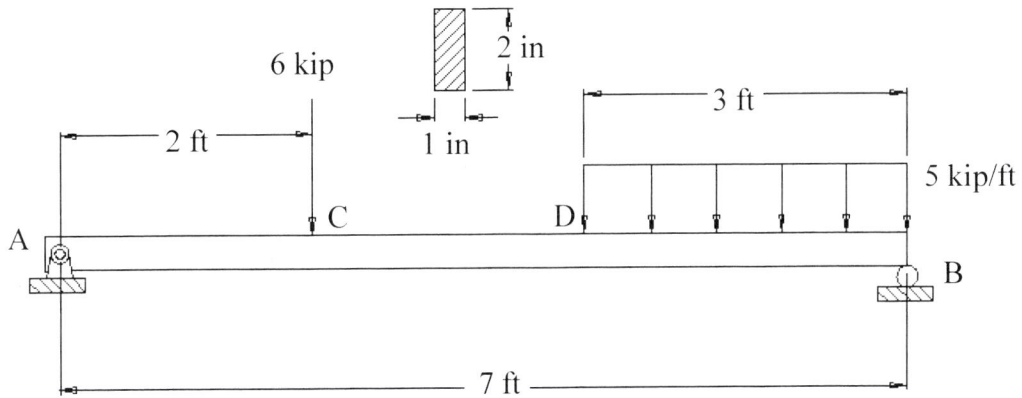

6.4 Preliminary Analysis

Free Body Diagram of the member:

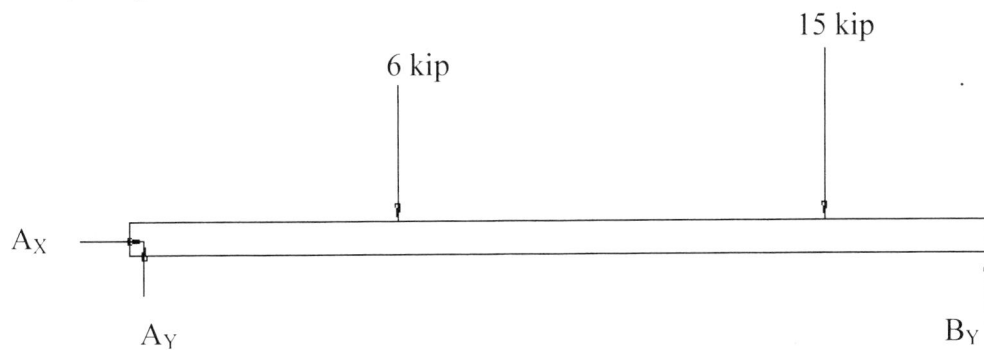

Applying the equations of equilibrium:

$$\Sigma M_{@A} = 0 = B_Y \times 7 - 6 \times 2 - 15 \times 5.5$$
$$\Sigma F_X = 0 = A_X$$
$$\Sigma F_Y = 0 = A_Y + B_Y - 6 - 15$$

Therefore,

$$B_Y = 13.5 \text{ kip}, \quad A_X = 0 \text{ and } A_Y = 7.5 \text{ kip}$$

Next, construct the shear and moment diagrams for the beam. Although this is not necessary for most FEA analyses, we will use this example to build up our confidence with *I-DEAS'* analysis results.

♦ Between A and C, 0 ft. $< X_1 < 2$ ft.:

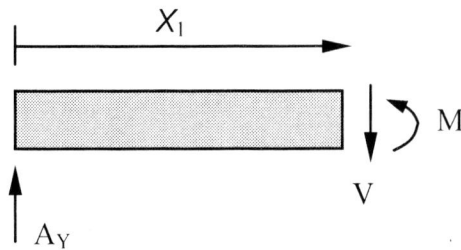

$$\sum F_Y = 0, \qquad A_Y - V = 0, \qquad V = 7.5 \text{ kip}$$

$$\sum M_{@X1} = 0, \quad -A_Y X_1 + M = 0, \qquad M = 7.5 \, X_1$$

♦ Between A and D, 2 ft. $< X_2 < 4$ ft.:

6 kip

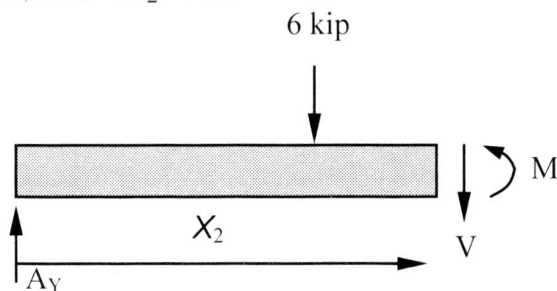

$$\sum F_Y = 0, \qquad A_Y - V - 6 = 0, \qquad V = 1.5 \text{ kip}$$

$$\sum M_{@X2} = 0, \quad -A_Y X_2 + M + 6 \, (X_2 - 2) = 0, \quad M = 1.5 \, X_2 + 12$$

♦ Between A and B, 4 ft. $< X_3 < 7$ ft.:

6 kip

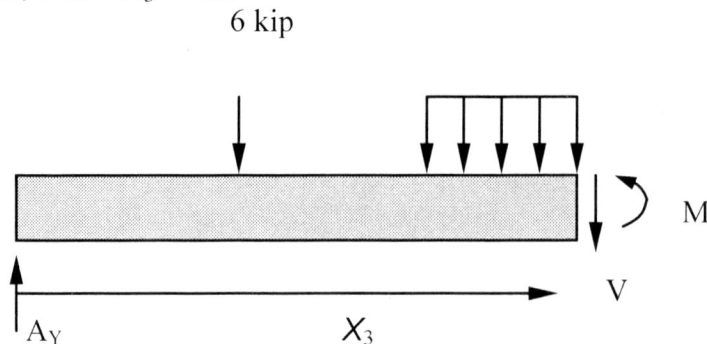

$$\sum F_Y = 0, \quad A_Y - V - 6 - 5 \, (X_3 - 4) = 0, \qquad V = (-5 \, X_3 + 21.5) \text{ kip}$$

$$\sum M_{@X3} = 0, \quad -A_Y X_3 + M + 6(X_3 - 2) + 5(X_3 - 4)(X_3 - 4)/2 = 0$$

$$M = -2.5 X_3^2 + 21.5 X_3 - 28$$

Shear and Moment diagrams:

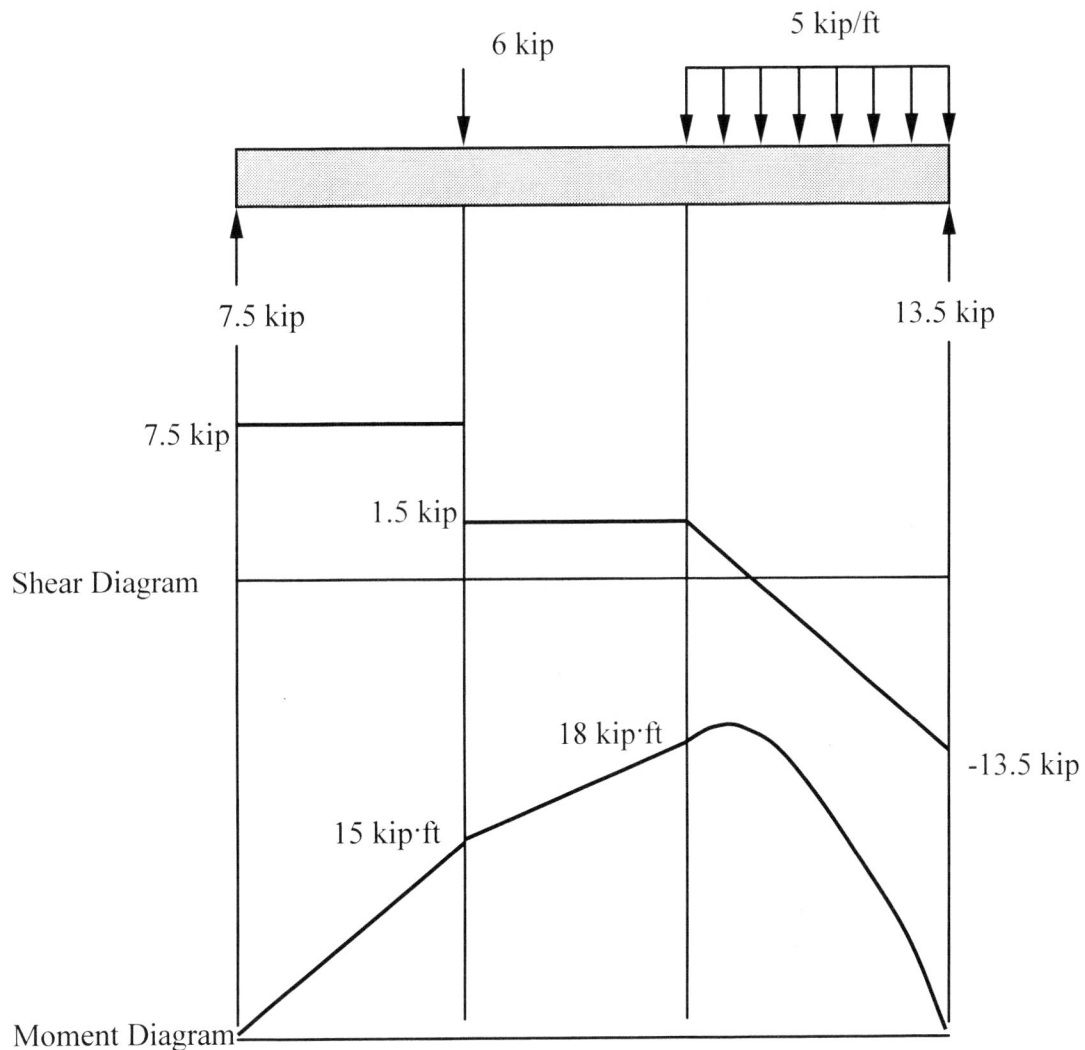

Determine the maximum normal stress developed in the beam:

$$V = -5 X_3 + 21.5 = 0, \quad X_3 = 4.3 \text{ ft.}$$

$$M = -2.5 X_3^2 + 21.5 X_3 - 28 = 18.225 \text{ kip-ft} = 218700 \text{ lb-in}$$

Therefore,

$$\sigma_{max} = \frac{MC}{I} = 18.255 \, (h/2)/(bh^3/12) = 47239 \text{ kip/ft}^2 = 3.28E+5 \text{ psi}$$

6.5 Starting *I-DEAS*

1. Login to the computer and bring up *I-DEAS 9*. In the *I-DEAS Start* window, start a new model file by filling in and selecting the items as shown below:

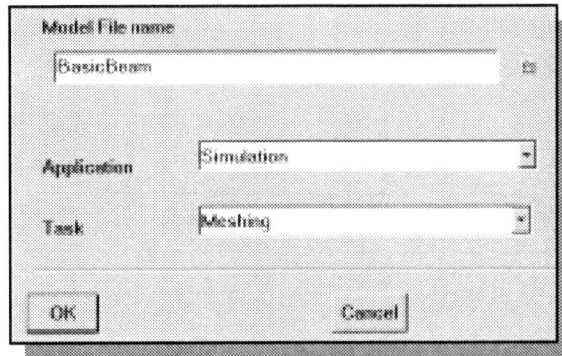

2. After you click **OK**, a *warning window* will appear indicating a new model file will be created. Click **OK** to exit the window and proceed to create the new file.

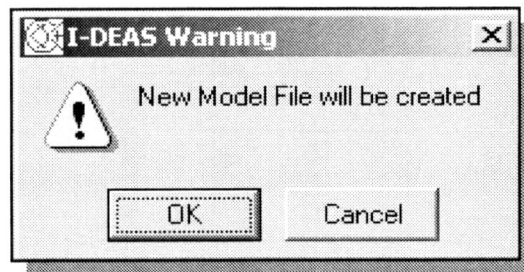

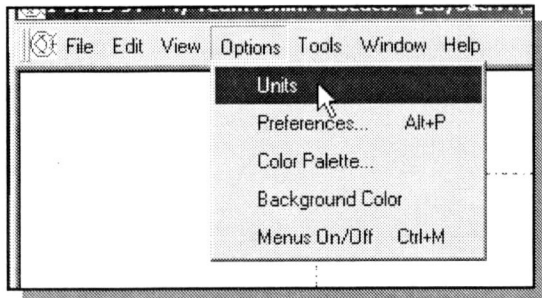

3. Use the left-mouse-button and select the **Options** menu in the icon panel.

4. Select the **Units** option.

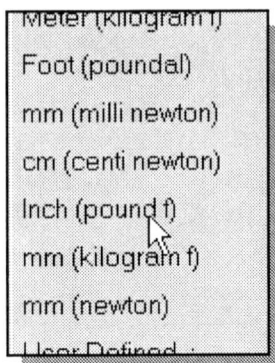

5. Set the units to **Inch (pound f)** by selecting from the menu of choices. This should be the first thing you set up when doing finite element analysis; the units you use <u>MUST</u> be consistent throughout the analysis.

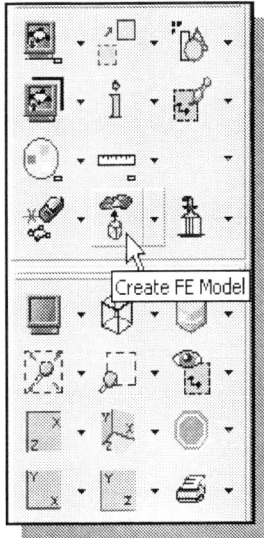

6. Choose **Create FE Model** in the icon panel. (The icon is located in the last row of the application specific icon panel.) The *FE Model Create* window appears.

7. In the *FE Model Create* window, enter **BasicBeam** in the *FE Model Name* box.

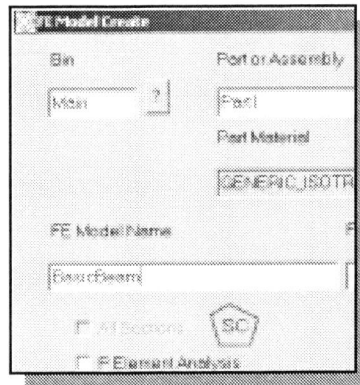

8. Click on the **OK** button to accept the settings.

9. In the *I-DEAS warning window*, click on the **OK** button to create a new part.

6.6 Workplane Appearance

The workplane is a construction tool; it is a coordinate system that can be moved in space. The size of the workplane display is only for our visual reference, since we can sketch on the entire plane, which extends to infinity.

1. Choose **Workplane Appearance** in the icon panel. (The icon is located in the second row of the application icon panel.) The *Workplane Attributes* window appears.

2. Toggle *on* the *Display Border* switch as shown.

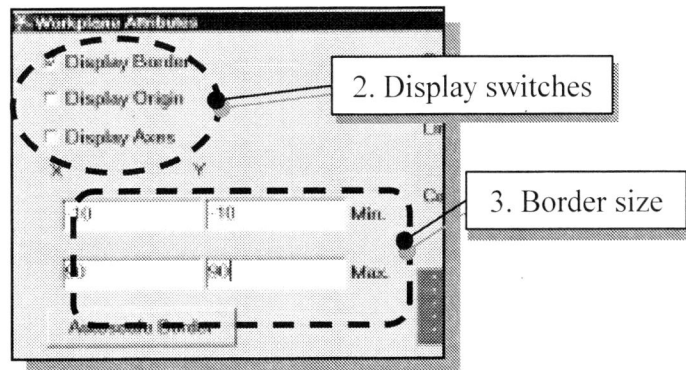

2. Display switches

3. Border size

3. Adjust the **workplane border size** by entering the *Min. & Max.* values as shown in the figure above.

4. Click on the **OK** button to exit the *Workplane Attributes* window.

6.7 Creating Node points

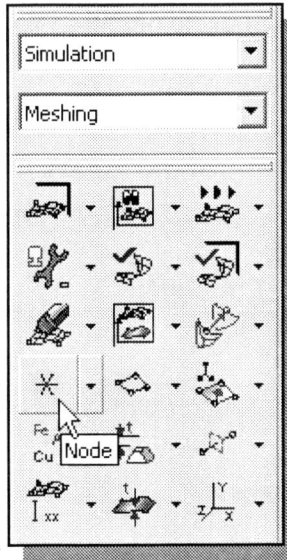

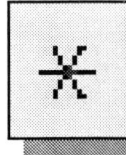

1. Choose **Node** in the icon panel. (The icon is located in the fourth row of the application icon panel.) The *Node* window appears.

2. Click on the **OK** button to accept the default settings, which will begin with *Node Number 1* and automatically increment the node numbers.

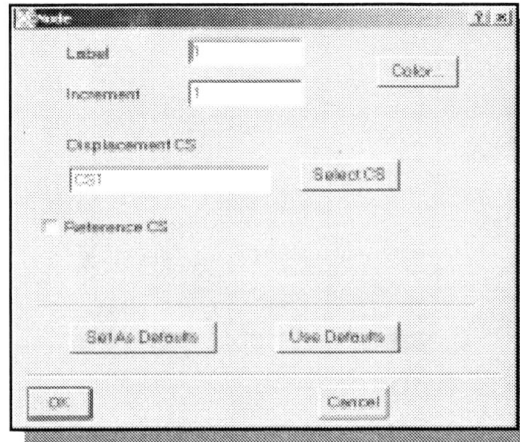

3. In the prompt window, the messages "*Enter Node 1 location*" and "*Enter X, Y, Z (0.0,0.0,0.0)*" are displayed. Click inside the prompt window and press the **ENTER** key once to place *Node 1* at the origin.

4. For *Node 2*, enter **2*12,0,0** (location of point C). *I-DEAS* will automatically do the calculations for us.

5. For *Node 3*, enter **4*12,0,0** (location of point D).

6. Pick **Done** in the popup menu.

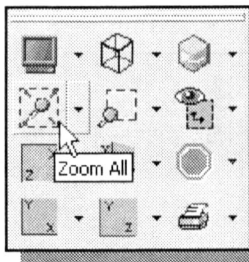

7. Choose **Zoom-All** in the display viewing icon panel.

8. Choose **Redisplay** in the display viewing icon panel.

6.8 Making Copies of Node Points

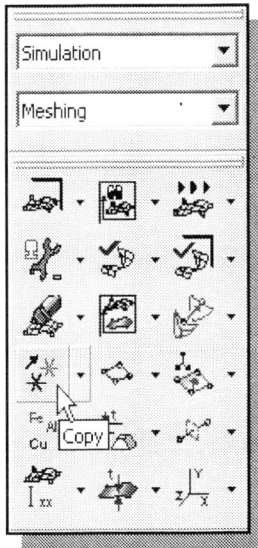

1. Choose **Copy Node** in the icon panel. (The icon is located in the same location as the **Node** icon.) The message "*Pick Nodes*" is displayed in the prompt window.

2. Pick **Node 3**. The message "*Pick Nodes (Done)*" is displayed in the prompt window.

3. Press the **ENTER** key to accept the selection and continue with the *Copy* option.

4. The message "*Enter number of copies (1)*" is displayed in the prompt window. Enter **10** to make ten copies of *Node Point 1*. The message "*Enter node start label,inc (4,1)*" is displayed in the prompt window.

5. Press the **ENTER** key to accept the default setting and continue with the *Copy* option.

6. The message "*Enter delta X, Y, Z (0.0,0.0,0.0)*" is displayed in the prompt window. Enter **0.3*12,0,0** to make the copies along the X-axis.

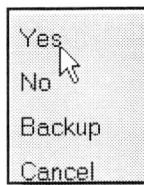

7. The message "*OK to keep these additions (Yes)*" is displayed in the prompt window. Press the **ENTER** key or select **Yes** in the popup menu.

8. Choose **Zoom-All** in the display viewing icon panel.

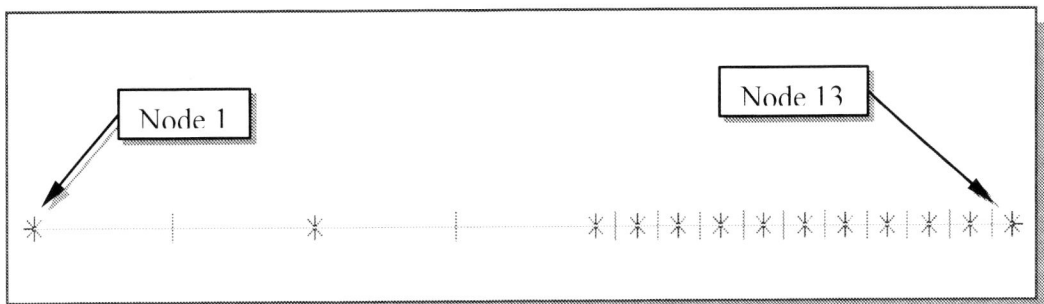

6.9 Material Property Table

❖ Before creating elements, we will create a *Material Property* table. The *Material Property* table contains general material information, such as *Modulus of Elasticity*, *Poisson's Ratio*, etc.

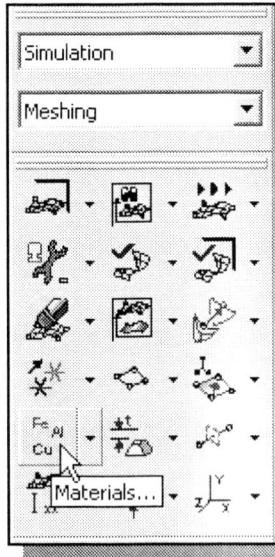

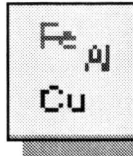

1. Choose **Materials** in the icon panel. (The icon is located in the fifth row of the task icon panel.) The *Materials* window appears.

2. Choose **Create** in the *Materials* window. The *Create Material* window appears.

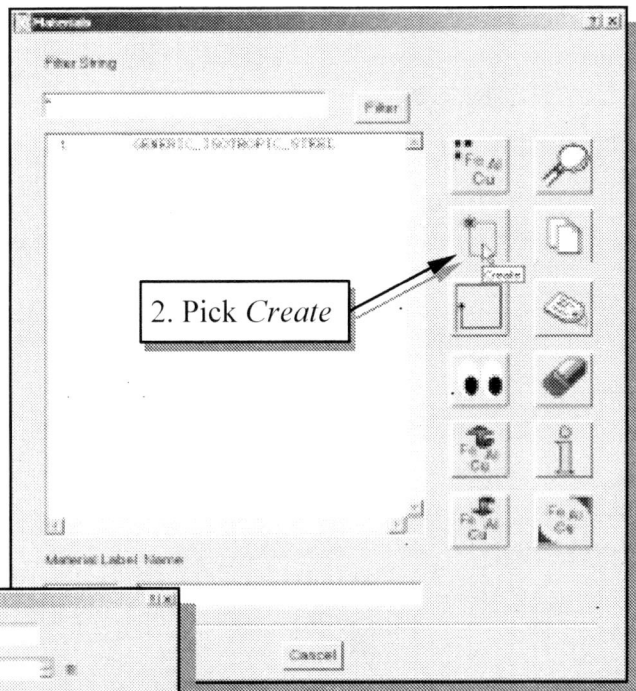

2. Pick *Create*

3. Type **STEEL** in the *Material Name* box.

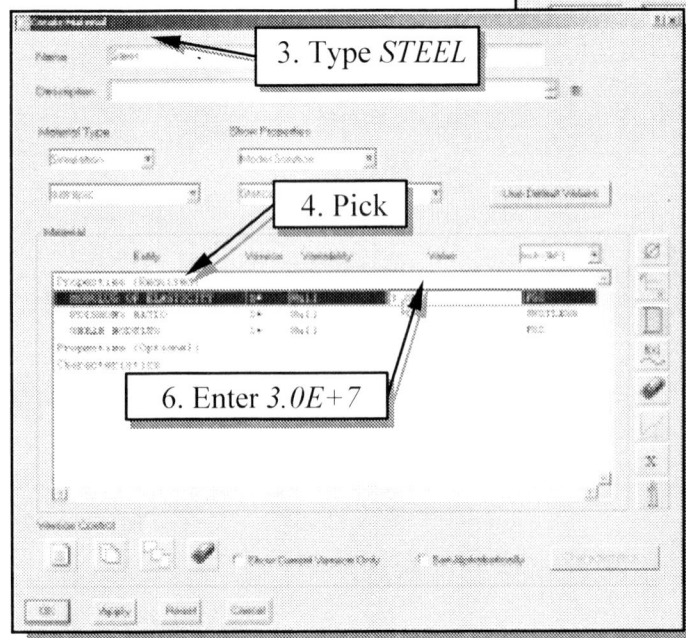

3. Type *STEEL*

4. Pick

6. Enter *3.0E+7*

4. Pick **Modulus of Elasticity** in the *Properties* list.

5. Click the *Value* box of the selected property.

6. Type **3.0E+7** in the *Properties Value* box.

7. Repeat the above steps and pick **Poisson's Ratio** in the *Properties* list.

8. Enter **0.3** in the *Properties Value* box.

Material				
Entity	Version	Variability	Value	Inch (lbF) ▼
Properties (Required)				
MODULUS OF ELASTICITY	1*	Constant	3.0e7	PSI
POISSONS RATIO	1*	Constant	0.3	UNITLESS
SHEAR MODULUS	1*	Null		PSI
Properties (Optional)...				
Characteristics...				

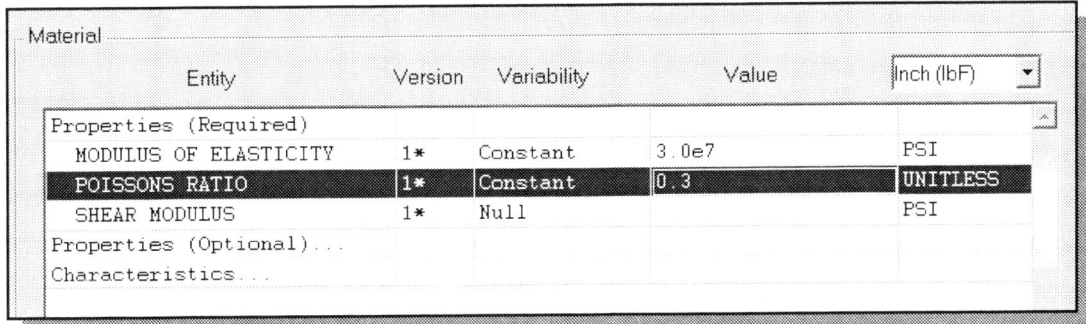

9. Click on the **OK** button to exit the *Create Material* window.

10. Click on the **OK** button to exit the *Materials* window.

6.10 Setting Up an Element Cross Section

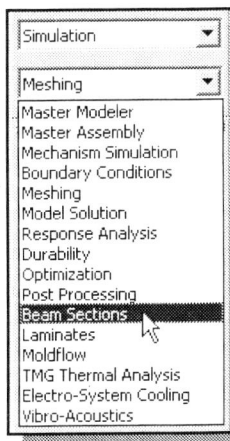

1. Switch to the **Beam Sections** task by selecting the task menu in the icon panel as shown.

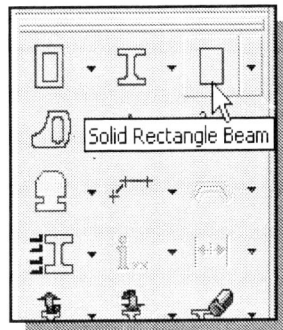

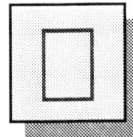

2. Choose **Solid Rectangular Beam** in the icon panel. (The icon is located in the first row of the task icon panel.) The message "*ENTER base*" is displayed.

3. Enter **7.5E-3** for the *base dimension*.

4. Press the **ENTER** key to make the height dimension the same as the base.

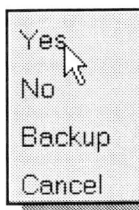

5. Pick **Yes** to complete creating the cross section.

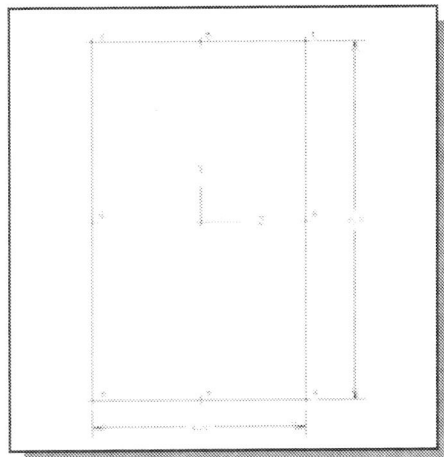

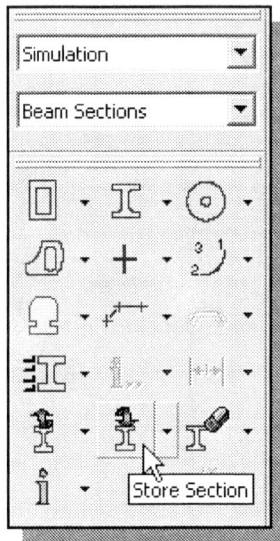

6. Choose **Store Section** in the icon panel. (The icon is located in the fifth row of the task icon panel.) The message "*Enter beam cross sect prop name or no (1-Rectangle 0.0075 x 0.0075).*" is displayed in the prompt window.

7. Press the **ENTER** key to accept the default name *(1-Rectangle 0.0075 x 0.0075).*

6.11 Creating Elements

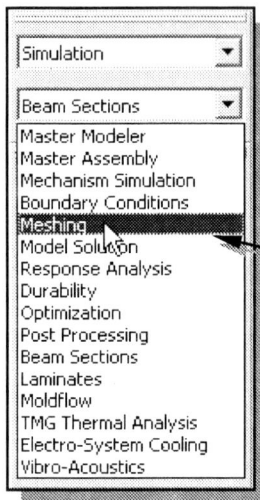

1. Switch back to the **Meshing** task by selecting the Meshing task in the task menu as shown.

Pick Meshing

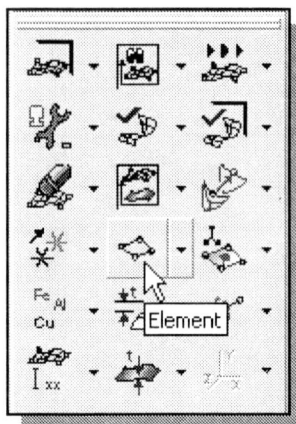

2. Choose **Element** in the icon panel. (The icon is located in the fourth row of the task specific icon panel. If the icon is not on top of the stack, press and hold down the left-mouse-button on the displayed icon to display all the choices.) The *Element* window appears.

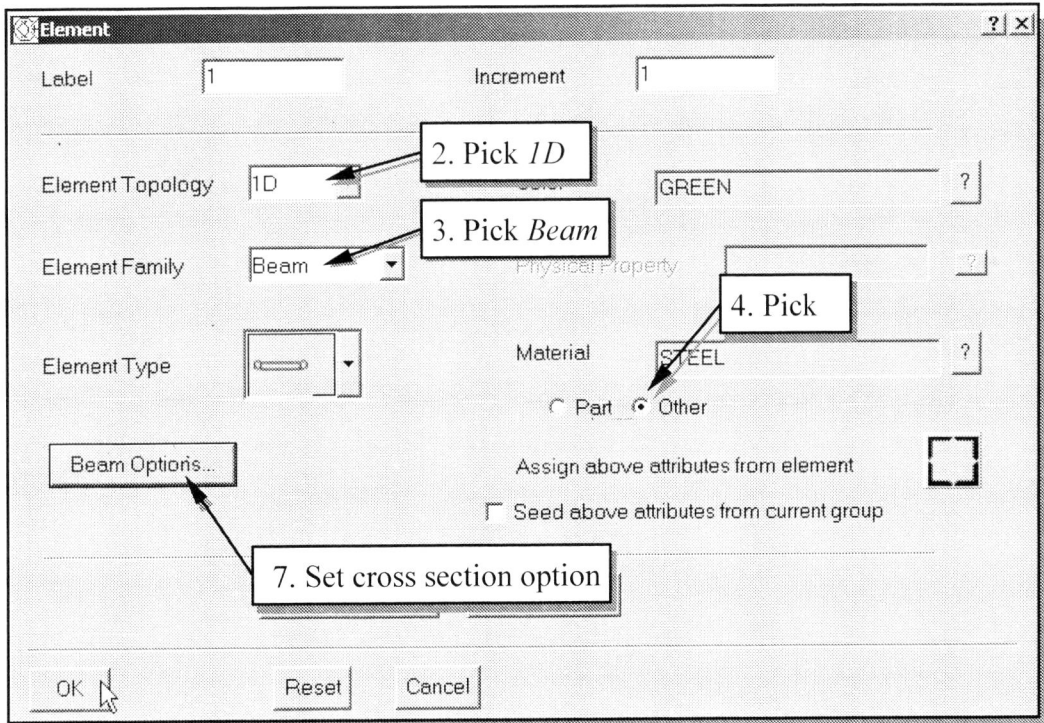

3. In the *Element* window, select **1D** element.

4. Pick **Beam** from the *Element Family* list.

5. Switch to the *Other* option and click on the [?] to set the material property.

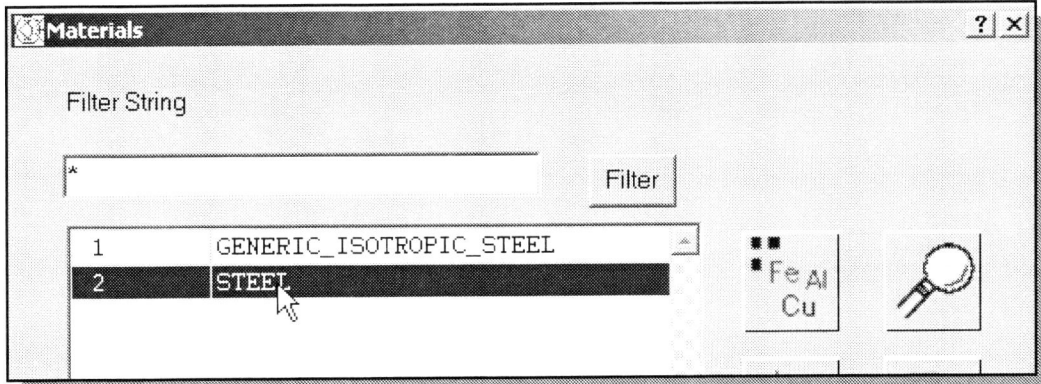

6. Choose **STEEL** in the *Materials* window.

7. Click on the **OK** button to accept the selection.

8. Click on the **Beam Options** button; the *Beam Options* window appears.

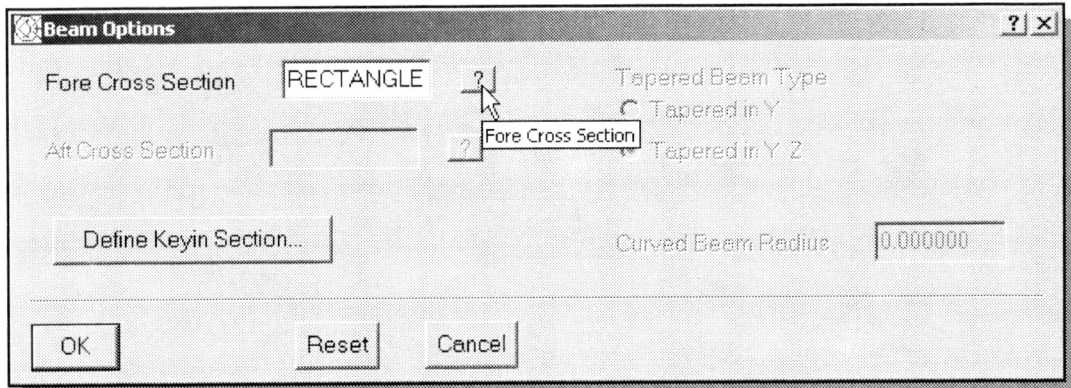

Beam Options

Fore Cross Section	RECTANGLE ?	Tapered Beam Type
		Tapered in Y
Aft Cross Section	? Fore Cross Section	Tapered in Y Z
Define Keyin Section...		Curved Beam Radius 0.000000

OK   Reset   Cancel

9. Click on the [**?**] button and confirm the rectangular section we just created is selected.

10. Click on the **OK** button to exit the *Fore Cross Section* window.

11. Click on the **OK** button to exit the *Beam Options* window.

12. Click on the **OK** button to exit the *Element* window. The message "*Pick Nodes*" is displayed in the prompt window.

13. *I-DEAS* expects us to select two node points to create an element in between. Pick **Node 1** and **Node 2** to create **Element 1** as shown.

14. Create the 12 elements as shown below. Note the **Nx** symbol as the cursor is moved next to a node point.

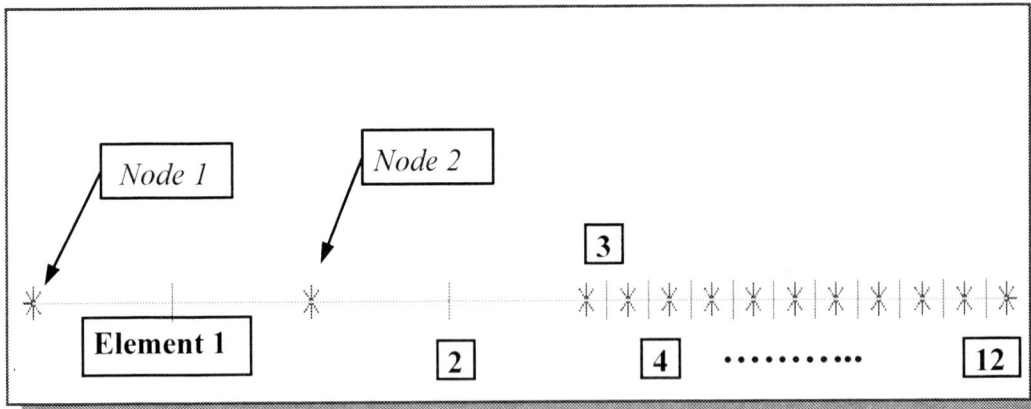

15. Press the **ENTER** key to exit the *Create Element* command.

6.12 Apply Boundary Conditions

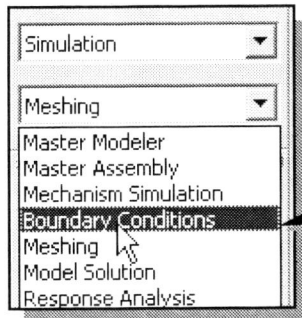

1. Switch to the **Boundary Conditions** task by selecting it in the task menu as shown.

   Pick *Boundary Conditions*

2. Choose ***Displacement Restraint*** in the icon panel. (The icon is located in the fourth row of the task icon panel.)

   Displacement Restraint...

3. The message "*Pick Nodes/Centerpoints/Vertices*" is displayed in the prompt window. Pick ***Node 1***. (If necessary, use the *Dynamic Zoom* function to assist the selection.)

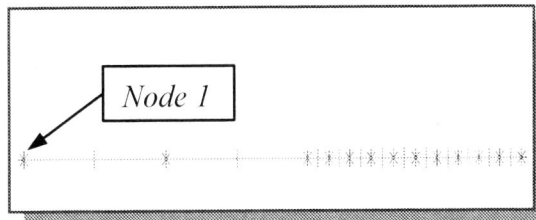

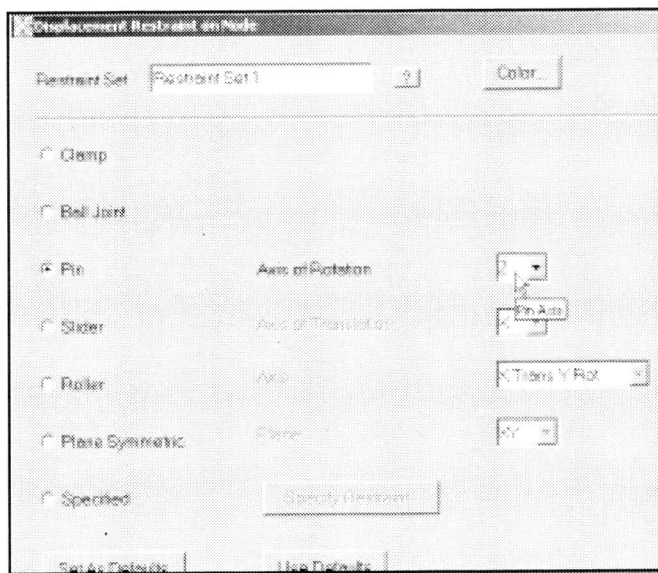

   Node 1

4. Press the **ENTER** key to continue. The *Displacement Restraint on Node* window appears.

5. Pick ***Pin Joint*** and set *Axis of Rotation* to **Z-axis**.

6. Click on the **OK** button to exit the *Displacement Restraint on Node* window.

7. Choose **Displacement Restraint** in the icon panel. (The icon is located in the fourth row of the task icon panel.)

8. The message *"Pick Nodes/Centerpoints/Vertices"* is displayed in the prompt window. Pick **Node 13** as shown below.

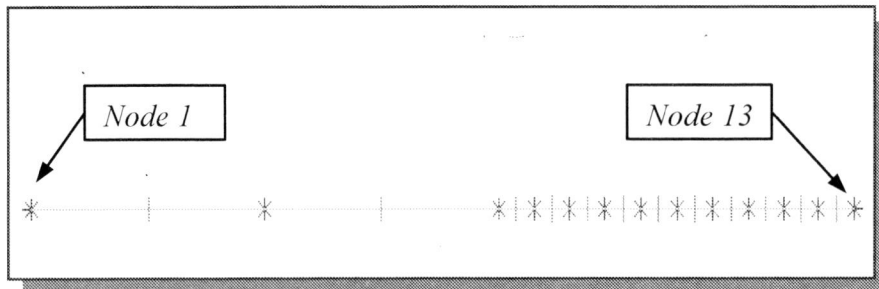

9. Press the **ENTER** key to accept the selection. The *Displacement Restraint on Node* window appears.

10. Pick **Roller** and set *Axis of Rotation* to **X Trans Z Rot**.

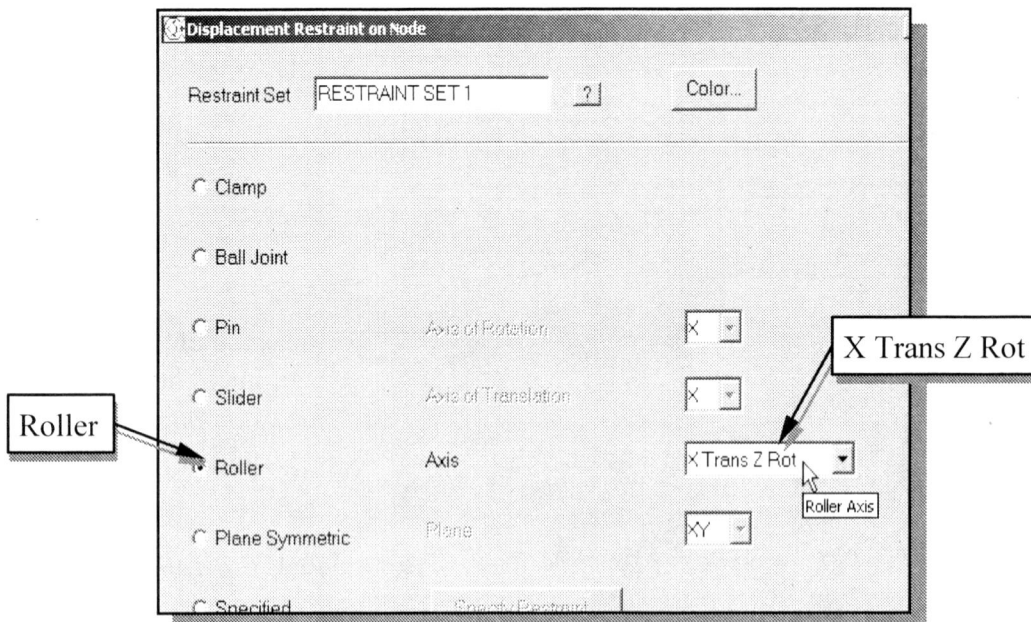

11. Click on the **OK** button to exit the *Displacement Restraint on Node* window.

➢ On your own, confirm that proper boundary conditions are applied to the *I-DEAS* FE model by examining the original structure on page 6-3.

6.13 Concentrated Force

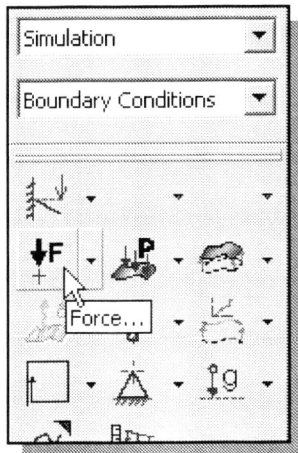

1. Choose **Force** in the icon panel. (The icon is located in the second row of the task icon panel.)

2. The message "*Pick entities*" is displayed in the prompt window. Pick **Node 2** as shown below.

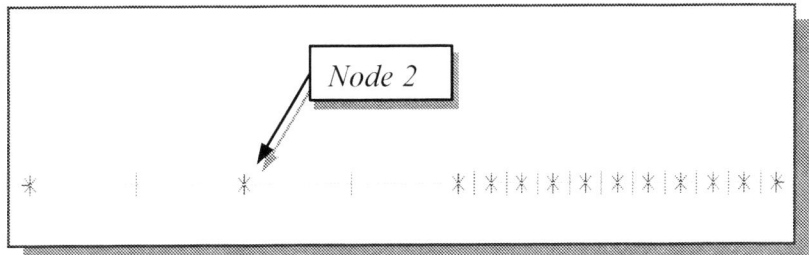

   Node 2

3. Press the **ENTER** key to accept the selection. The *Force on Node* window appears.

4. The load at *Node 2* is 6000 lb. in the negative Y direction. Enter **-6000** in the *Y Force* box.

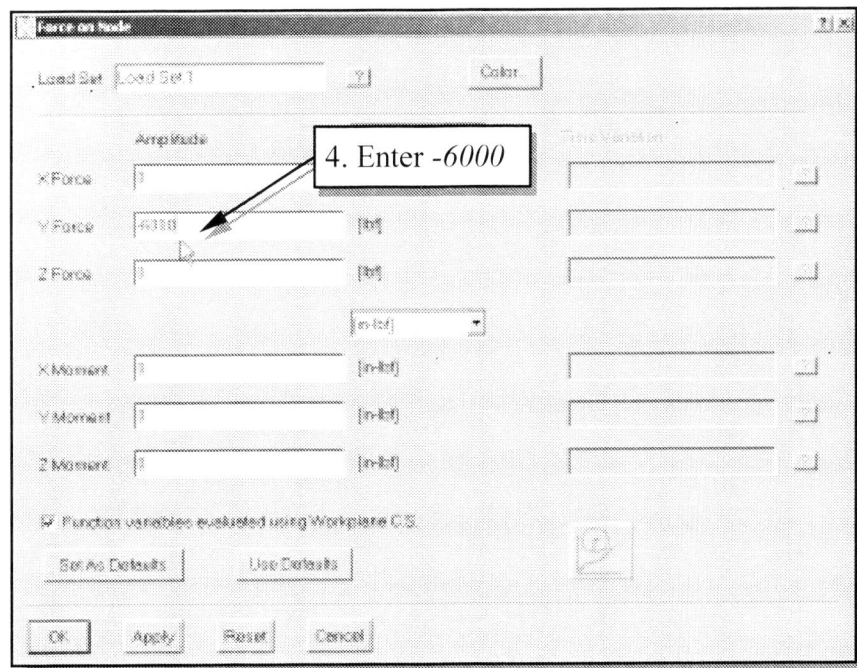

   4. Enter *-6000*

5. Click on the **OK** button to accept the settings and end the *Force on Node* command.

6.14 Distributed Load

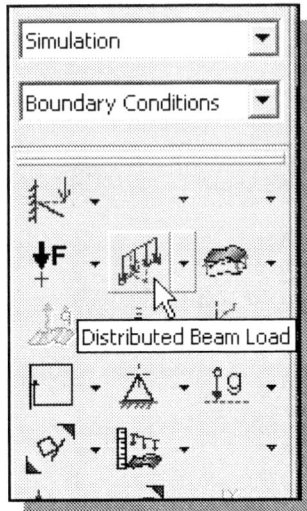

1. Choose **Distributed Load** in the icon panel. (The icon is located in the second row of the task icon panel.) The message "*Select beams to be loaded, Pick Elements*" is displayed in the prompt window.

2. Pick **Element 3** through **Element 12** by enclosing the elements inside a selection window.

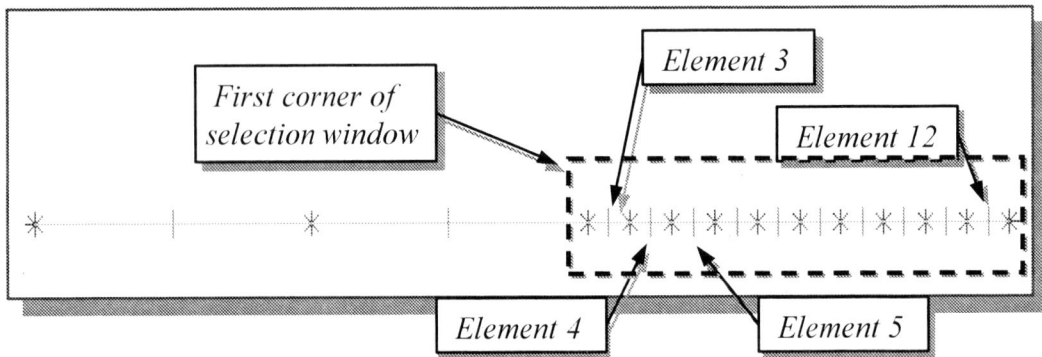

3. Press the **ENTER** key to accept the selections.

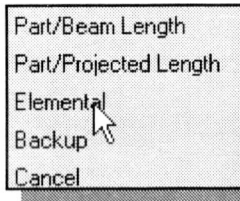

4. The message "*Select coordinate system of forces (Part/Beam length)*" is displayed in the prompt window. Choose **Elemental** in the popup menu.

5. The message "*Select type of load (Constant)*" is displayed in the prompt window. Press the **ENTER** key to apply a uniformly distributed load.

6. The message "*Enter distributed Axial force (0.0)*" is displayed in the prompt window. Press the **ENTER** key to accept the default setting.

7. The message "*Enter distributed Y force (0.0)*" is displayed in the prompt window. Enter the value of **-5000**. The minus sign indicates the direction of the load.

8. The message "*Enter distributed Z force (0.0)*" is displayed in the prompt window. Press the **ENTER** key to accept the default setting.

9. The message "*Enter distributed X moment (0.0)*" is displayed in the prompt window. Press the **ENTER** key to accept the default setting.

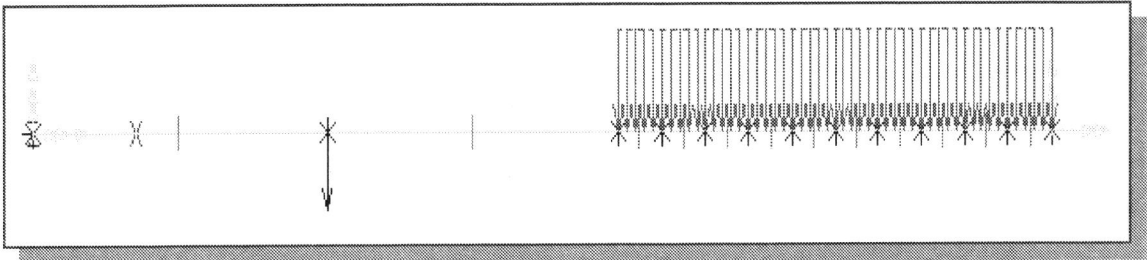

10. The message "*Enter distributed Y moment (0.0)*" is displayed in the prompt window. Press the **ENTER** key to accept the default setting.

11. The message "*Enter distributed Z moment (0.0)*" is displayed in the prompt window. Press the **ENTER** key to accept the default setting.

12. The message "*Enter color name or no. (10-ORANGE)*" is displayed in the prompt window. Press the **ENTER** key to accept the default setting.

6.15 Running the Solver

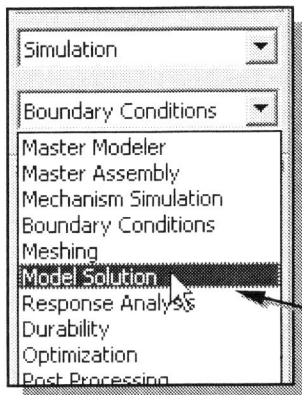

1. Switch to the **Model Solution** task by selecting the task in the task menu as shown.

    Pick *Model Solution*

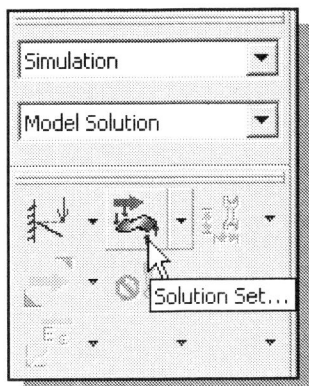

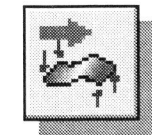

2. Choose **Solution Set** in the icon panel. (The icon is located in the first row of the task icon panel.) The *Manage Solution Sets* window appears.

3. Click on the **Create** button. The *Solution Set* window appears.

4. Click on the **Output Selection** button. The *Output Selection* window appears.

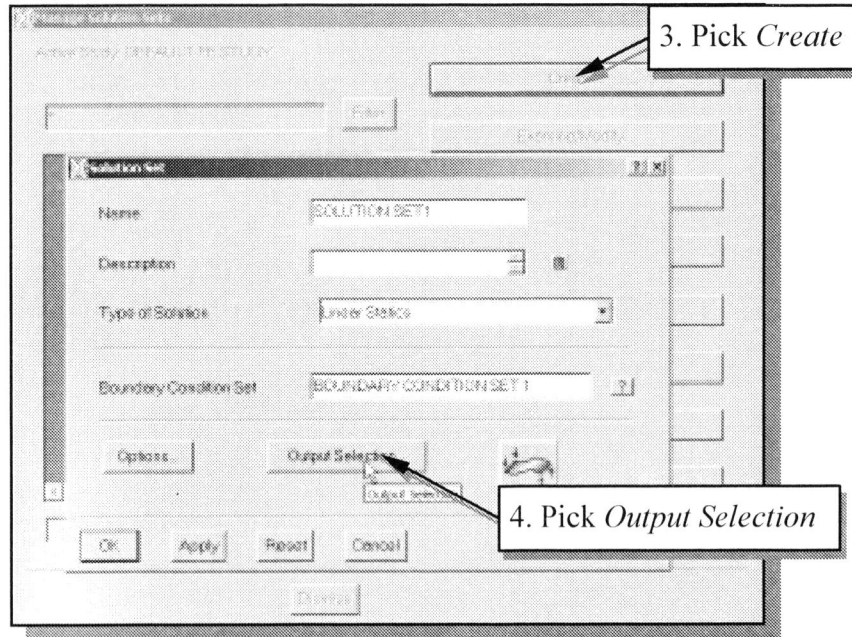

3. Pick *Create*

4. Pick *Output Selection*

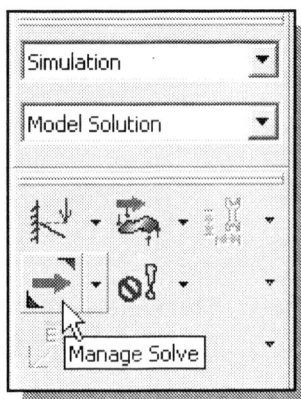

5. On your own, set the **Element Forces** output option to **Store** in the *Output* list.

6. Click on the **OK** button to exit the *Output Selection* window.

7. Click on the **OK** button to exit the *Solution Set* window.

8. Pick **Dismiss** to exit the *Manage Solution Sets* window.

9. Choose **Manage Solve** in the icon panel. (The icon is located in the second row of the task icon panel.) The *Solve* window appears.

10. In the *Solve* window, pick the **Solve** icon to find the solutions.

➤ I-DEAS will begin the solving process. <u>DO NOT</u> close any windows. Any errors or warnings are displayed in the list window.

6.16 Viewing the results

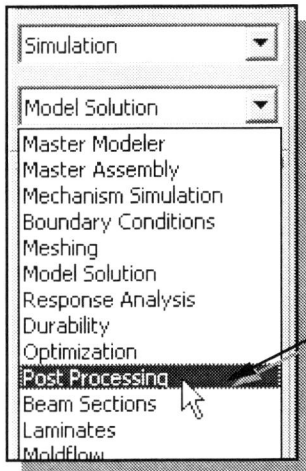

1. Switch to the **Post Processing** task by selecting it in the task menu as shown.

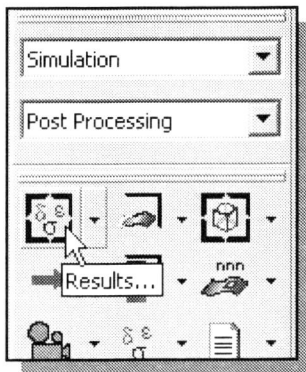

Pick *Post Processing*

2. Choose **Results Selection** in the icon panel. (The icon is located in the first row of the task icon panel.)

3. The *Results Selection* window appears. Pick **Reaction Force** from the list.

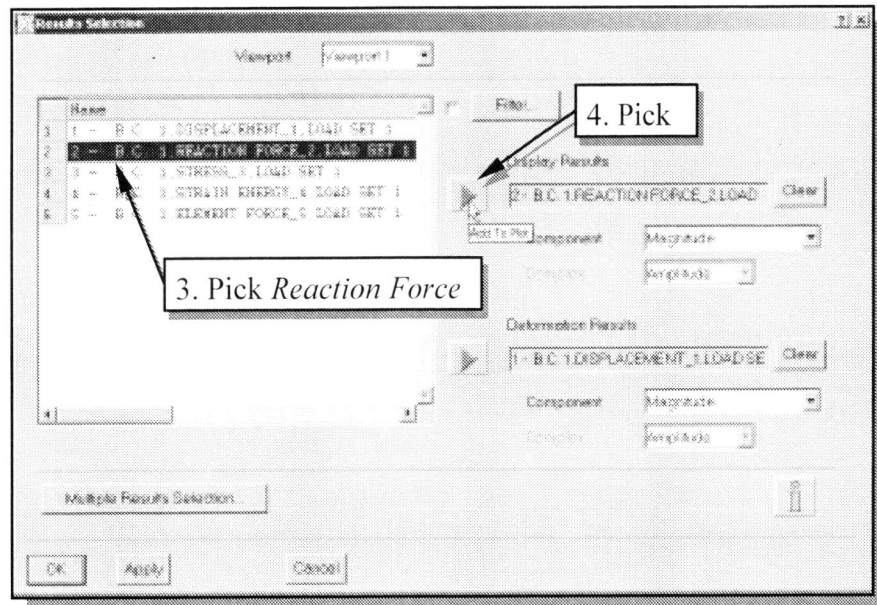

4. Pick

3. Pick *Reaction Force*

4. Click on the **Triangle** button to set the *Reaction Force* as the *Display Results*.

5. Click on the **OK** button to exit the *Results Selection* window.

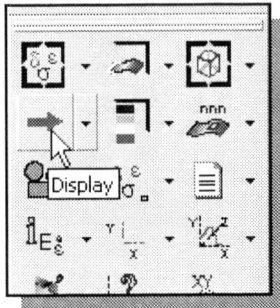

6. Choose **Results Display** in the icon panel. (The icon is located in the second row of the task icon panel.)

7. The message "*Pick elements*" is displayed in the prompt window. Pick **Element 12** as shown below, note the E12 symbol displayed near the center of the element.

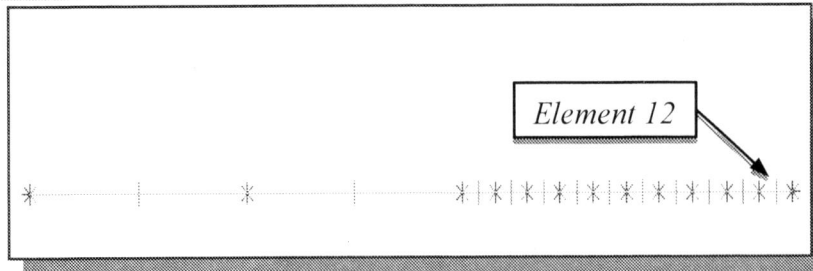

Element 12

8. Press the **ENTER** key or the middle-mouse-button accept the selection and display the reaction force associated with this element.

➤ The computer solution is **1.43E+05 lb**. Our hand calculation was **13.5E+03 lb**. (see section 6.4 Preliminary Analysis). What went wrong?

Redisplay	F12
Refresh	Alt+R
Stereo Viewing	

9. Choose **Refresh** in the icon panel. (The icon is located in the first row of the display icon panel.)

10. Choose **Results Display** in the icon panel. (The icon is located in the second row of the task icon panel.)

11. The message "*Pick elements*" is displayed in the prompt window. Pick **Element 1** as shown.

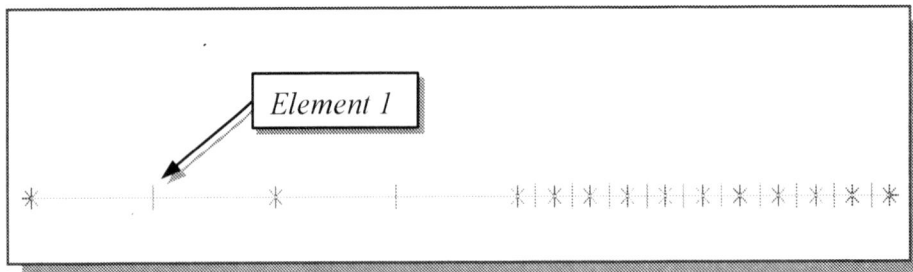

Element 1

12. Press the **ENTER** key or the middle-mouse-button to display the reaction force associated with this element.

➤ The computer solution is **4.29E+04 lbs**., and our preliminary calculation (see section 6.4 Preliminary Analysis) was **7.5E+03 lbs**. The computer solution is about five times more than our preliminary calculation.

6.17 What Went Wrong?

❖ The computer solution indicates the total external load is **18.59 E+4 lb** (186 kips). It should be **21 kips** (6 + 5 x 3 = 21 kips). Examining the FE model, it becomes clear that the problem occurred when we applied the distributed load. The distributed load should be **5kip/ft**, but the system of units we are using is *inches (pounds)*. **The units used __MUST__ be consistent throughout the analysis.**

Converting the distributed load to lb/in:

$$5\text{kip/ft} = (5000/12) \text{ lb/in}$$

6.18 Correct the Mistake

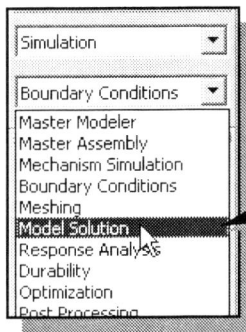

1. Switch to the **Model Solution** task by selecting the task in the task menu as shown.

Pick *Model Solution*

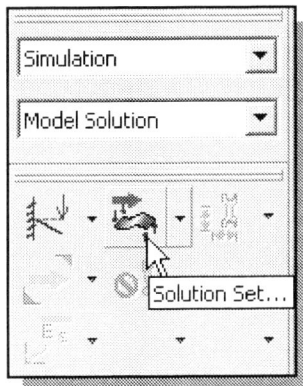

2. Choose **Solution Set** in the icon panel. (The icon is located in the first row of the task icon panel.)

3. The *Manage Solution Sets* window appears. Select **Solution SET1**.

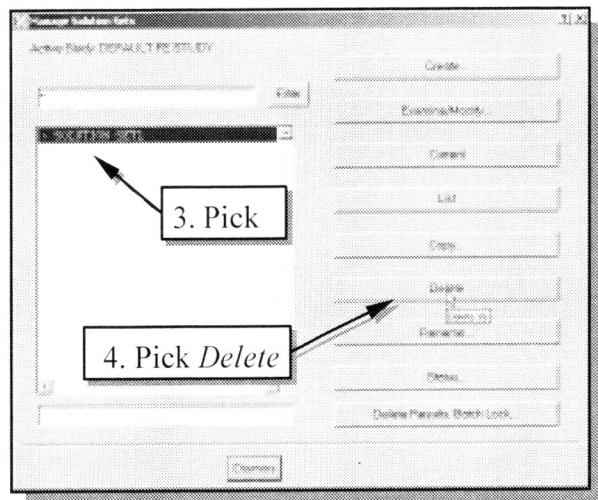

4. Click on the **Delete** button to remove the solution set.

5. Click on the **OK** button to confirm deleting *Solution Set1*.

6. Click on the **Dismiss** button to exit the *Manage Solution Sets* window.

3. Pick

4. Pick *Delete*

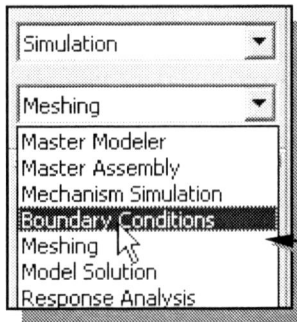

7. Switch to the **Boundary Conditions** task by selecting it in the task menu as shown.

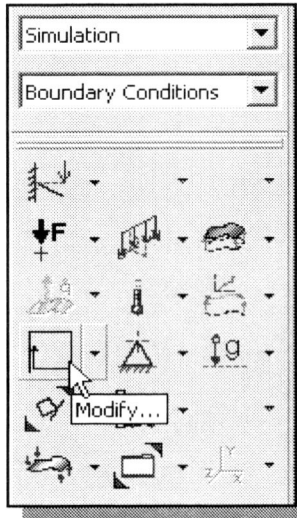

Pick *Boundary Conditions*

8. Choose **Modify** in the icon panel. (The icon is located in the fourth row of the task icon panel.)

9. The message "*Pick entities*" is displayed in the prompt window. Pick the distributed loads by enclosing the elements inside a selection window.

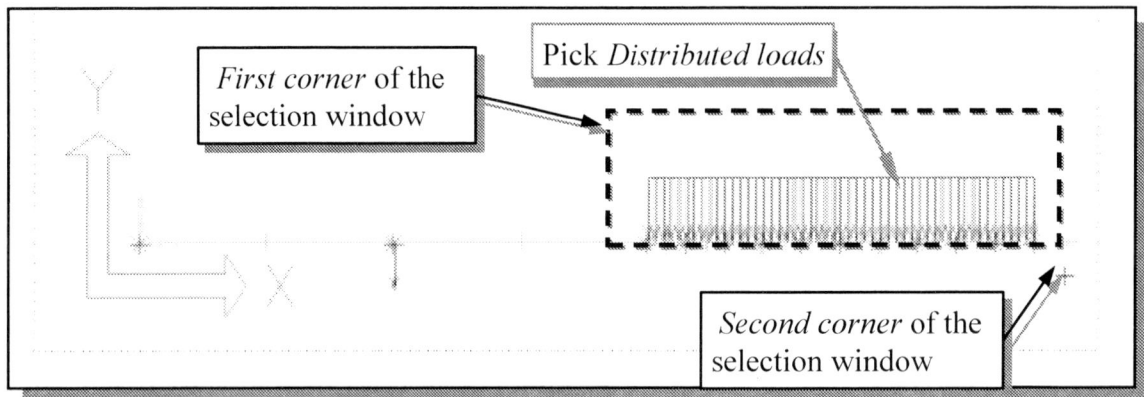

First corner of the selection window

Pick *Distributed loads*

Second corner of the selection window

10. Press the **ENTER** key or the middle-mouse-button to accept the selections and proceed to modify the selected distributed loads.

➤   On your own, re-enter the information regarding the distributed load. At the prompt "*Enter distributed Y force (0.0)*" is displayed in the prompt window. Enter the value of **-5000/12**. The minus sign indicates the direction of the load. Refer to page 6-18 if you are uncertain about any of the specific definitions of the loads.

6.19 Run the Solver

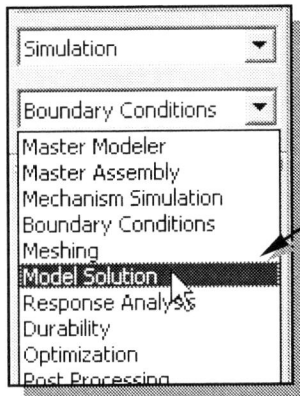

1. Switch to the **Model Solution** task by selecting the task in the task menu as shown.

Pick *Model Solution*

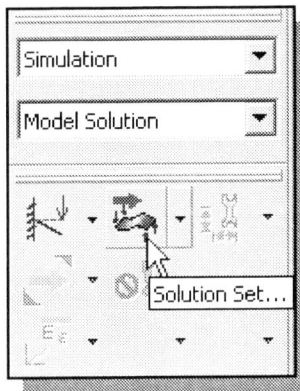

2. Choose **Solution Set** in the icon panel. (The icon is located in the first row of the task icon panel.) The *Manage Solution Sets* window appears.

3. Click on the **Create** button. The *Solution Set* window appears.

4. Click on the **Output Selection** button. The *Output Selection* window appears.

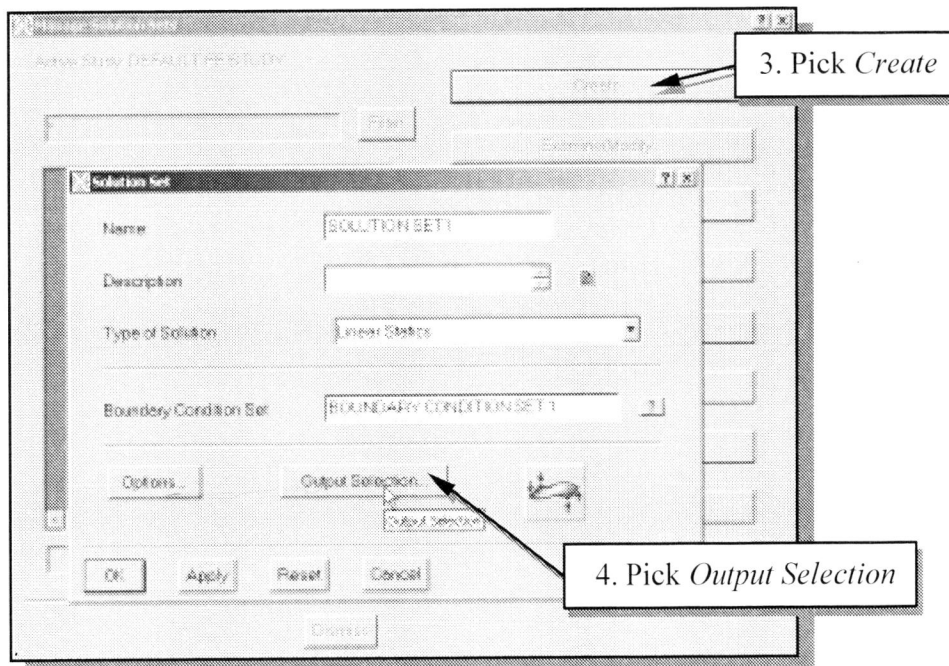

3. Pick *Create*

4. Pick *Output Selection*

5. On your own, set the **Element Forces** output option to **Store** in the *Output* list.

6. Click on the **OK** button to exit the *Output Selection* window.

7. Click on the **OK** button to exit the *Solution Set* window.

8. Pick **Dismiss** to exit the *Manage Solution Sets* window.

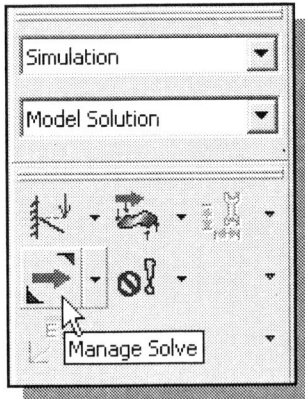

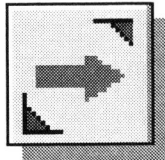

9. Choose ***Manage Solve*** in the icon panel. (The icon is located in the second row of the task icon panel.) The *Solve* window appears.

10. In the *Solve* window, pick the ***Solve*** icon to find the solutions.

➤ *I-DEAS* will begin the solving process. <u>DO NOT</u> close any windows. Any errors or warnings are displayed in the list window.

6.20 View the Results – Reaction Forces

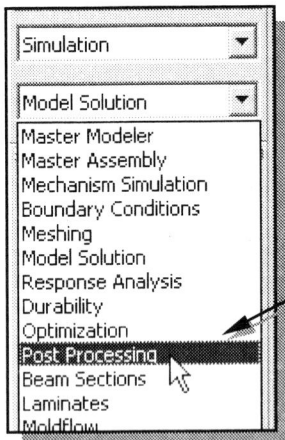

1. Switch to the **Post Processing** task by selecting it in the task menu as shown.

    Pick *Post Processing*

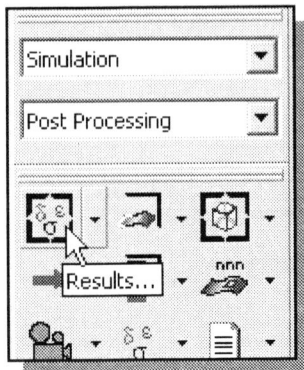

2. Choose ***Results Selection*** in the icon panel. (The icon is located in the first row of the task icon panel.)

3. The *Results Selection* window appears. Pick **Reaction Force** from the list.

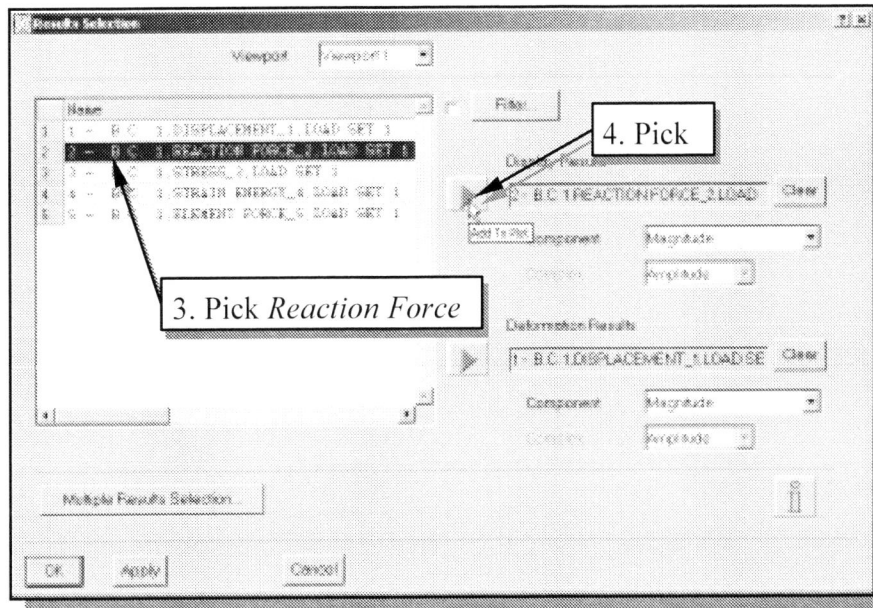

4. Click on the **Triangle** button to set the *Reaction Force* as *the Display Results*.

5. Click on the **OK** button to exit the *Results Selection* window.

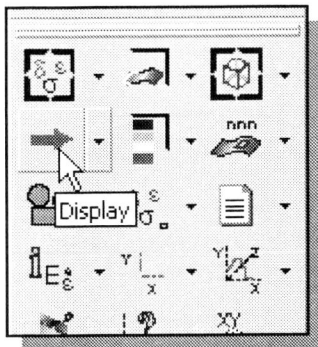

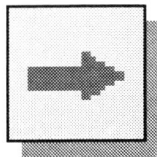

6. Choose **Results Display** in the icon panel. (The icon is located in the second row of the task icon panel.)

7. The message "*Pick elements*" is displayed in the prompt window. Pick **Element 12** as shown below, note the E12 symbol displayed near the center of the element.

8. Press the **ENTER** key or the middle-mouse-button accept the selection and display the reaction force associated with this element.

➤ The computer solution is **1.35E+04 lb.**, which matches our hand calculation **13.5 kips** (see section 6.4 Preliminary Analysis).

> ➤ On your own, examine the reaction at the pin joint (left-end of the structure). Does the computer solution match with our preliminary calculation?

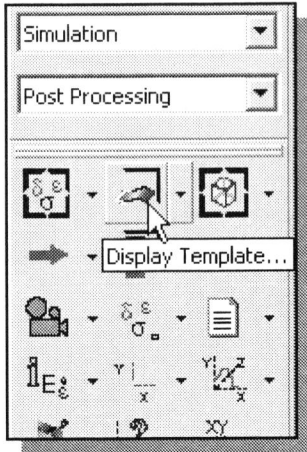

9. Choose **Display Template** in the icon panel. (The icon is located in the first row of the task icon panel.)

10. Click on the **Contour** button.

11. Click on **Fast Display** to switch off the *Fast Display* option as shown.

12. Click the **OK** button to exit the *Contour Options* window.

13. Click the **OK** button to exit the *Display Template* window.

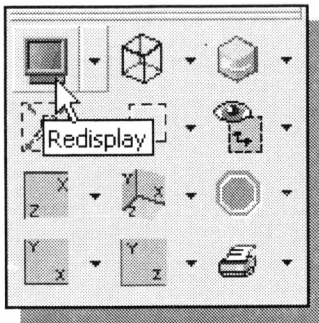

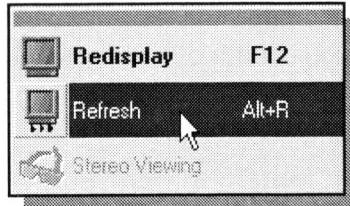

14. Before continuing to the next section, choose **Refresh,** in the icon panel, to reset the screen display.

6.21 Shear Diagram

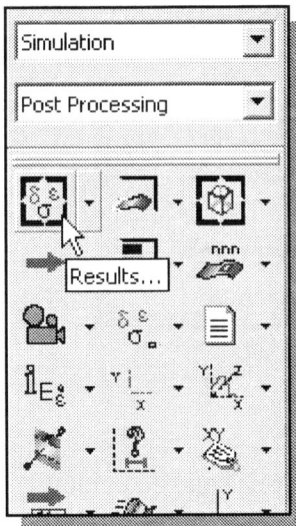

1. Choose **Results Selection** in the icon panel. (The icon is located in the first row of the task icon panel.)

2. The *Results Selection* window appears. Pick **Element Force** from the list.

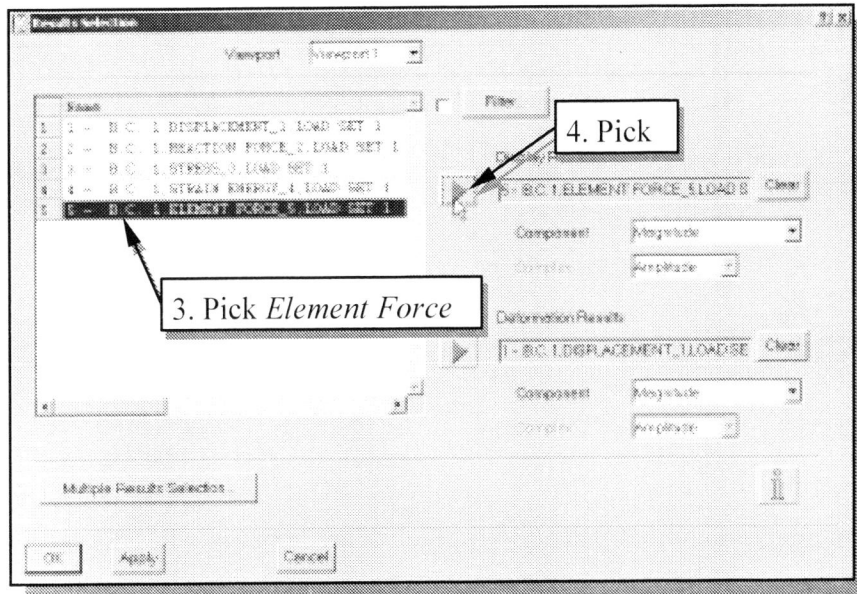

3. Click on the **Triangle** button to set the *Element Force* as *the Display Results*.

4. Click on the **OK** button to exit the *Results Selection* window.

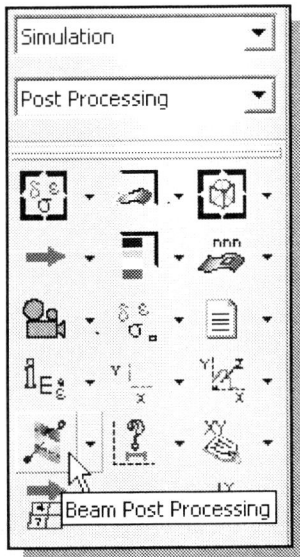

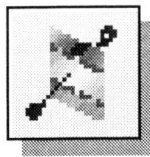

5. Choose **Beam Post Processing** in the icon panel. (The icon is located in the fifth row of the task icon panel.) A cascading menu appears in the graphics window.

6. Select **Force & Stress** in the cascading menu.

7. Select **Data Component** in the popup menu.

8. Select **Force** in the popup menu.

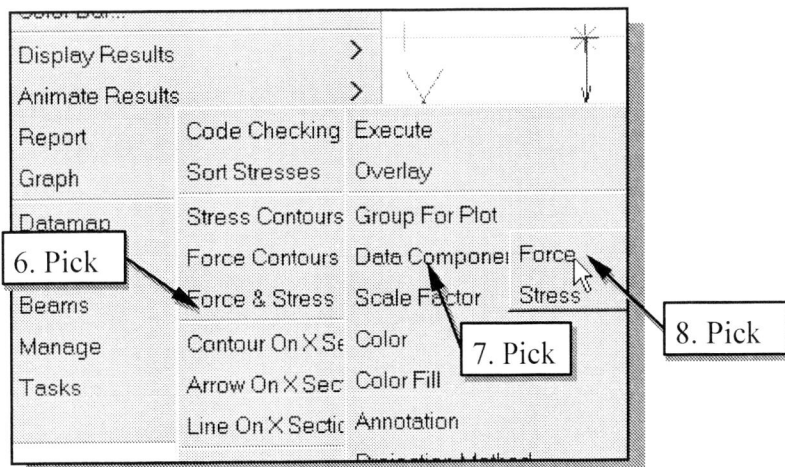

Axial Force
Shear Force In Y
Shear Force In Z
Torque
Moment About Y
Moment About Z
SRSS Shear Force
SRSS Bending Mome

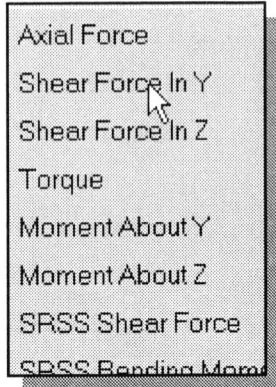

9.   In the popup menu, select **Shear Force in Y**.

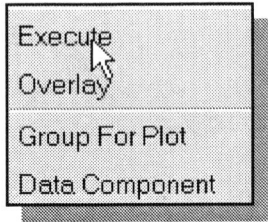

Execute
Overlay
Group For Plot
Data Component

10. In the cascading menu, select
Execute.

11. Press down the **CTRL** key and hit the **M** key to switch **off** the cascading menu.

❖   The shear diagram should appear as shown below. Does the diagram match
with the preliminary analysis we did at the beginning of the chapter?

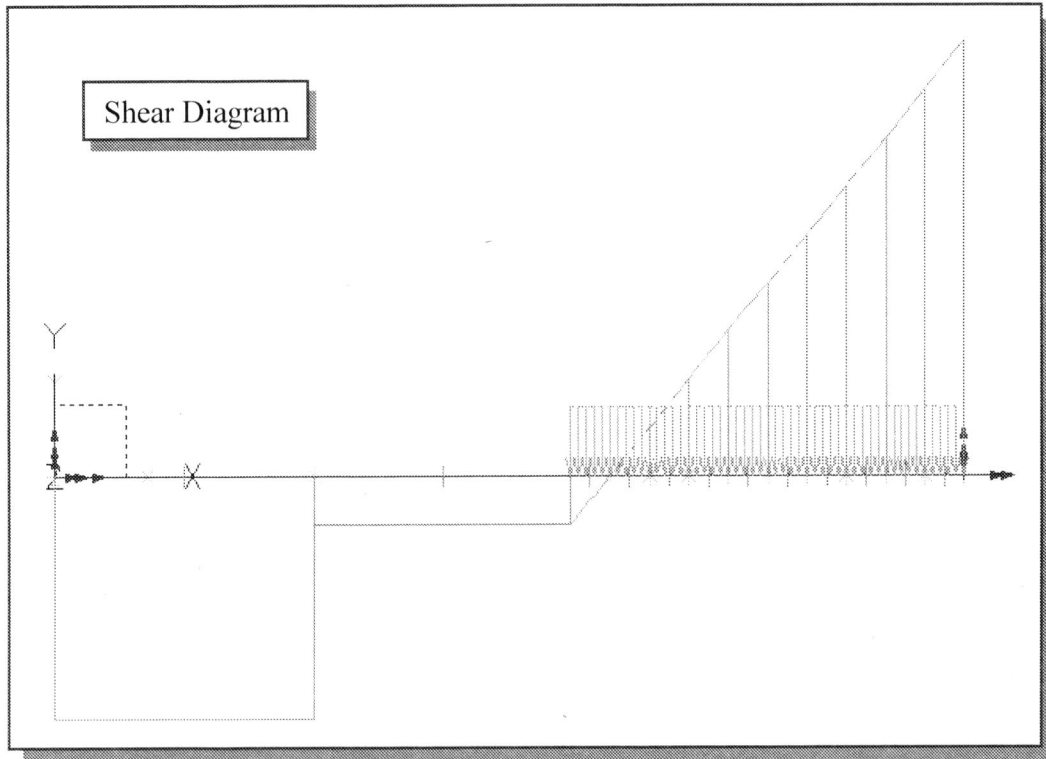

Shear Diagram

➢   Note the computer calculation of the maximum shear force is **13500 lb**.

Report   Code Checking Execute
Graph   Sort Stresses Overlay
Datamap   Stress Contours Group For Plot
Path   Force Contours Data Component
Beams   Force & Stress Scale Factor
Manage   Contour On X Se Color
Tasks   Arrow On X Sec Color Fill

❖   Note that choosing **Beam Post
Processing → Force & Stress → Scale
Factor**, you can enter a scale factor to adjust
the scale of the displayed diagram.

6.22 Moment Diagram

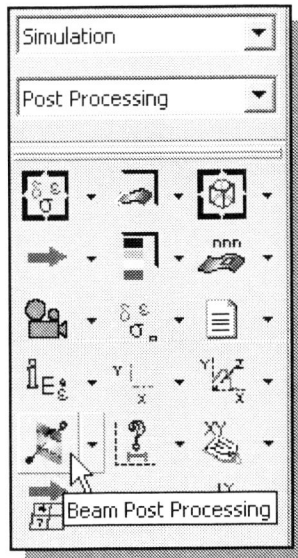

1. Choose **Beam Post Processing** in the icon panel. (The icon is located in the fifth row of the task icon panel.) The cascading menu appears in the graphics window.

2. Select **Force & Stress** in the cascading menu.

3. Select **Data Component** in the popup menu.

4. Select **Force** in the popup menu.

2. Pick
3. Pick
4. Pick

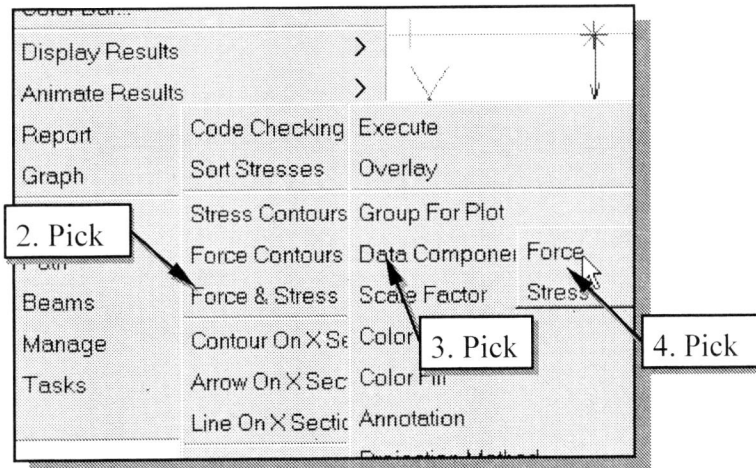

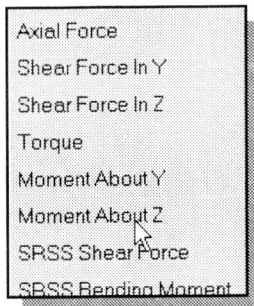

5. In the popup menu, select **Moment About Z.**

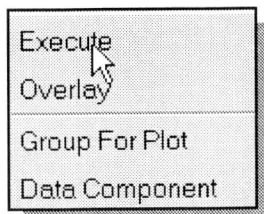

6. In the cascading menu, select **Execute**.

7. Press down the **CTRL** key and hit the **M** key to switch off the cascading menu.

❖   The shear diagram should appear as shown below. Does the diagram match with the preliminary analysis we did at the beginning of the chapter?

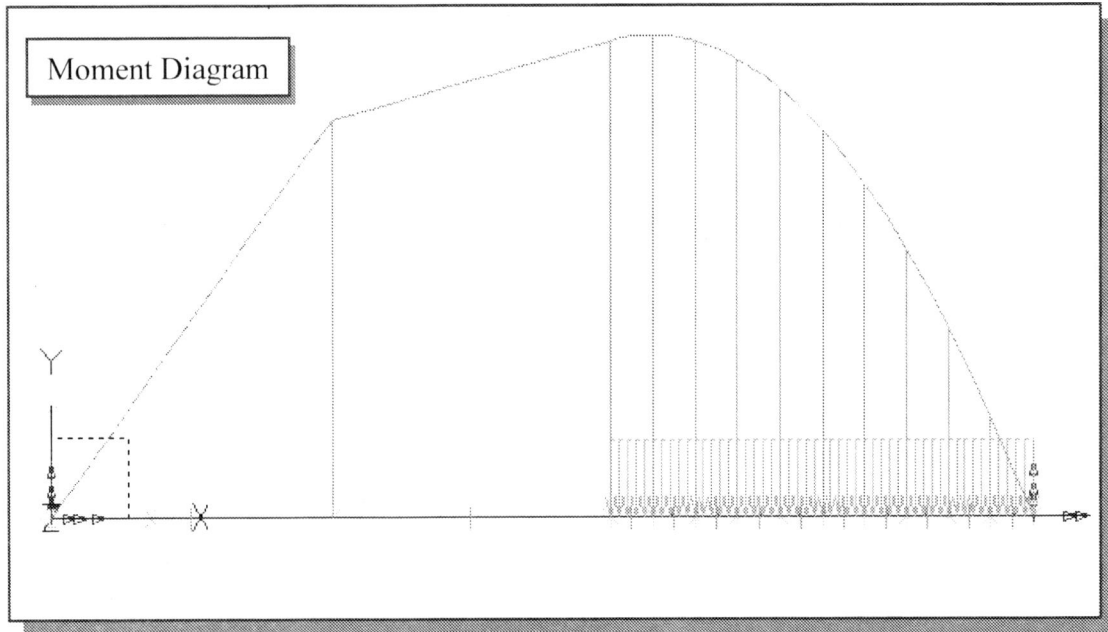

Moment Diagram

➢   Note the computer calculated the maximum value of **218700 in-lb**. Does this value match with the preliminary analysis we did at the beginning of the chapter?

Questions:

1. For a *beam element* in three-dimensional space, what is the number of degrees of freedom it possesses?

2. What are the assumptions for the beam element?

3. What are the differences between *Truss Element* and *Beam Element*?

4. What are the relationships between the shear diagram and the moment diagram?

5. Identify and describe the following commands:

   (a)

   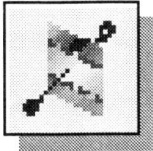

   (b)

   (c)

   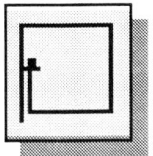

   (d)

Exercises:

Determine the maximum stress produced by the loads and create the shear and moment diagram.

1.   Material: Steel, Diamater 2.5 in.

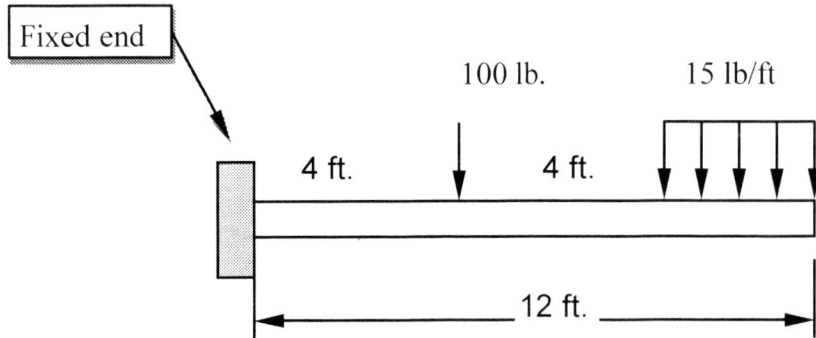

| Fixed end |

100 lb.   15 lb/ft

4 ft.   4 ft.

12 ft.

2.   Material: Aluminum

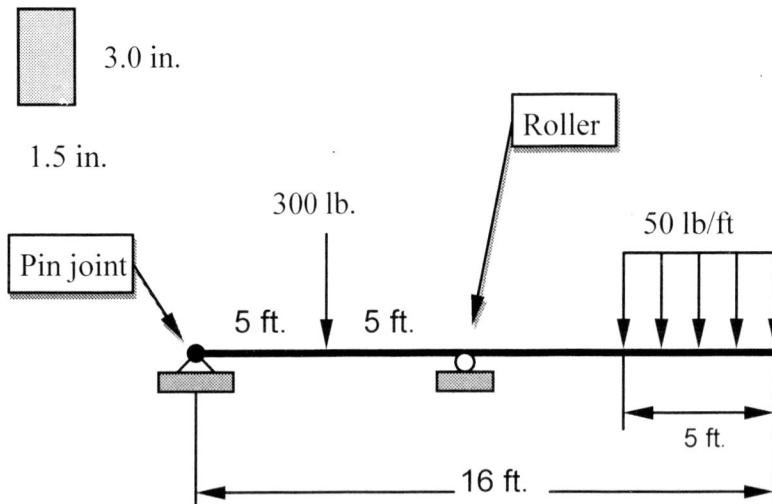

3.0 in.

1.5 in.

| Roller |

300 lb.   50 lb/ft

| Pin joint |

5 ft.   5 ft.

5 ft.

16 ft.

Chapter 7
Beam Analysis Tools

```
RESULTS: 5- B.C. 1,ELEMENT FORCE_5,LOAD SET 1
Data component: Z BENDING MOMENT
Maximum amplitude =  95232.3
```

Shear Diagram

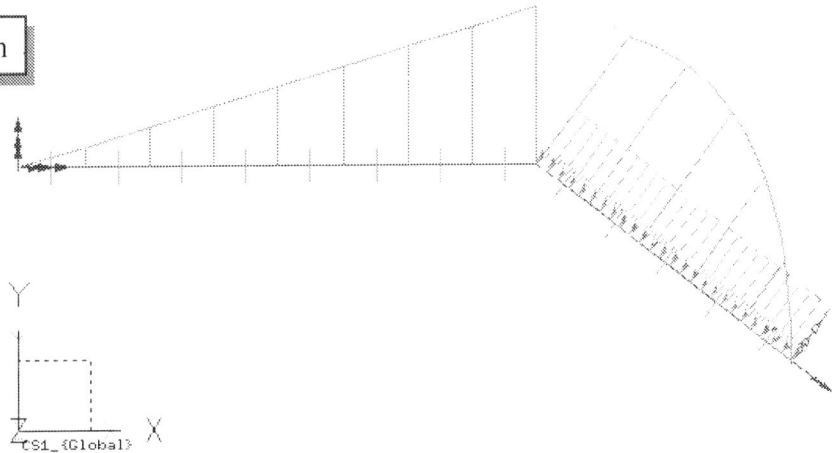

Y

X

CS1_{Global}

Learning Objectives

When you have completed this lesson, you will be able to:
- ◆ **Create FE models using the Auto-Mesh command.**
- ◆ **Create User Coordinate Systems.**
- ◆ **Apply Elemental loads on Beam Elements.**
- ◆ **Create Inclined Supports.**
- ◆ **Use the Modify Node command.**
- ◆ **Analyze Structures with Combined Stresses.**

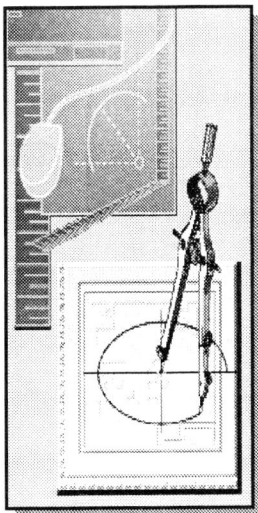

7.1 Introduction

The beam element is the most common of all structural elements as it is commonly used in buildings, towers, bridges and many other structures. The use of beam elements in finite element models provides the engineer with the capability to solve rather complex structures, which could not be easily done with conventional approaches. In many cases, the use of beam elements also provides a very good approximation of the actual three-dimensional members, and there is no need to do a three-dimensional analysis. *I-DEAS* provides an assortment of tools to make the creation of finite element models easier and more efficient. This chapter demonstrates the use of *automatic mesh-generation* to help generate nodes and elements, the use of a *displacement coordinate system* to account for inclined supports, and the application of *elemental loads* along individual elements. The effects of several internal loads that occur simultaneously on a member's cross-section, such as axial load and bending, are considered in the illustrated example. Although the illustrated example is a two-dimensional problem, the principle can also be applied to three-dimensional beam problems as well as other types of elements.

7.2 Problem Statement

Determine the state of stress at *point C* (measured 1.5 m from *point A*), and the maximum normal stress that the load produces in the member. (Structural Steel A36)

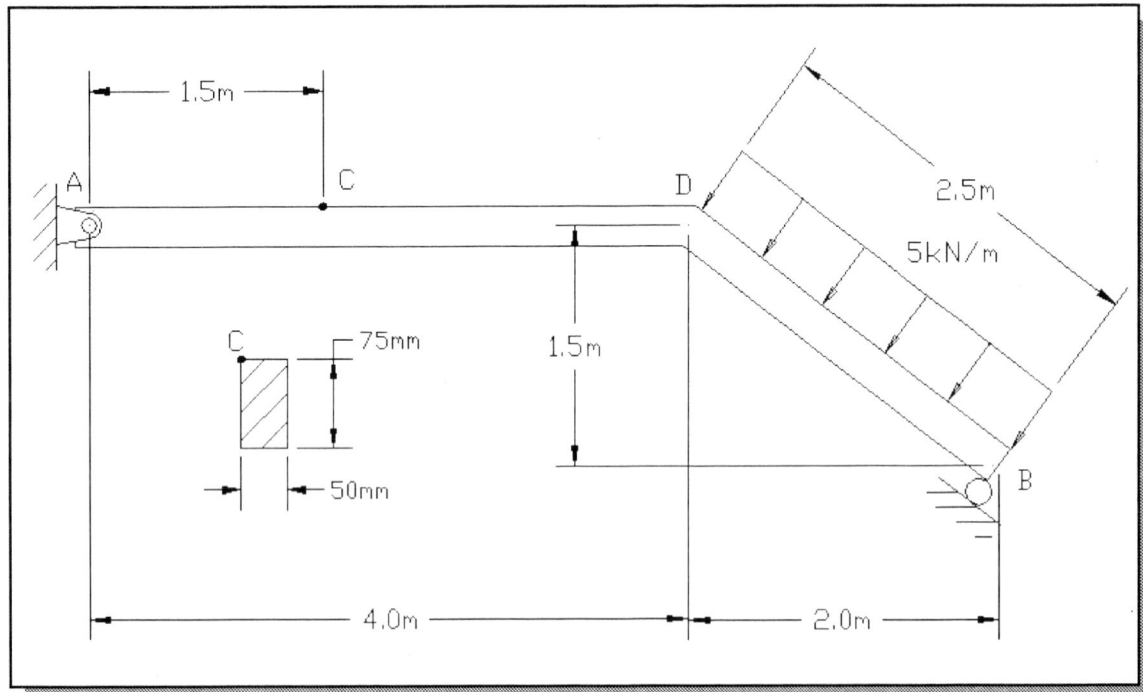

7.3 Preliminary Analysis

Free Body Diagram of the member:

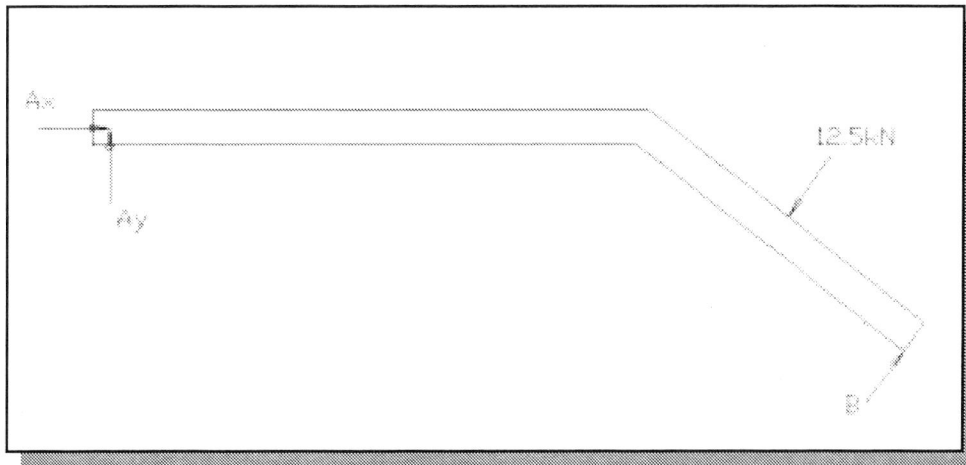

Applying the equations of equilibrium:

$$\Sigma M_{@A} = 0, \ \Sigma F_X = 0, \ \Sigma F_Y = 0,$$

Therefore $B = 9.76$ kN, $A_X = 1.64$ kN and $A_Y = 2.19$ kN

Consider a segment of the horizontal portion of the member:

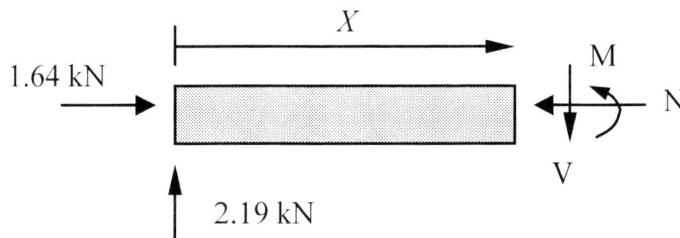

$\Sigma F_X = 0, \quad 1.64 - N = 0, N = 1.64$ kN
$\Sigma F_Y = 0, \quad 2.19 - V = 0, V = 2.19$ kN
$\Sigma M_{@X} = 0, \ - A_Y X + M = 0, \quad M = 2.19 \ X$ kN-m

At *point C*, X = 1.5 m and

$$N = 1.64 \text{ kN}, \ V = 2.19 \text{ kN}, \ M = 3.29 \text{ kN-m}$$

➤ The state of stress at *point C* can be determined by using the principle of superposition. The stress distribution due to each loading is first determined, and then the distributions are superimposed to determine the resultant stress distribution. The principle of superposition can be used for this purpose provided a linear relationship exists between the stress and the loads.

7.4 Stress Components

Normal Force:
   The normal stress at C is a compressive uniform stress.

$$\sigma_{normal\_force} = 1.64kN/(0.075 \times 0.05)m^2 = 0.437 \text{ MPa}$$

Shear Force:
   Point C is located at the top of the member. No shear stress existed at *point C*.

$$\tau_{shear\_force} = 0$$

Bending Moment:
   Point C is located at 37.5 mm from the neutral axis. The normal stress at C is a compressive uniform stress.

$$\sigma_{bending\_moment} = (3.29kN\text{-}m \times 0.0375m)/(1/12 \times 0.05 \times (0.075)^3)m^4$$
$$= 70.16 \text{ MPa}$$

Superposition:
   The shear stress is zero and combining the normal stresses gives a compressive stress at *point C*:

$$\sigma_C = 0.437 \text{ MPa} + 70.16 \text{ MPa} = 70.6 \text{ MPa}$$

Examine the horizontal segment of the member:

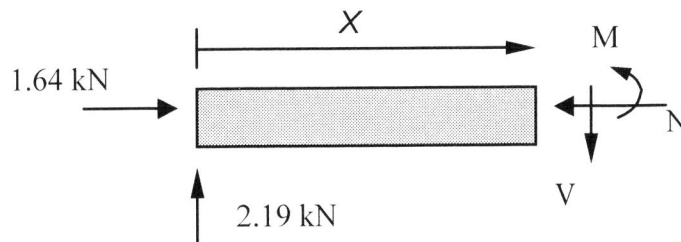

$\sum F_X = 0, \quad 1.64 - N = 0, N = 1.64$ kN

$\sum F_Y = 0, \quad 2.19 - V = 0, V = 2.19$ kN

$\sum M_{@X} = 0, \quad -A_Y X + M = 0, \quad M = 2.19 \, X$ kN-m

The maximum normal stress for the horizontal segment will occur at *point D*, where X = 4m.

$$\sigma_{normal\_force} = 1.64kN/(0.05 \, X \, 0.075)m^2 = 0.437 \text{ MPa}$$

$$\sigma_{bending\_moment} = (8.76kN\text{-}m \, X \, 0.0375m)/(1/12 \, X \, 0.05 \, X \, (0.075)^3)m^4$$
$$= 186.9MPa$$

$$\sigma_{max@D} = 0.437 \text{ MPa} + 186.9 \text{ MPa} = 187.34 \text{ MPa}$$

➢ Does the above calculation provide us the maximum normal stress developed in the structure? To be sure, it would be necessary to check the stress distribution along the inclined segment of the structure. We will rely on the *I-DEAS* solutions to find the maximum stress developed. The above calculation (the state of stress at *point C* and *point D*) will serve as a check to the *I-DEAS* FEA solutions.

7.5 Starting *I-DEAS*

1. Login to the computer and bring up *I-DEAS*. Start a new model file by filling in the items as shown below in the *I-DEAS Start* window:

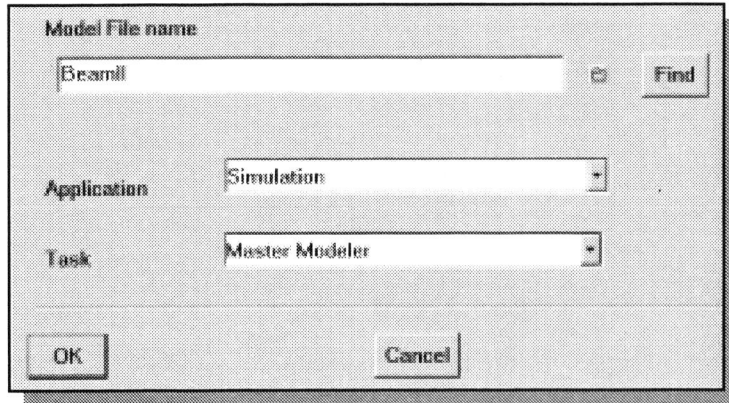

2. After you click **OK**, a *warning window* will appear indicating a new model file will be created. Click **OK** to continue.

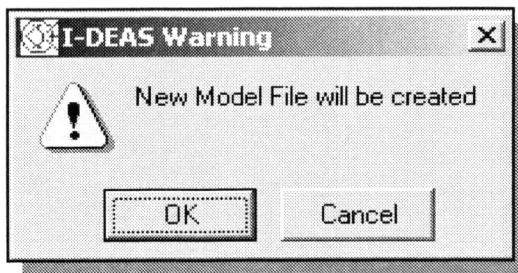

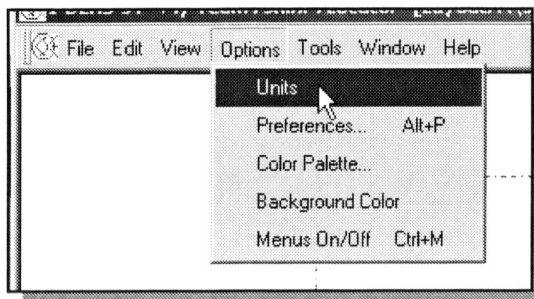

3. Use the left-mouse-button and select the **Options** menu in the icon panel.

4. Select the **Units** option.

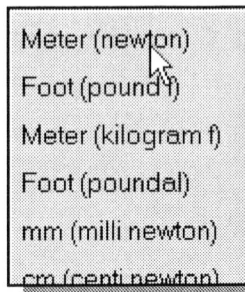

5. Set the units to **Meter (newton)** by selecting it from the menu of choices.

❖ The units used in an FEA analysis <u>MUST</u> be consistent throughout the analysis.

7.6 Workplane Appearance

❖ The workplane is a construction tool; it is a coordinate system that can be moved in space. The size of the workplane display is only for our visual reference, since we can sketch on the entire plane, which extends to infinity.

1. Choose **Workplane Appearance** in the icon panel. (The icon is located in the second row of the application icon panel.) The *Workplane Attributes* window appears.

2. Toggle **on** the *Display Border* switch as shown.

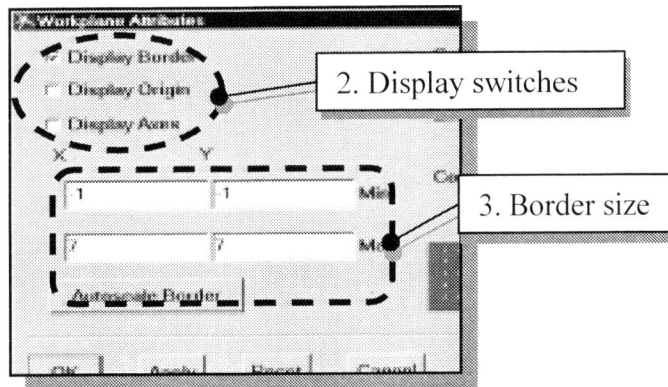

2. Display switches

3. Border size

3. Adjust the **workplane border size** by entering the *Min. & Max.* values as shown in the figure above.

4. Click on the **OK** button to exit the *Workplane Attributes* window.

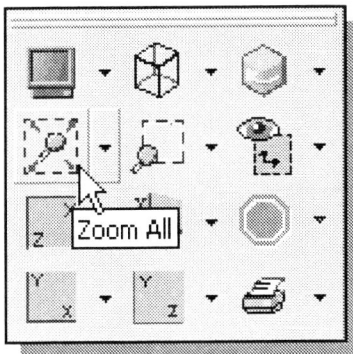

5. Choose **Zoom-All** in the display viewing icon panel.

6. Choose **Redisplay** in the display viewing icon panel

7.7 Using the Create Part option

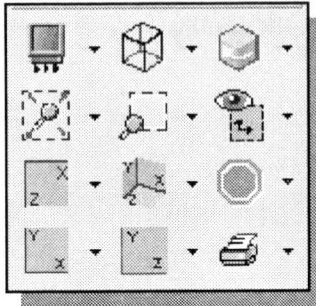

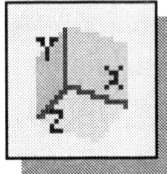

1. Choose **Isometric View** in the display viewing icon panel.

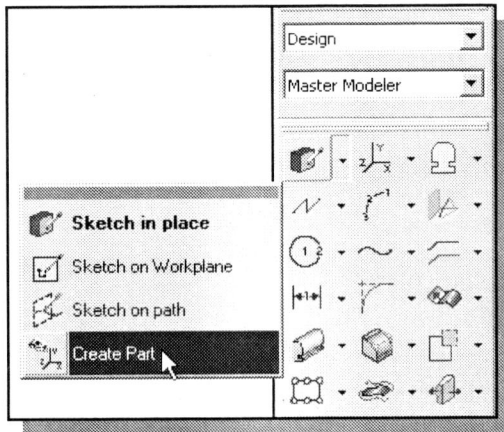

2. Choose **Create Part** in the icon panel.

   ➢ The icon is located in the first row of the task specific icon panel. The icon is located in the same stack as the *Sketch In Place* icon. Press and hold down the left-mouse-button on the icon stack to display the choice menu.

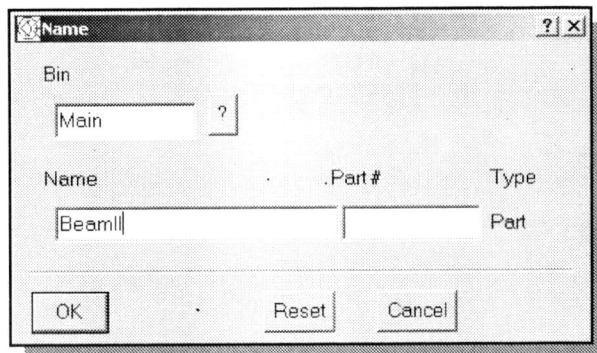

3. The *Name Part* window appears on the screen, enter **BeamII** as the name of the part as shown.

4. Click on the **OK** button to proceed with the **Create Part** command.

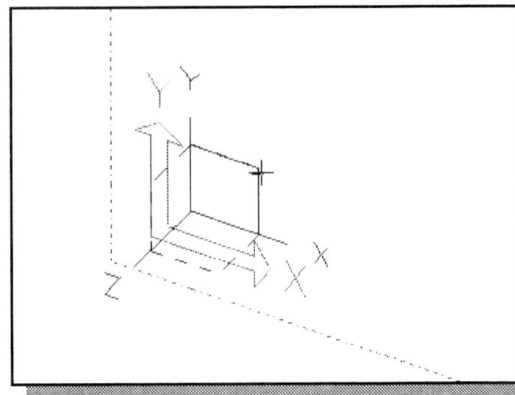

5. In the prompt window, the message *"Pick plane to sketch on"* is displayed. Pick the **XY** plane of the newly created coordinate system as shown. (Note that the default work plane, **blue** color, is still aligned to the XY plane of the world coordinate system. Aligning the sketch plane to the newly created coordinate system assures the proper association of the features to the part.)

6. Press the **ENTER** key once, or click once with the right-mouse-button to accept the placement of the workplane.

7.8 Creating a Wireframe Model of the System

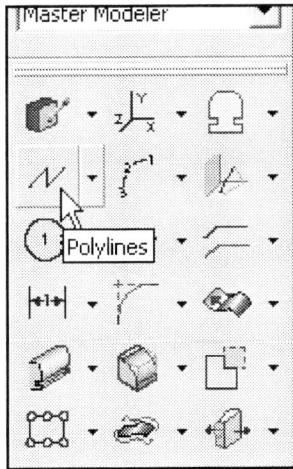

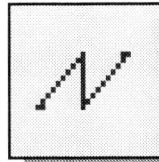

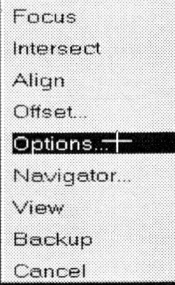

1. Pick **Polylines** in the icon panel. (The icon is located in the second row of the task specific icon panel.) The message *"Locate start"* is displayed in the prompt window.

2. Move the cursor inside the graphics window. Press and hold down the right-mouse-button to display the popup option menu. Slide the cursor up and down in the popup menu and select **Options...**

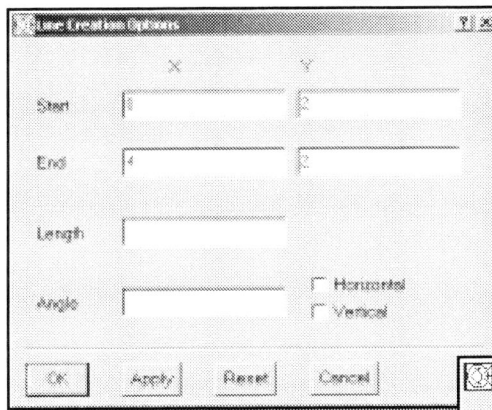

3. Fill in the **Start** and **End** point coordinates as shown.

4. Click the **Apply** button to accept the settings and create a line.

5. Fill in the **End** point coordinates as shown in the below figure.

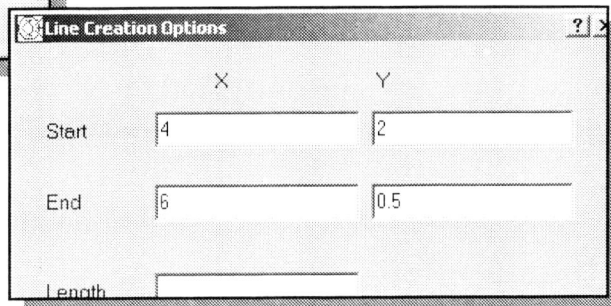

6. Click the **OK** button to accept the settings and create a second line.

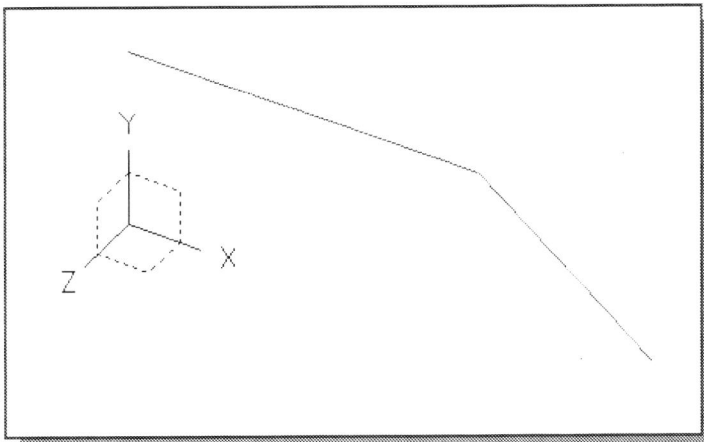

Line Creation Options		? >
	X	Y
Start	4	2
End	6	0.5
Length		

7. Press the **ENTER** key or the middle-mouse-button to end the current line segments.

8. Press the **ENTER** key or the middle-mouse-button to end the *Polylines* command.

7.9 A CAD Model is NOT an FEA Model

The two lines we just created represent the system that we will be analyzing. This model is created under the *Master Modeler*, and it is a CAD model that contains geometric information about the system. The CAD model can be used to provide geometric information needed in the finite element model. However, the CAD model is not an FEA model. The CAD model is typically developed to provide geometric information necessary for manufacturing. All details must be specified and all dimensions are required. The manufactured part and the CAD model are identical in terms of geometric information. The FEA model uses the geometric information of the CAD model as the starting point, but the FEA model usually will adjust some of the basic geometric information of the CAD model. The FEA model usually will contain additional nodes and elements. Idealized boundary conditions and external loads are also required in the model. The goal of finite element analysis is to gain sufficient reliable insights into the behaviors of the real-life system. Many assumptions are made in the finite element analysis procedure to simplify the analysis, since it is not possible or practical to simulate all factors involved in real-life systems. The finite element analysis procedure provides an idealized approximation of the real-life system. It is therefore not practical to include all details of the system in the FEA model; the associated computational cost cannot be justified in doing so. It is a common practice to begin with a more simplified FEA model. Once the model has been solved accurately and the result has been interpreted, it is feasible to consider a more refined model in order to increase the accuracy of the prediction of the actual system. We will next create the FEA model for the steel member by using the geometric information of the wireframe model.

7.10 Creating an Element Cross-Section

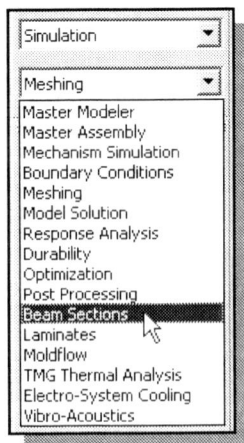

1. Switch to the **Beam Sections** task by selecting the task menu in the icon panel as shown.

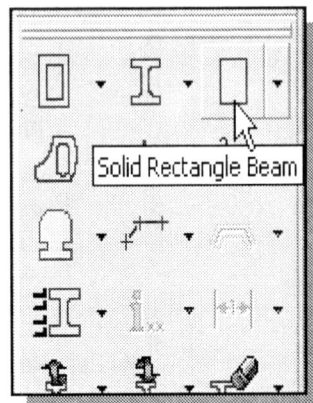

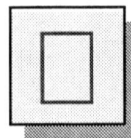

2. Choose **Solid Rectangular Beam** in the icon panel. (The icon is located in the first row of the task icon panel.) The message "*ENTER base*" is displayed.

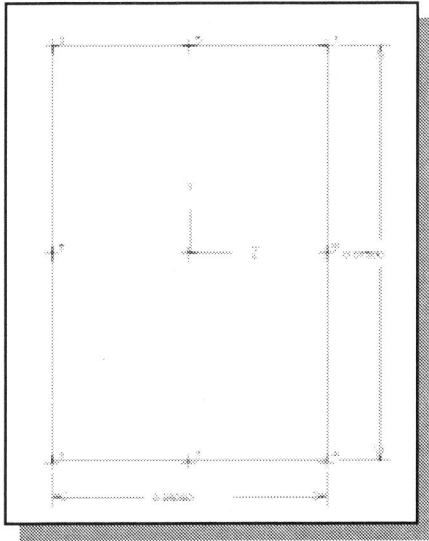

3. Enter **0.05** for the *base dimension*.

4. Enter **0.075** for the *height dimension*.

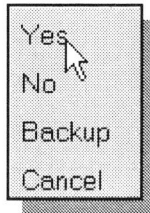

5. In the popup menu, pick **Yes** to complete creating the cross section.

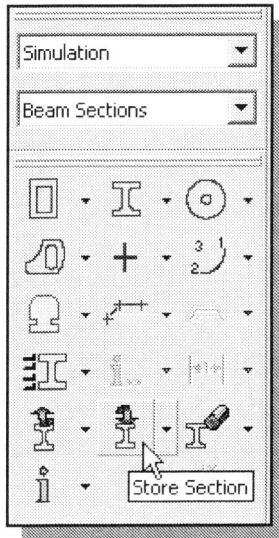

6. Choose ***Store Section*** in the icon panel. (The icon is located in the fifth row of the task icon panel.) The message *"Enter beam cross sect prop name or no (1-Rectangle 0.05 x 0.075)."* is displayed in the prompt window.

7. Press the **ENTER** key to accept the default name *(1-Rectangle 0.05 x 0.075)*.

7.11 Setup of an FEA Model

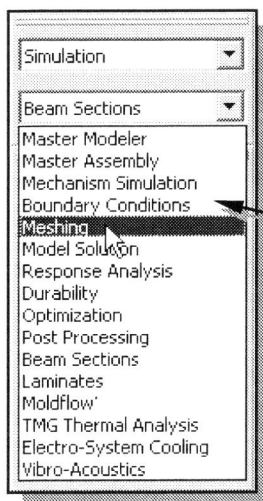

1. Switch back to the **Meshing** task by selecting the Meshing task in the task menu as shown.

Pick Meshing

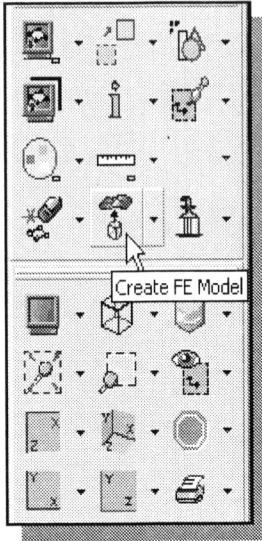

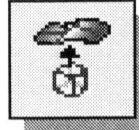

2. Choose **Create FE Model** in the icon panel. (The icon is located in the last row of the application specific icon panel.) The *FE Model Create* window appears.

3. In the *FE Model Create* window, enter **BeamII** in the *FE Model Name* box.

4. Click on the **OK** button to accept the settings.

❖ Note the *Part* name is displayed in the *FE Model Create* window.

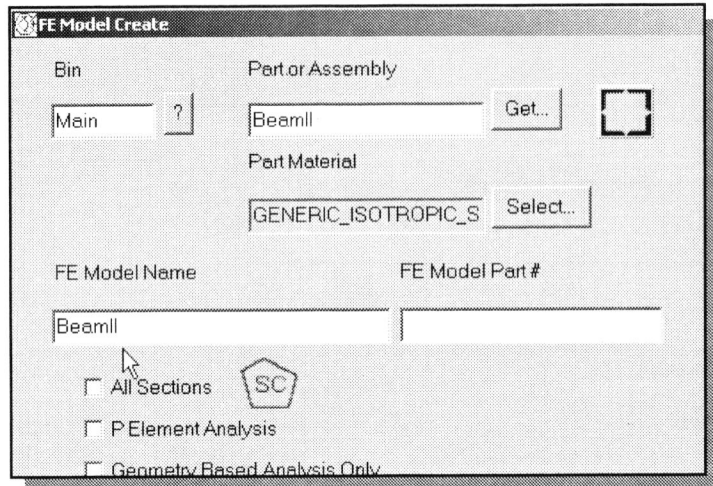

FE Model Create

Bin	Part or Assembly
Main ?	BeamII Get...
	Part Material
	GENERIC_ISOTROPIC_S Select...

| FE Model Name | FE Model Part # |
| BeamII | |

☐ All Sections   SC
☐ P Element Analysis
☐ Geometry Based Analysis Only

7.12 Material Property Table

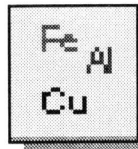

1. Choose **Materials** in the icon panel. (The icon is located in the fourth row of the task icon panel.) The *Materials* window appears.

❖ In this example, we will use the default material property set, *Generic_Isotropic_Steel*, for our analysis.

2. Choose the default material, **Generic_Isotropic_Steel**, in the list window.

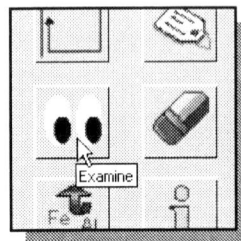

3. Click on the **Examine** icon and review the material properties.

4. Click on the **OK** button to exit the *Examine* window.

5. Click on the **OK** button to exit the *Materials* window.

7.13 Define Beam Auto-Mesh

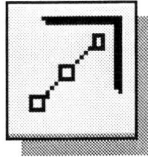

1. Choose **Define Beam Mesh** in the icon panel. (The icon is located in the first row of the task icon panel. Press and hold down the left-mouse-button on the displayed icon and select the *Define Beam Mesh* command by releasing the left-mouse-button when the option is highlighted.)

2. The message "*Pick edges/Curves*" is displayed in the prompt window. Pick both lines by pressing down the **SHIFT** key and left-clicking each line.

3. Press the **ENTER** key or the middle-mouse-button to accept the selection.

4. The *Define Mesh* window appears. Enter **0.5** in the *Element Length* box.

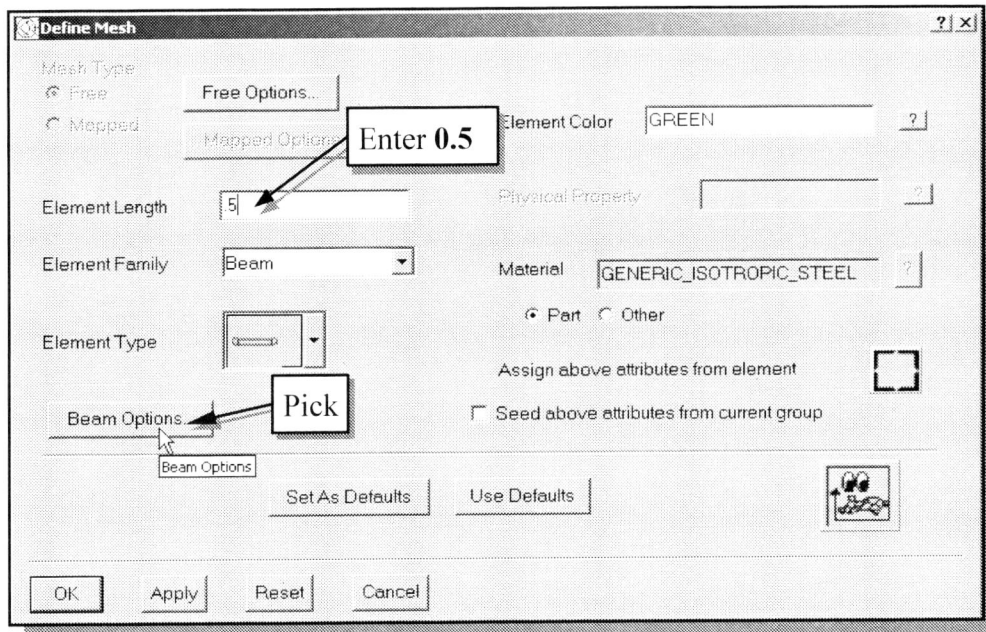

5. Pick **Beam Options** and confirm the cross section is set to *1-Rectangular 0.05x0.075.*

6. Click on the **OK** button to exit the *Beam Options* window.

7. Click on the **OK** button to exit the *Define Beam Mesh* window.

7.14 Create Beam Auto-Mesh

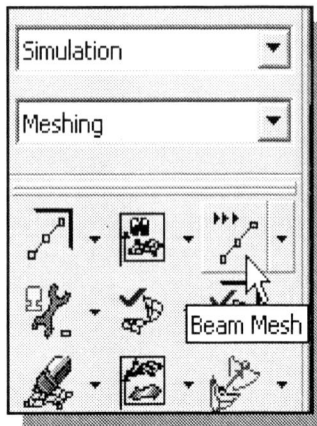

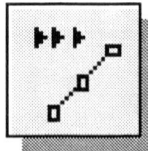

1. Choose **Create Beam Mesh** in the icon panel. (The icon is located in the first row of the task icon panel. Press and hold down the left-mouse-button on the displayed icon and select the *Create Beam Mesh* command by releasing the left-mouse-button when the option is highlighted.) The message "*Pick edges/Curves*" is displayed in the prompt window.

2. Pick both lines by pressing down the **SHIFT** key and left-clicking each line.

3. Press the **ENTER** key or the middle-mouse-button to accept the selections. New nodes and elements are created.

4. Press the **ENTER** key or the middle-mouse-button to accept the additions.

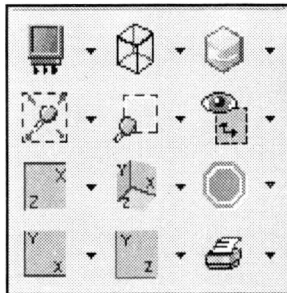

5. Pick **Isometric View** in the icon panel.

6. Pick **Shaded Software** or **Hidden Lines Removal** in the icon panel.

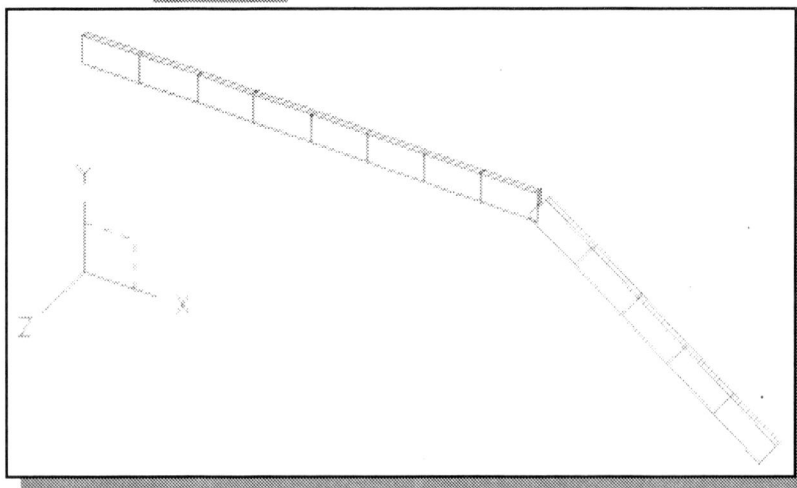

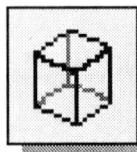

7. Pick **Line** in the icon panel to reset the display back to *wireframe display*.

7.15 Create a Coordinate System for the Inclined Support

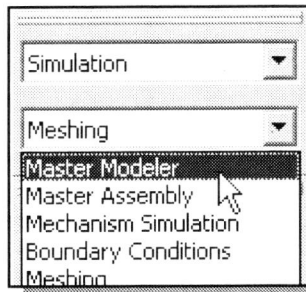

1. Switch to the **Master Modeler** task by selecting the task in the task menu as shown.

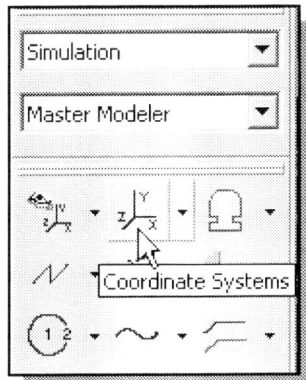

2. Choose **Coordinate Systems** in the icon panel. (The icon is located in the sixth row of the task icon panel.)

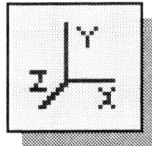

3. The message "*Pick entity for coordinate system to reference*" is displayed in the prompt window. Pick one of the edges of the *3D part coordinate system.*

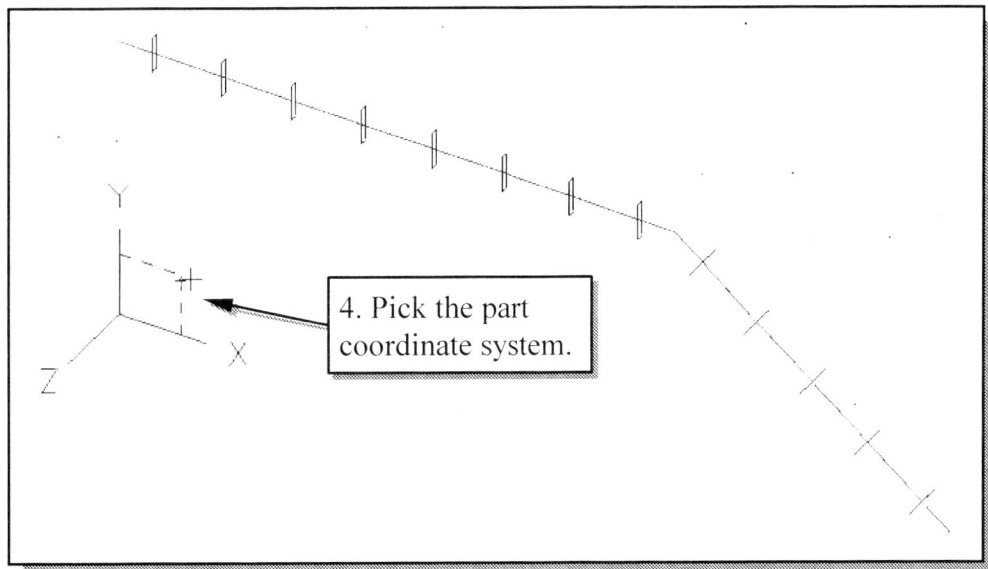

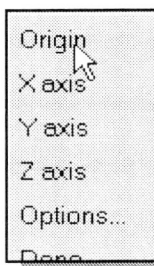

4. Pick the part coordinate system.

4. In the popup menu, pick **Origin** to define the alignment of the new coordinate system.

5. The message "*Pick origin definition (Done)*" is displayed in the prompt window. Select the **right endpoint** of the model as shown below.

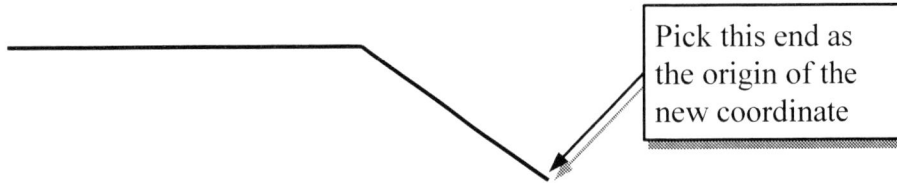

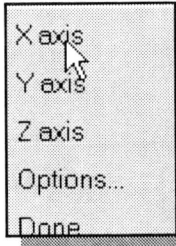

Pick this end as the origin of the new coordinate

X axis
Y axis
Z axis
Options...
Done

6. Pick **X axis** in the popup menu.

7. The message "*Pick X axis definition*" is displayed in the prompt window. Pick the **inclined line** to define the new X-axis direction. (Select the wireframe geometry.)

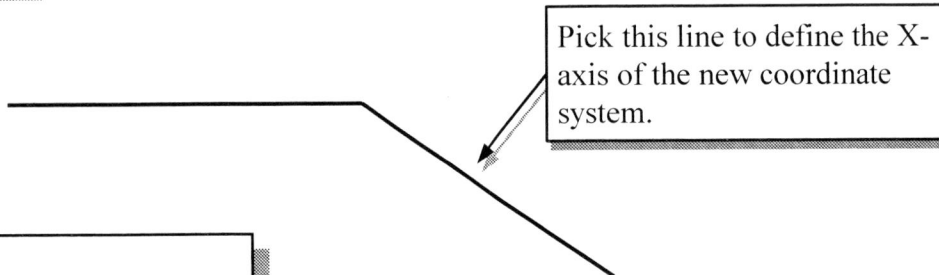

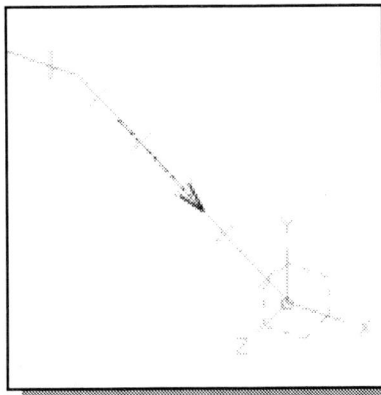

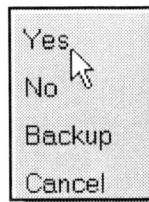

Pick this line to define the X-axis of the new coordinate system.

Yes
No
Backup
Cancel

8. Pick **Yes** to accept the direction for the new X-axis.

9. Pick **Done** to complete the creation of the new coordinate system.

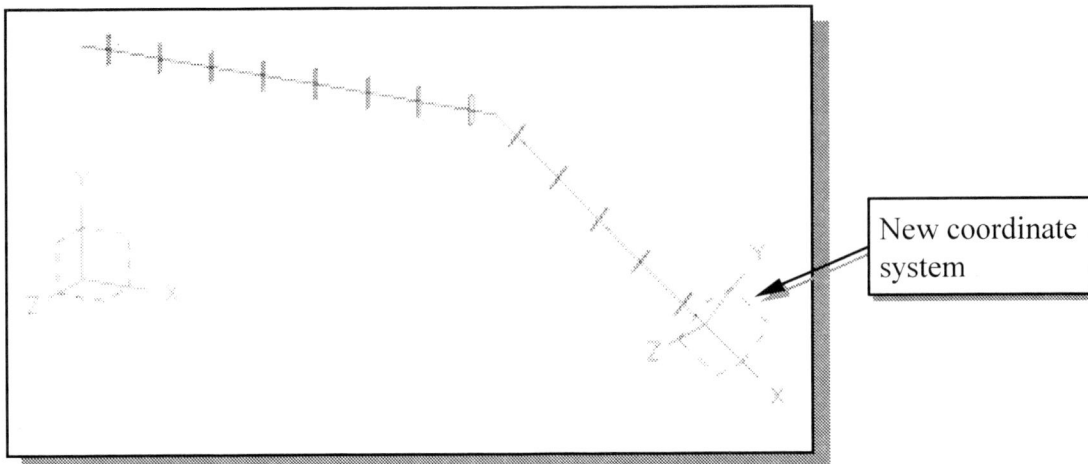

New coordinate system

7.16 Node Orientation Alignment

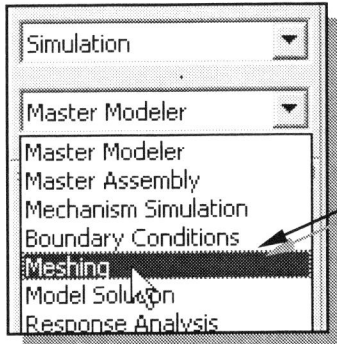

1. Switch back to the **Meshing** task by selecting the Meshing task in the task menu as shown.

Pick *Meshing*

2. Choose ***Modify Node*** in the icon panel. (The icon is located in the fourth row of the task icon panel. In the same stack as the *Node* icon.)

3. The message "*Pick Nodes*" is displayed in the prompt window. Select the node located at the right endpoint of the member as shown below.

Pick the end node.

4. Press the **ENTER** key or the middle-mouse-button to accept the selection. The message "*Select node attribute to modify*" is displayed in the prompt.

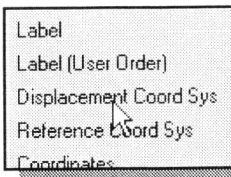

5. Select **Displacement Coord Sys** in the popup menu.

6. Pick the new coordinate system as shown below.

7. Pick **Yes** to accept the modification.

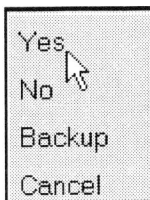

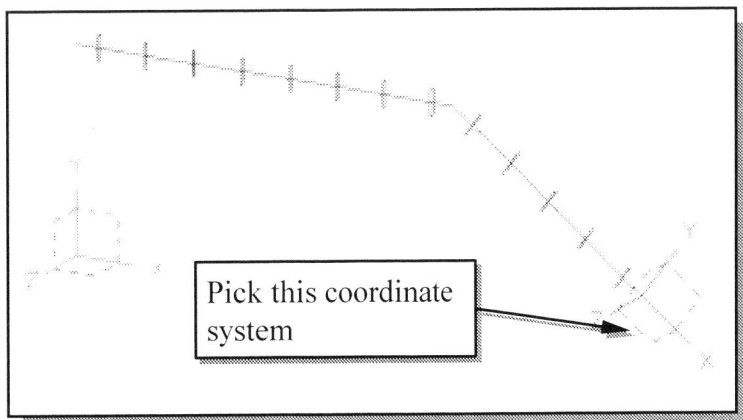

Pick this coordinate system

7.17 Apply Boundary Conditions

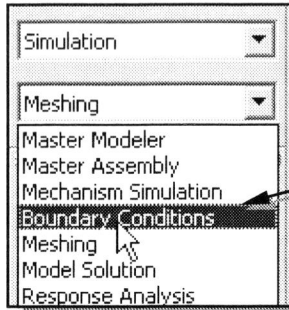

1. Switch to the **Boundary Conditions** task by selecting it in the task menu as shown.

Pick *Boundary Conditions*

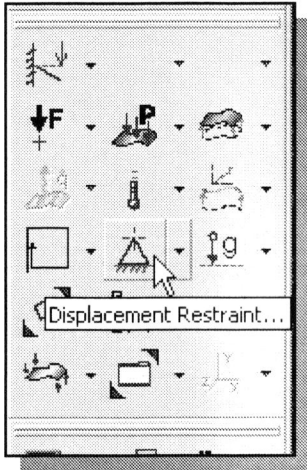

2. Choose ***Displacement Restraint*** in the icon panel. (The icon is located in the fourth row of the task icon panel.)

3. The message "*Pick Nodes/Centerpoints/Vertices*" is displayed in the prompt window. Pick ***Node 1*** (left endpoint of the horizontal member).

Pick the node located at this end of the member.

4. Press the **ENTER** key or the middle-mouse-button to accept the selection. The *Displacement Restraint on Node* window appears.

5. Pick ***Pin*** joint and set *Axis of Rotation* to **Z-axis**.

6. Click on the **OK** button to exit the *Displacement Restraint on Node* window.

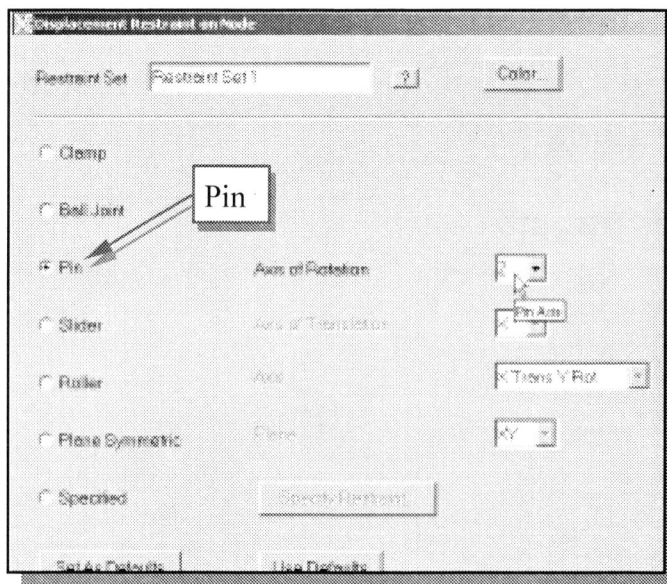

7.18 Apply Boundary Condition for the Inclined Support

1. Choose **Displacement Restraint** in the icon panel. (The icon is located in the fourth row of the task icon panel.)

2. The message "*Pick Nodes/Centerpoints/Vertices*" is displayed in the prompt window. Pick the node that the new coordinate system is aligned to (right endpoint of the inclined member).

Pick the node located at this end of the member.

3. Press the **ENTER** key or the middle-mouse-button to accept the selection. The *Displacement Restraint on Node* window appears.

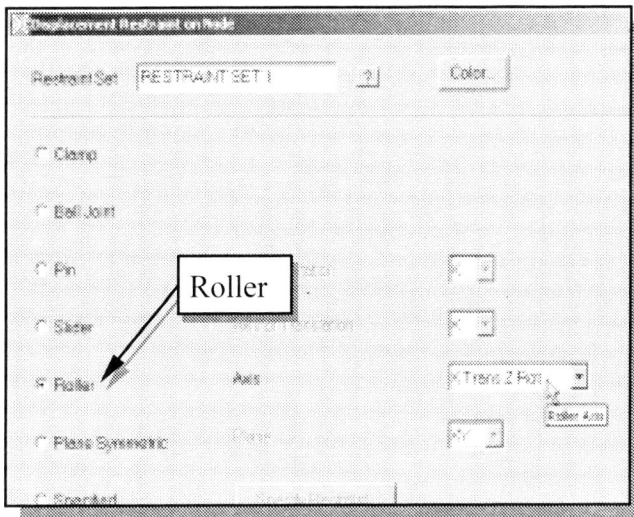

4. Pick **Roller** and set the *Axis* option to **X Trans Z Rot**.

Roller

5. Click on the **OK** button to exit the *Displacement Restraint on Node* window. Your screen should appear as shown.

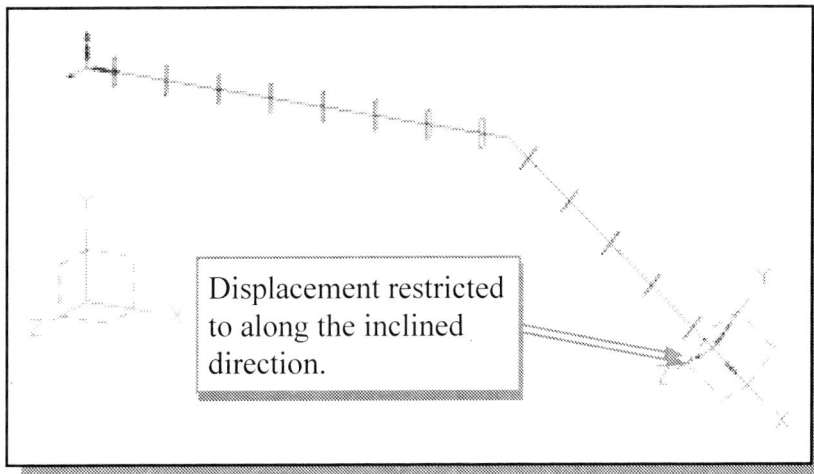

Displacement restricted to along the inclined direction.

7.19 Distributed Load

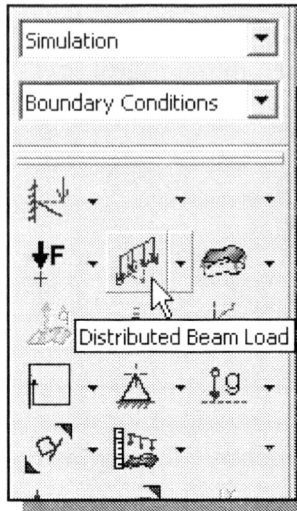

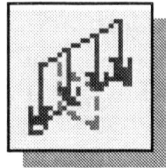

1. Choose **Distributed Load** in the icon panel. (The icon is located in the second row of the task icon panel.)

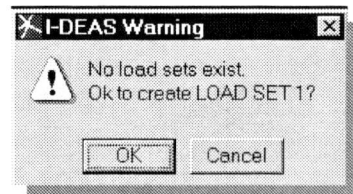

2. Click on the **OK** button in the *warning window* to create LOAD SET 1.

3. The message "*Select beams to be loaded, Pick Elements*" is displayed in the prompt window. Pick the **inclined elements** as shown below. (Press down the **SHIFT** key and left-click on each element or use a selection window.)

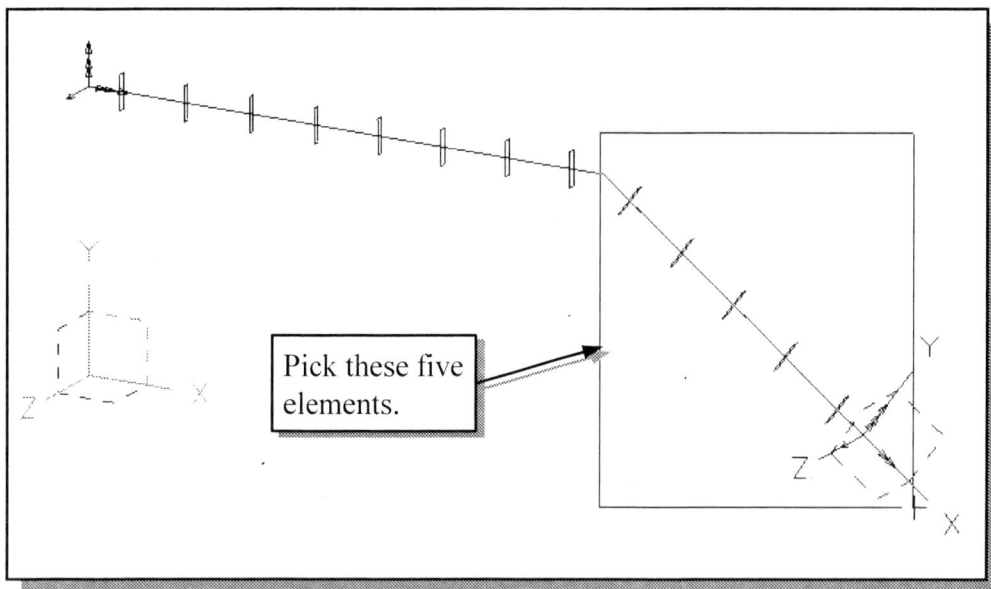

Pick these five elements.

4. Press the **ENTER** key or the middle-mouse-button to accept the selection and continue.

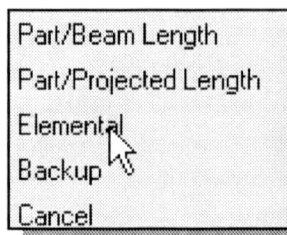

5. The message "*Select coordinate system of forces (Part/Beam length)*" is displayed in the prompt window. Pick **Elemental** in the popup menu.

6. The message *"Select type of load (Constant)"* is displayed in the prompt window. Press the **ENTER** key to apply a uniformly distributed load.

7. The message *"Enter distributed axial force (0.0)"* is displayed in the prompt window. Press the **ENTER** key to accept the default setting.

8. The message *"Enter distributed Y shear force (0.0)"* is displayed in the prompt window. Enter the value of **-5000**. The minus sign indicates the direction of the load.

9. The message *"Enter distributed Z shear force (0.0)"* is displayed in the prompt window. Press the **ENTER** key to accept the default setting.

10. The message *"Enter distributed torque (0.0)"* is displayed in the prompt window. Press the **ENTER** key to accept the default setting.

11. The message *"Enter distributed Y bending moment (0.0)"* is displayed in the prompt window. Press the **ENTER** key to accept the default.

12. The message *"Enter distributed Z bending moment (0.0)"* is displayed in the prompt window. Press the **ENTER** key to accept the default.

13. The message *"Enter color name or no. (10-ORANGE)"* is displayed in the prompt window. Press the **ENTER** key to accept the default.

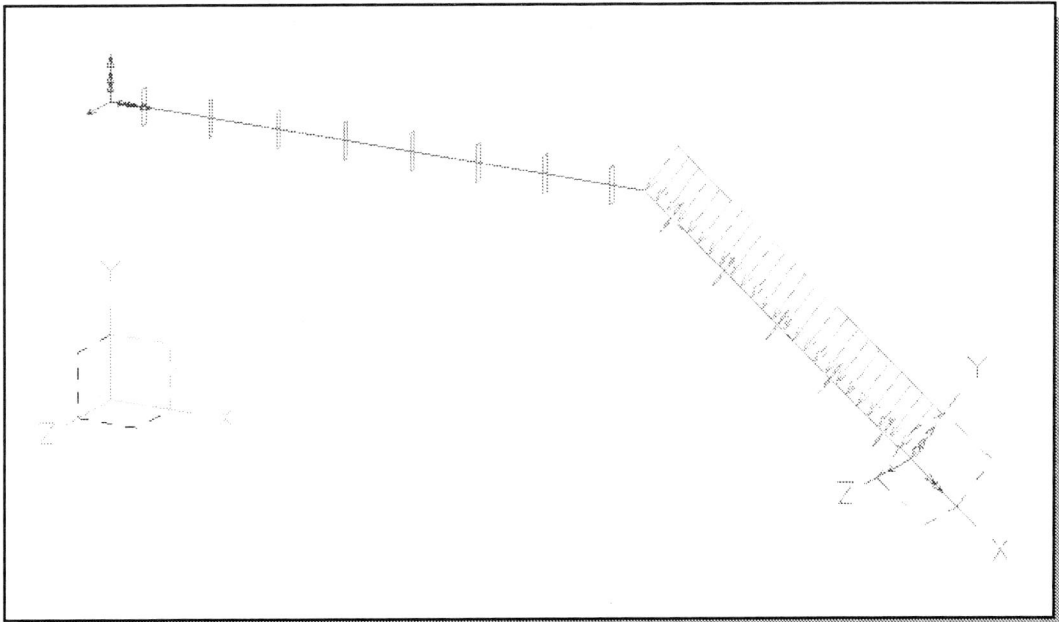

❖ Note that when using the **Elemental** coordinate, the X direction is always aligned to the beam direction.

7.20 Running the Solver

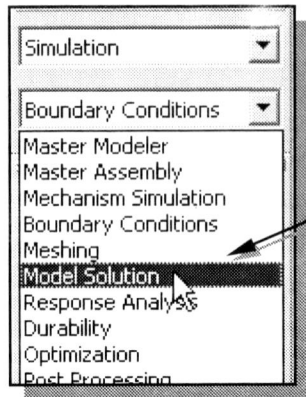

1. Switch to the **Model Solution** task by selecting the task in the task menu as shown.

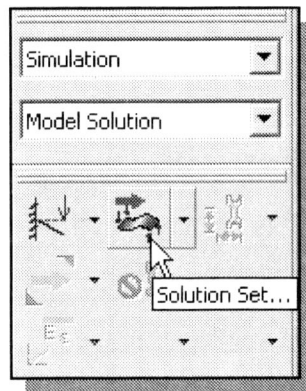

   Pick *Model Solution*

2. Choose **Solution Set** in the icon panel. (The icon is located in the first row of the task icon panel.) The *Manage Solution Sets* window appears.

3. Click on the **Create** button. The *Solution Set* window appears.

4. Click on the **Output Selection** button. The *Output Selection* window appears.

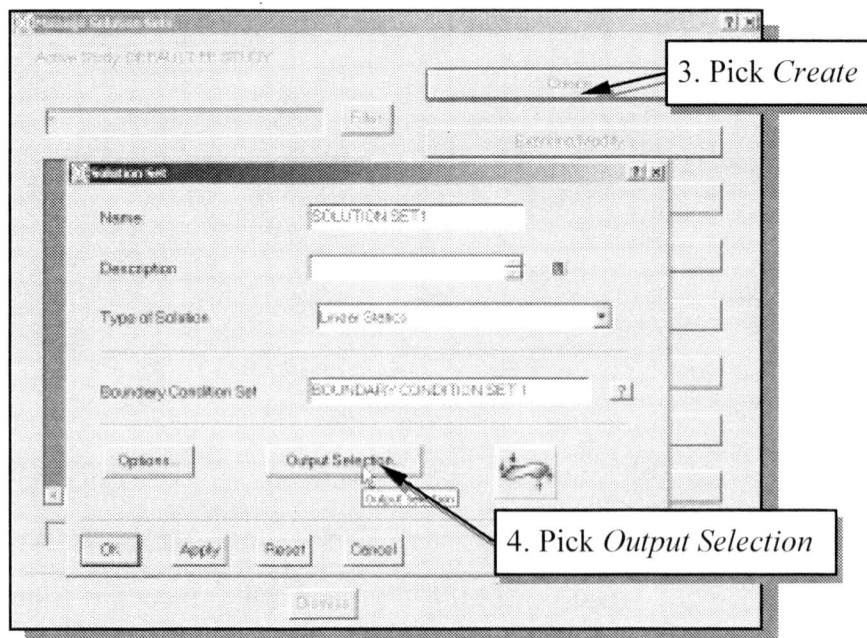

3. Pick *Create*

4. Pick *Output Selection*

5. On your own, set the **Element Forces** output option to **Store** in the *Output* list.

6. Click on the **OK** button to exit the *Output Selection* window.

7. Click on the **OK** button to exit the *Solution Set* window.

8. Pick **Dismiss** to exit the *Manage Solution Sets* window.

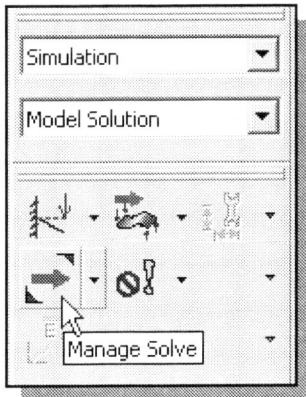

9. Choose **Manage Solve** in the icon panel. (The icon is located in the second row of the task icon panel.) The *Solve* window appears.

10. In the *Solve* window, pick the **Solve** icon to find the solutions.

 ➢ *I-DEAS* will begin the solving process. <u>DO NOT</u> close any windows. Any errors or warnings are displayed in the list window.

7.21 Viewing the results – Reaction Forces

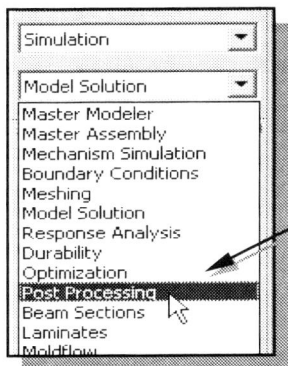

1. Switch to the **Post Processing** task by selecting it in the task menu as shown.

Pick *Post Processing*

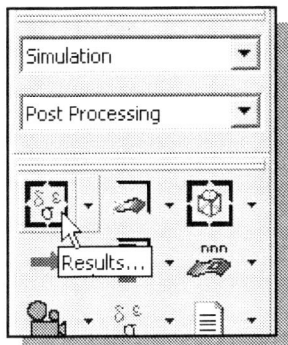

2. Choose **Results Selection** in the icon panel. (The icon is located in the first row of the task icon panel.)

3. The *Results Selection* window appears. Pick **Reaction Force** from the list.

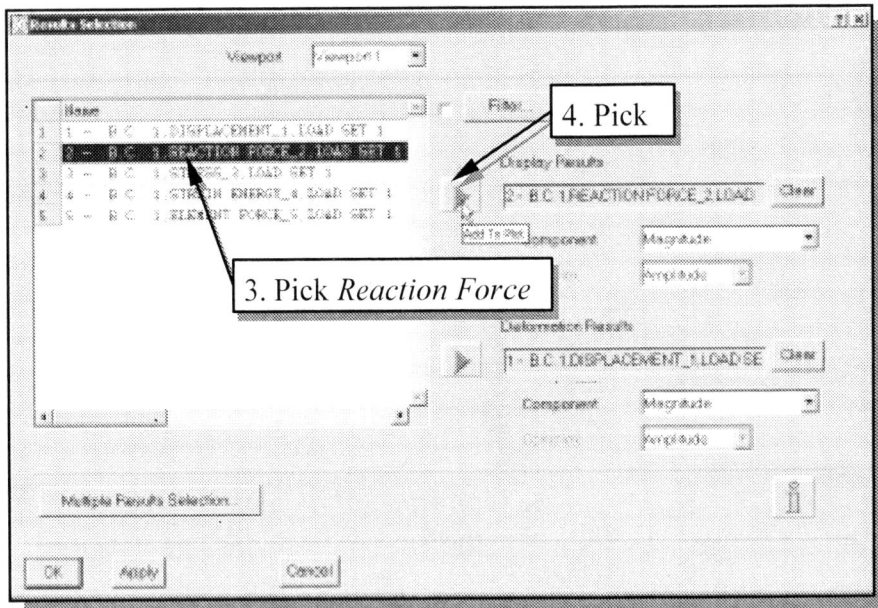

4. Click on the **Triangle** button to set the *Reaction Force* as the *Display Results*.

5. Click on the **OK** button to exit the *Results Selection* window.

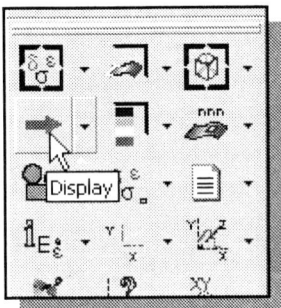

6. Choose **Results Display** in the icon panel. (The icon is located in the second row of the task icon panel.)

7. Pick the last inclined element, the element associated with the inclined support.

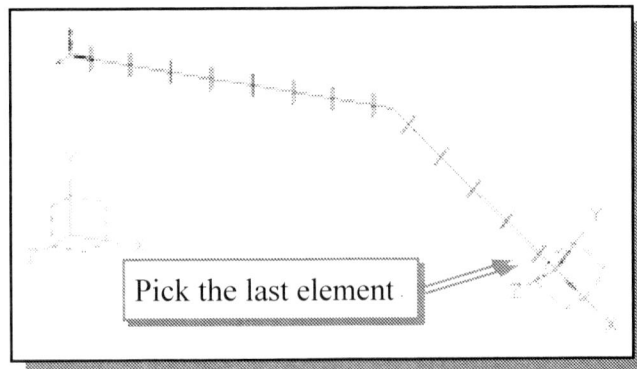

Pick the last element

8. Press the **ENTER** key or the middle-mouse-button to accept the selection and display the reaction force associated with this element.

➢ The computer solution is **9.76 kN**, which matches our hand calculation 9.76 kN (see section 7.3 **Preliminary Analysis.**)

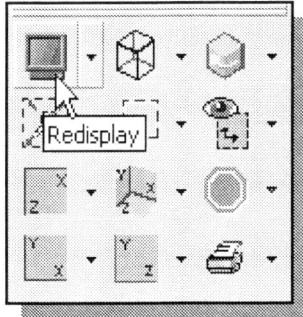

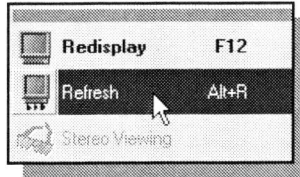

9. Choose **Refresh** in the icon panel. (The icon is located in the first row of the display icon panel.)

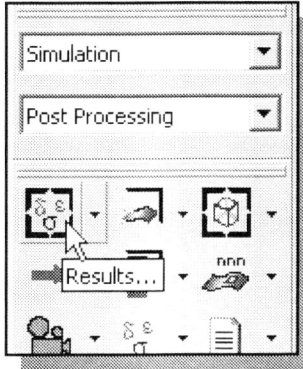

10. Choose **Results Selection** in the icon panel. (The icon is located in the first row of the task icon panel.) The *Results Selection* window appears.

11. Select the **X component** under the *Display Results Component*.

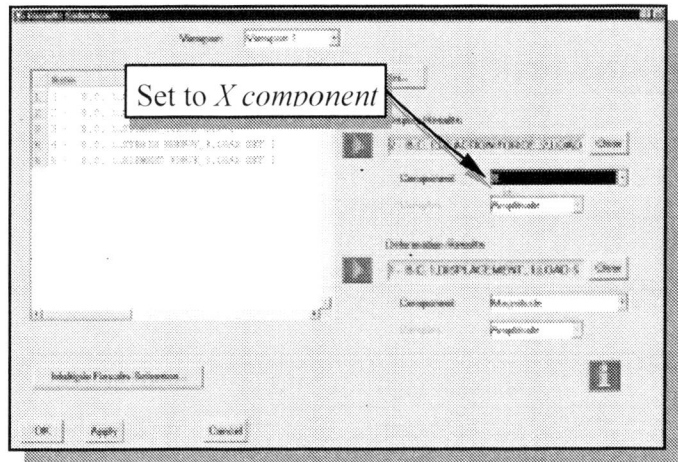

12. Click on the **OK** button to exit the *Results Selection* window.

13. Choose **Results Display** in the icon panel.

Set to *X component*

14. Pick the first left element as shown in the below figure.

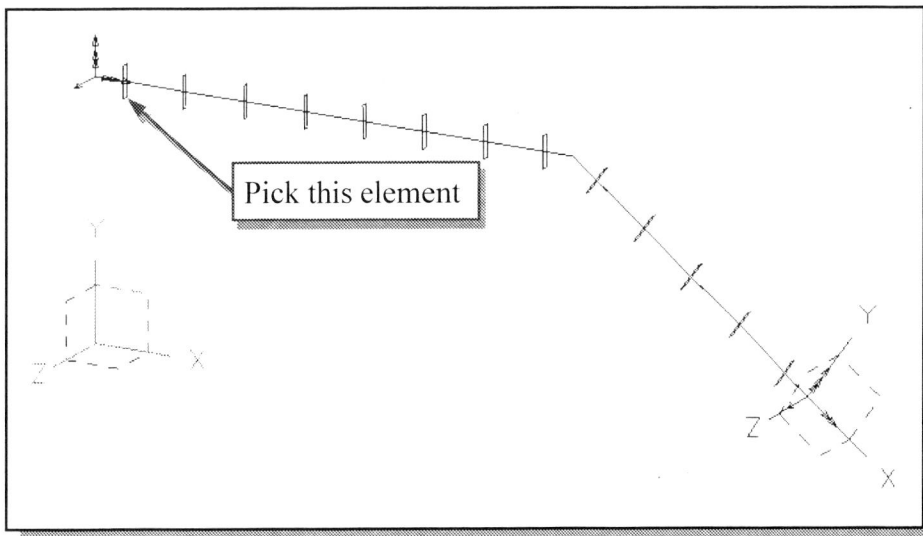

Pick this element

15. Press the **ENTER** key or the middle-mouse-button to accept the selection and display the reaction force associated with the selected element.

➤ The computer solution is **1.64 kN**, which matches our hand calculation 1.64 kN (see section 7.3 Preliminary Analysis.)

16. Choose **Refresh** in the icon panel. (The icon is located in the first row of the display icon panel.)

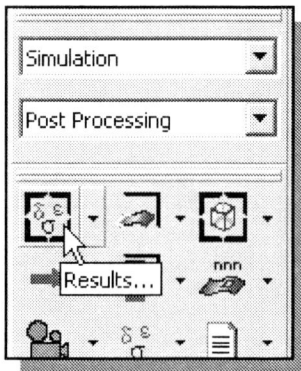

17. Choose **Results Selection** in the icon panel. (The icon is located in the first row of the task icon panel.) The *Results Selection* window appears.

18. Select **Y component** under the *Display Results Component*.

19. Click on the **OK** button to exit the *Results Selection* window.

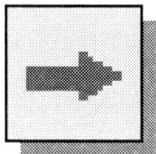

20. Choose **Results Display** in the icon panel.

21. Pick the element as shown.

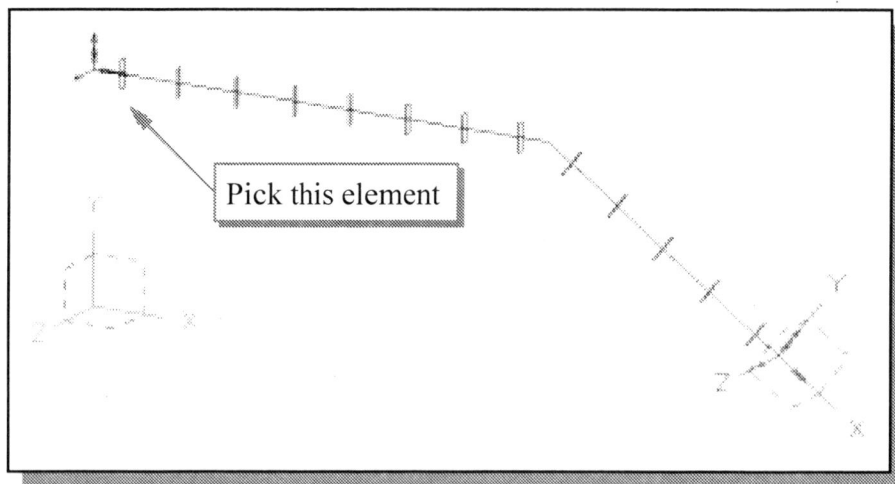

Pick this element

22. Press the **ENTER** key to display the reaction force associated with this element.

➤ The computer solution is **2.19 kN**, which matches our hand calculation 2.19 kN (see section 7.3 Preliminary Analysis.)

7.22 Z-Bending Moment Diagram

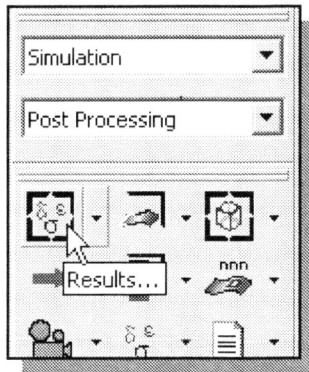

1. Choose **Results Selection** in the icon panel. (The icon is located in the first row of the task icon panel.)

2. The *Results Selection* window appears. Pick **Element Force** from the list.

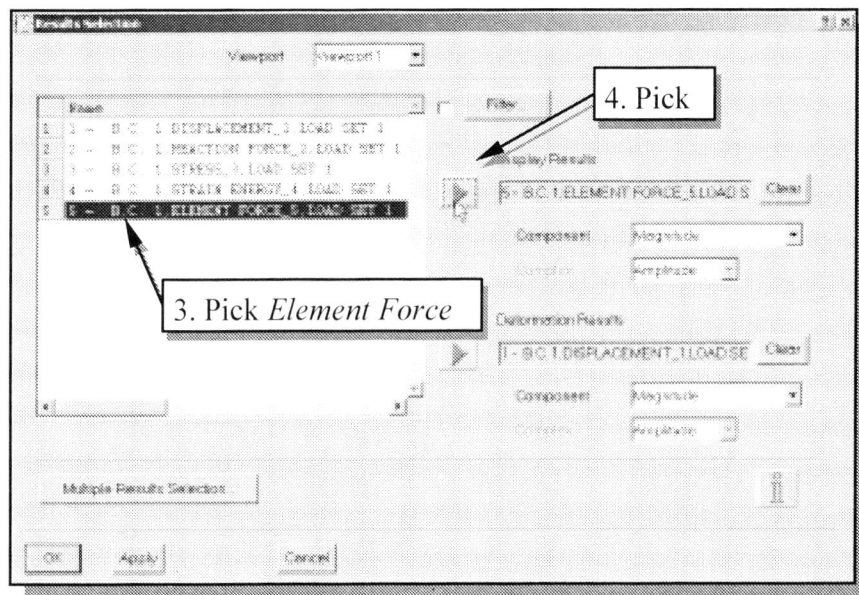

4. Pick

3. Pick *Element Force*

3. Click on the **Triangle** button to set the *Element Force* as the *Display Results*.

4. Click on the **OK** button to exit the *Results Selection* window.

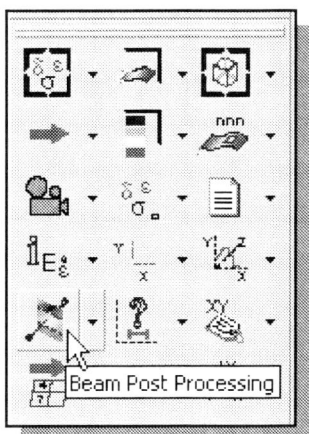

5. Choose **Beam Post Processing** in the icon panel. (The icon is located in the fifth row of the task icon panel.) The cascading menu appears in the graphics window.

6. Select **Force & Stress** in the cascading menu.

7. Select **Data Component** in the popup menu.

8. Select **Force** in the popup menu as shown on the next page.

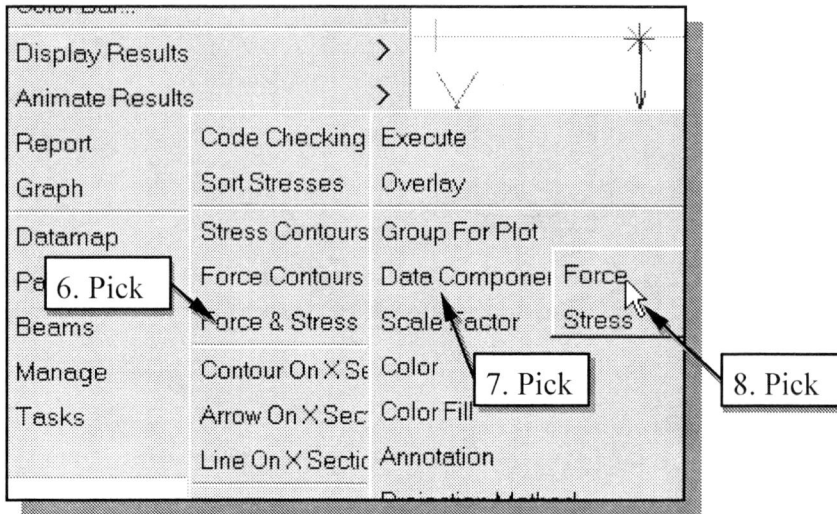

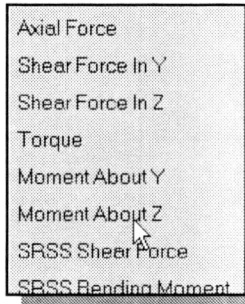

9. In the popup menu, select **Moment About Z.**

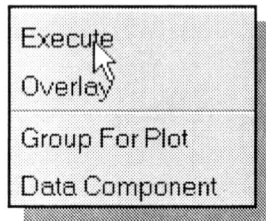

10. In the cascading menu, select **Execute**.

11. Press down the **CTRL** key and hit the **M** key to switch off the cascading menu.

❖   The bending diagram should appear as shown. Does the diagram match with the preliminary analysis we did at the beginning of the chapter?

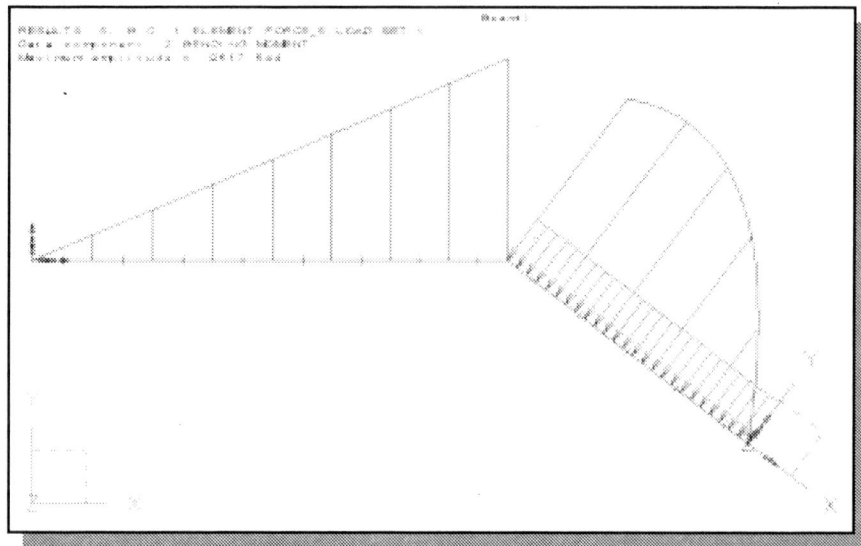

➢   The moment diagram shows us that the maximum moment in the structure occurs at a point that is located on the inclined member.

7.23 Axial Load Diagram

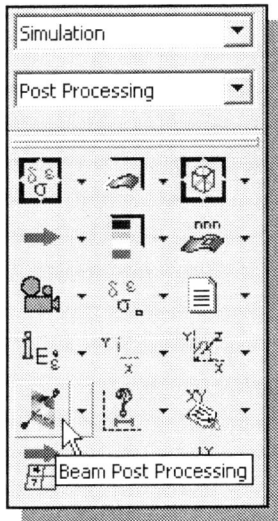

1. Choose **Beam Post Processing** in the icon panel. (The icon is located in the fifth row of the task icon panel.) The cascading menu appears in the graphics window.

2. Select **Force & Stress** in the cascading menu.

3. Select **Data Component** in the popup menu.

4. Select **Force** in the popup menu.

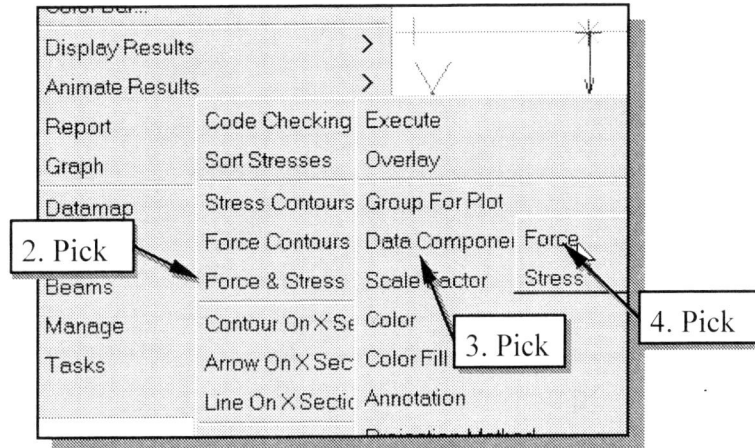

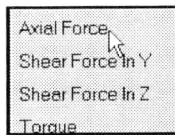

5. In the popup menu, select **Axial Force.**

6. In the cascading menu, select **Execute**.

7. Press down the **CTRL** key and hit the **M** key to switch off the cascading menu.

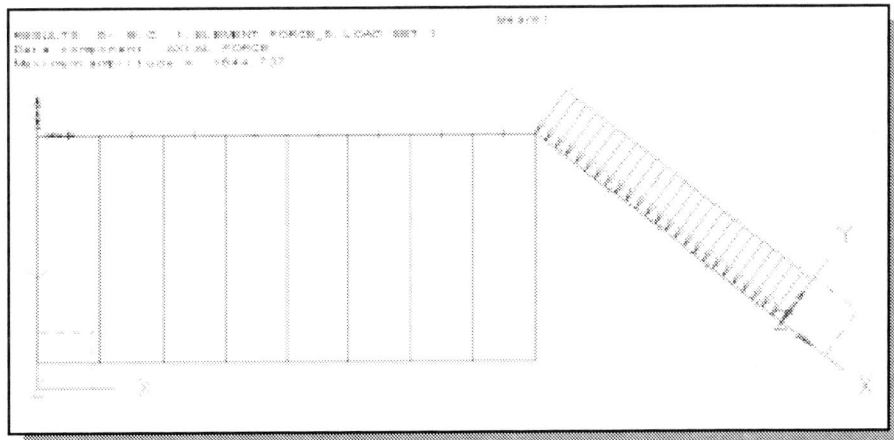

➤ The axial load diagram shows us that there is constant normal stress from the axial load along the horizontal segment and zero axial load along the inclined segment.

7.24 Shear Diagram

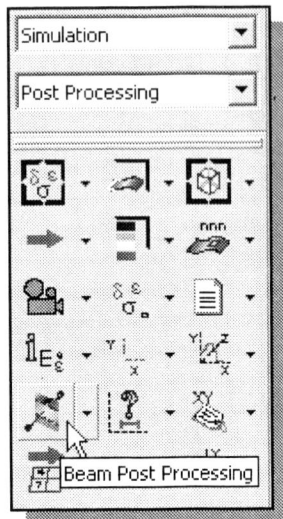

1. Choose **Beam Post Processing** in the icon panel. (The icon is located in the fifth row of the task icon panel.) The cascading menu appears in the graphics window.

2. Select **Force & Stress** in the cascading menu.

3. Select **Data Component** in the popup menu.

4. Select **Force** in the popup menu.

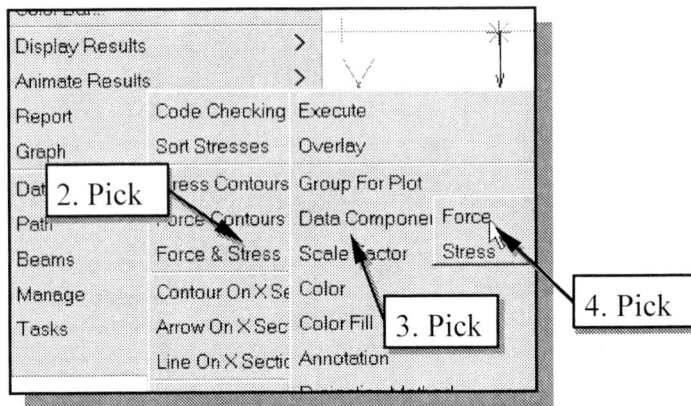

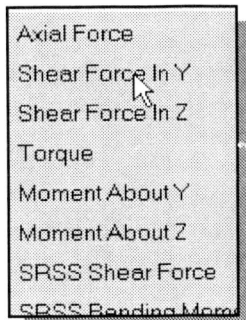

5. In the popup menu, select **Shear Force in Y**.

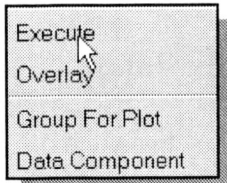

6. In the cascading menu, select **Execute**.

7. Press down the **CTRL** key and hit the **M** key to switch off the cascading menu.

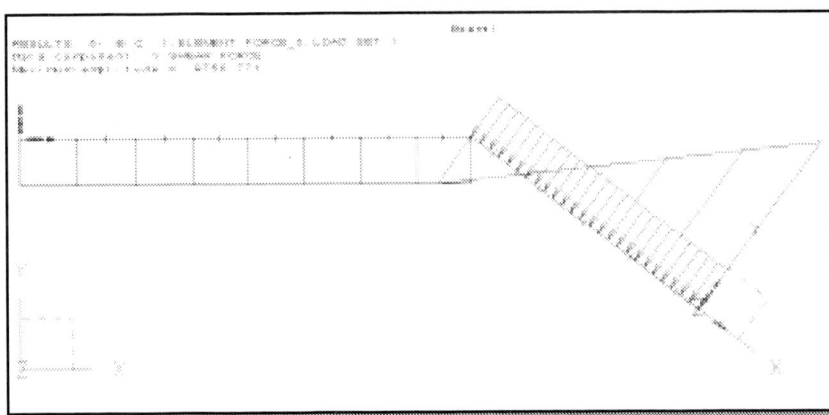

➤ The *Y shear diagram* shows a constant shear stress along the horizontal segment and a higher shear force at the right-end of the inclined segment.

7.25 Maximum Normal Stress in the Member

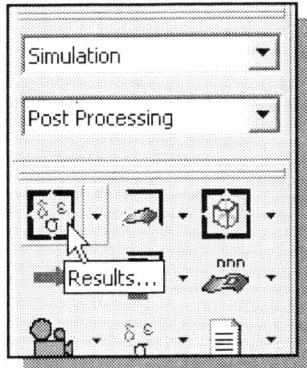

1. Choose **Results Selection** in the icon panel. (The icon is located in the first row of the task icon panel.) The *Results Selection* window appears.

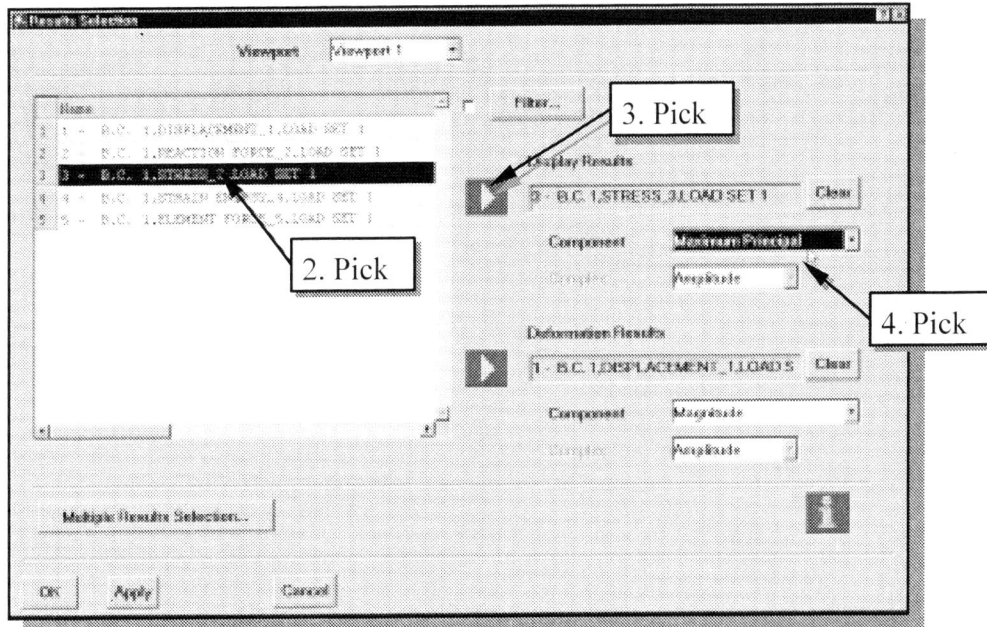

2. Pick **Stress** from the list.

3. Click on the **Triangle** button to set the display option.

4. Select the **Maximum Principal** *Component* option.

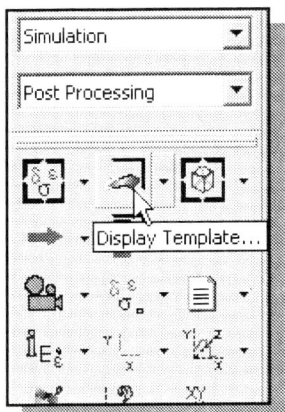

5. Click on the **OK** button to exit the *Results Selection* window.

6. Choose **Display Template** in the icon panel. (The icon is located in the first row of the task icon panel.)

7. Click on the **Contour** button.

8. Click on **Fast Display** to switch off the *Fast Display* option as shown.

9. Click the **OK** button to exit the *Contour Options* window.

10. Click the **OK** button to exit the *Display Template* window.

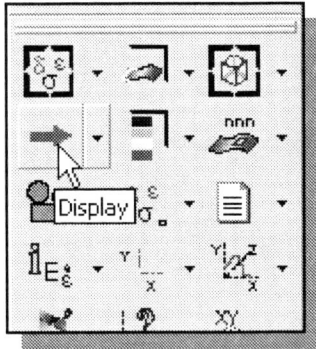

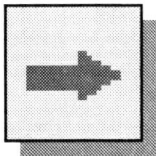

11. Choose **Results Display** in the icon panel. (The icon is located in the second row of the task icon panel.) The message "*Pick elements*" is displayed in the prompt window.

12. Press the **ENTER** key to continue. (Stresses on all elements will be processed)

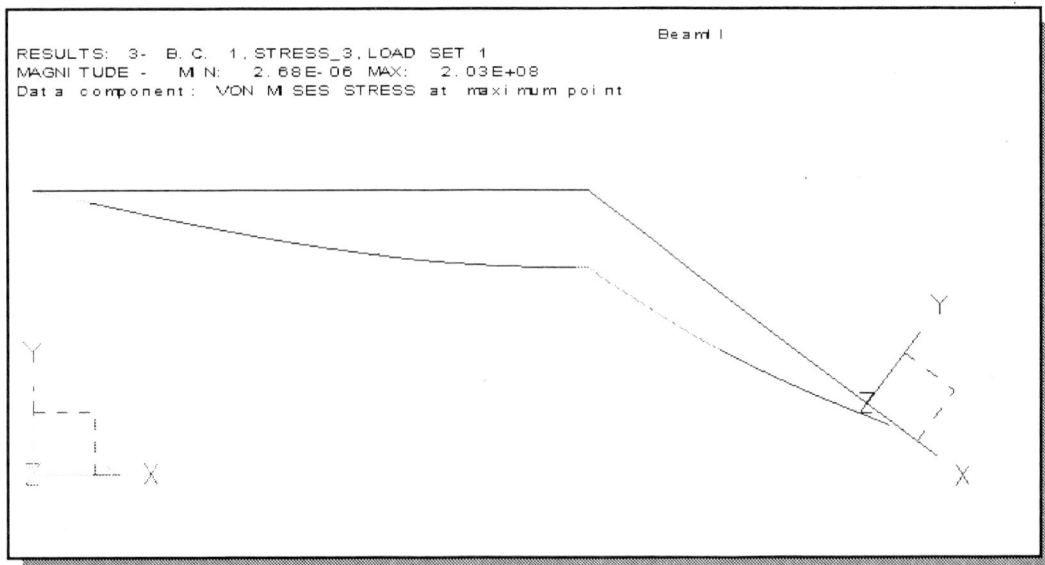

```
                                                    Beam I
RESULTS: 3-  B. C.  1, STRESS_3, LOAD SET 1
MAGNITUDE -    MIN:   2.68E-06 MAX:   2.03E+08
Data component:  VON MISES STRESS at maximum point
```

➤   The computer solution for the maximum stress is **203 MPa**, which occurs along the inclined segment of the member. On your own, examine the maximum normal stress developed in the structure and compare the *I-DEAS* solution to the calculation performed at the beginning of the chapter.

Questions:

1. What are the main differences between a CAD model and an FEA model?

2. How do we account for inclined supports and apply *elemental* loads on individual elements in *I-DEAS*?

3. Describe the method of *Superposition.*

4. Describe the general procedure in using the *Auto-Mesh* option.

5. Identify and describe the following commands:

(a)

(b)

(c)

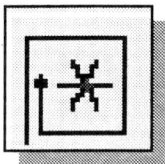

(d)

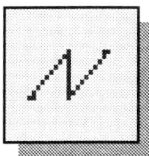

Exercises:

Determine the maximum stress produced by the loads and create the shear and moment diagrams for the structures.

1. Fixed-end support.
    Material: Steel rod,
        Diamater 1.0 in.

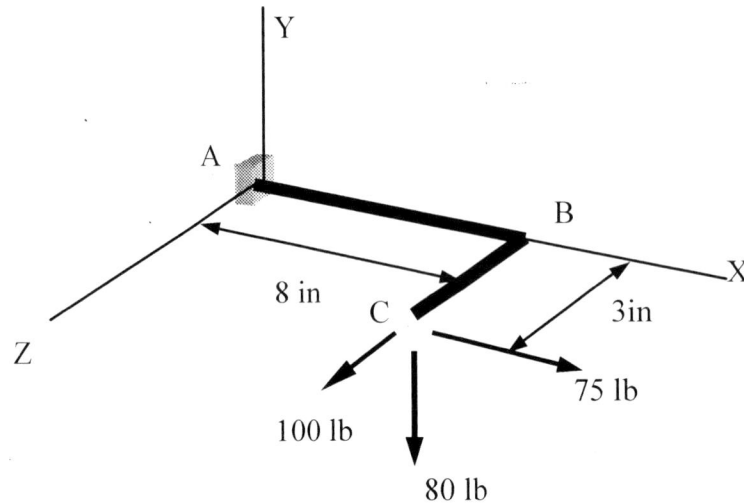

2. Pin-Joint and roller supports.
    Material: Aluminum Alloy 6061 T6

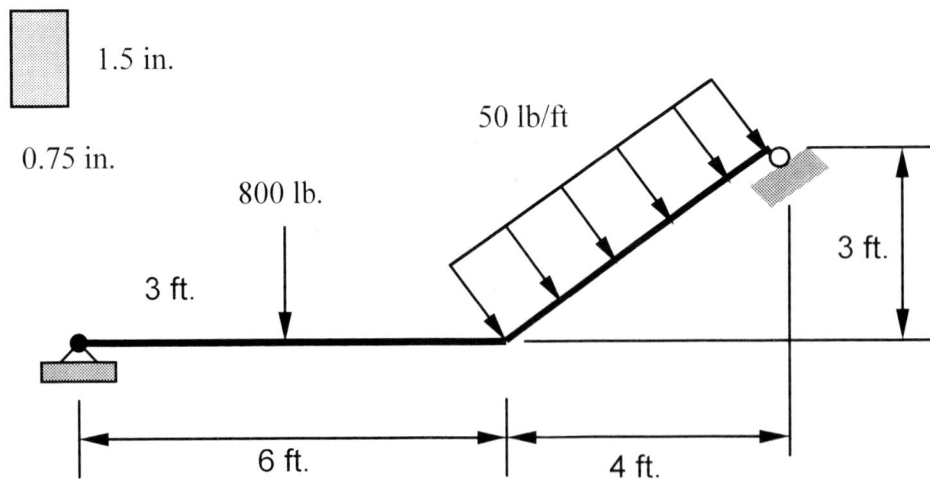

Chapter 8
Statically Indeterminate Structures

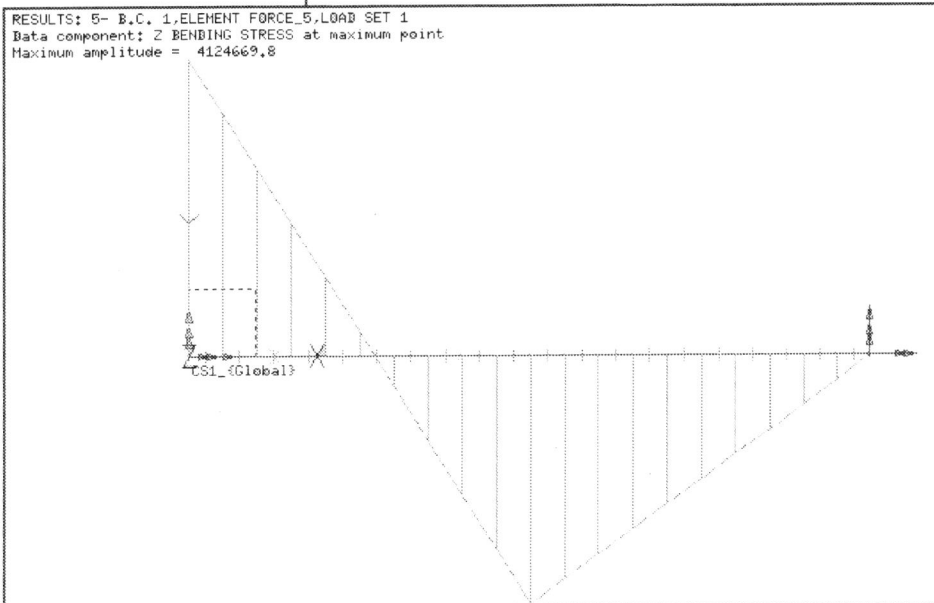

```
RESULTS: 5- B.C. 1,ELEMENT FORCE_5,LOAD SET 1
Data component: Z BENDING STRESS at maximum point
Maximum amplitude =   4124669.8
```

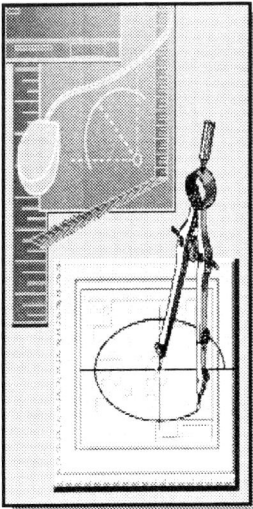

Learning Objectives

When you have completed this lesson, you will be able to:
- ◆ Perform Statically Indeterminate Beam Analysis.
- ◆ Understand and Apply the Principle of Superposition.
- ◆ Identify Statically Indeterminate Structures.
- ◆ Apply and Modify Boundary Conditions on Beams.
- ◆ Use the Check Coincident Nodes command.
- ◆ Generate Shear and Moment diagrams.

8.1 Introduction

Up to this point, we have dealt with a very convenient type of problem in which the external reactions can be solved by using the *equations of Statics* (Equations of Equilibrium). This type of problem is called *statically determinate*, where the *equations of Statics* are directly applicable to *determine* the loads in the system. Fortunately, many real problems are statically determinate, and our effort is not in vain. An additional large class of problems may be reasonably approximated as though they are *statically determinate*. However, there are those problems that are not statically determinate, and these are called *statically indeterminate structures*. Static indeterminacy can arise from redundant constraints, which means more supports than are necessary to maintain equilibrium. The reactions that can be removed leaving a stable system statically determinate are called *redundant*. For statically indeterminate structures, the number of unknowns exceeds the number of static equilibrium equations available. Additional equations, beyond those available from *Statics*, are obtained by considering the geometric constraints of the problem and the force-deflection relations that exist. Most of the derivations of the force-deflection equations are fairly complex; the equations can be found in design handbooks. There are various procedures for solving statically indeterminate problems. In recent years, with the improvements in computer technology, the finite element procedure has become the most effective and popular way to solving statically indeterminate structures.

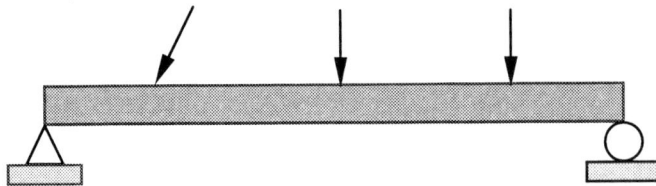

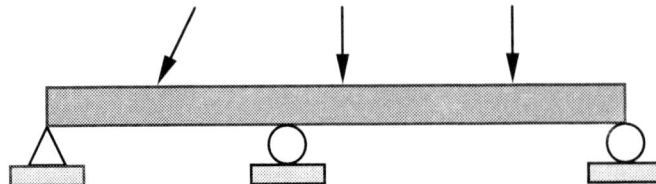

Equations of equilibrium:

$$\sum M = 0$$
$$\sum F = 0$$

Statically determinate structure:

Statically indeterminate structure:

8.2 Problem Statement

Determine the reactions at point A and point B. Also determine the maximum normal stress that the loading produces in the steel member (Diameter 2").

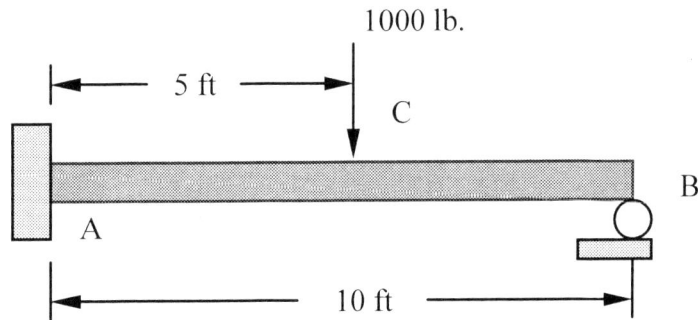

8.3 Preliminary Analysis

The reactions at point A and B can be determined by using the principle of superposition. By removing the roller support at point B, the displacement at point B due to the loading is first determined. Since the displacement at point B should be zero, the reaction at point B must cause an upward displacement of the same amount. The displacements are superimposed to determine the reaction. The *principle of superposition* can be used for this purpose, and once the reaction at B is determined, the other reactions are determined from the equations of equilibrium.

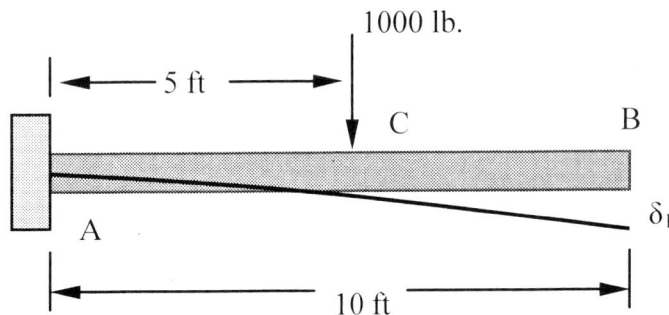

The displacement δ_1 can be obtained from most strength of materials textbooks and design handbooks:

$$\delta_1 = \frac{5\,PL^3}{48\,EI}$$

2

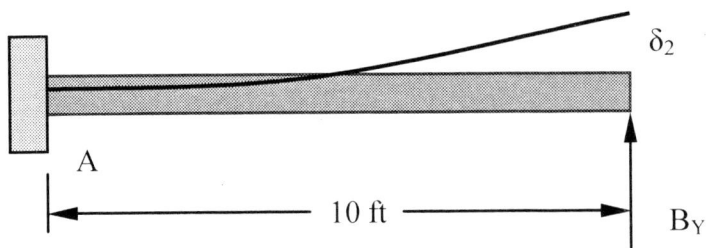

The displacement δ_2 due to the load can also be obtained from most strength of materials textbooks and design handbooks:

$$\delta_2 = \frac{B_Y L^3}{3\,E\,I}$$

3   (Superposition of 1 and 2)

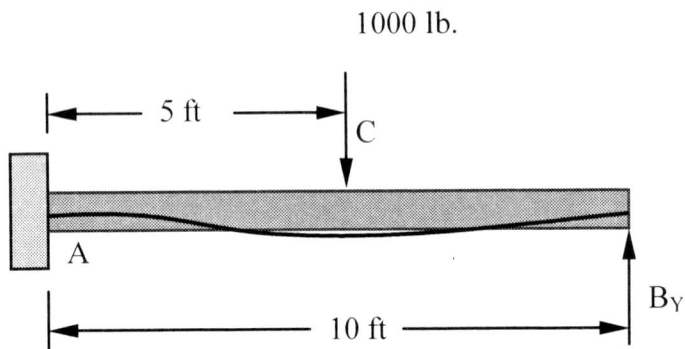

1000 lb.

$$\delta_B = -\delta_1 + \delta_2 = 0$$

$$\frac{-5\,PL^3}{48\,EI} + \frac{B_Y L^3}{3\,E\,I} = 0$$

$$B_Y = \frac{5\,P}{16} = 312.5\ \text{lb}$$

Free Body Diagram of the system:

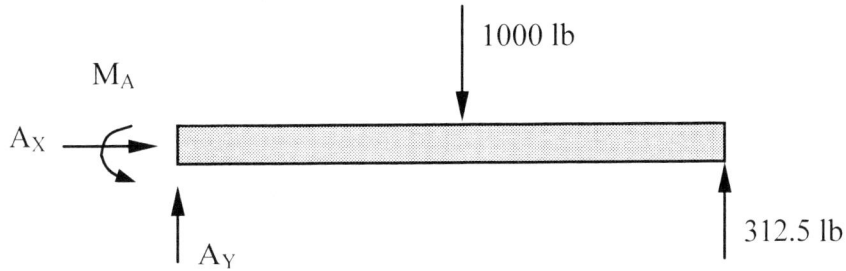

The reactions at A can now be solved:

$$\Sigma F_X = 0, \quad A_X = 0$$

$$\Sigma F_Y = 0, \quad 312.5 - 1000 + A_Y = 0, \quad A_Y = 687.5 \text{ lb.}$$

$$\Sigma M_{@A} = 0, \quad 312.5 \times 10 - 1000 \times 5 + M_A = 0, \quad M_A = 1875 \text{ ft-lb}$$

The maximum normal stress at point A is from the bending stress:

$$\sigma = \frac{MC}{I}$$

for the circular cross section:

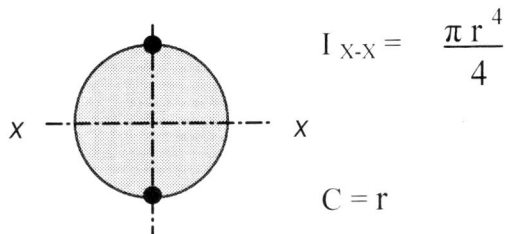

$$I_{X-X} = \frac{\pi r^4}{4}$$

$$C = r$$

Therefore,

$$\sigma_A = \frac{MC}{I} = \frac{4 M r}{\pi r^4} = 4.125 \times 10^6 \text{ lb/ft}^2$$

8.4 Starting *I-DEAS*

1. Login to the computer and bring up *I-DEAS*. Start a new model file by filling in the items as shown below in the *I-DEAS Start* window:

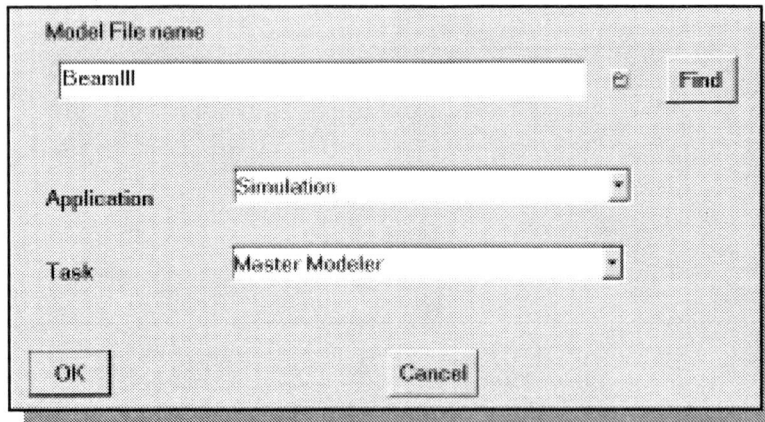

Model File name	
Beamlll	Find
Application	Simulation
Task	Master Modeler
OK	Cancel

2. After you click **OK**, a *warning window* will appear indicating a new model file will be created. Click **OK** to continue.

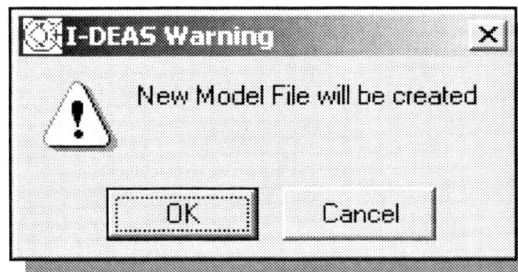

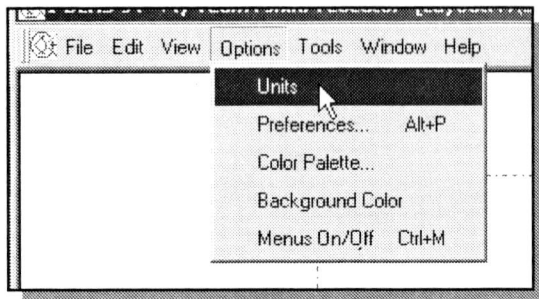

> **I-DEAS Warning**   ✕
> ⚠ New Model File will be created
> OK   Cancel

File Edit View Options Tools Window Help

Units
Preferences...   Alt+P
Color Palette...
Background Color
Menus On/Off   Ctrl+M

3. Use the left-mouse-button and select the **Options** menu in the icon panel.

4. Select the **Units** option.

Meter (newton)
Foot (pound f)
Meter (kilogram f)
Foot (poundal)
mm (milli newton)
cm (centi newton)

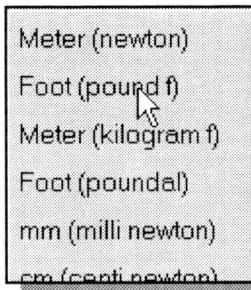

5. Set the units to **Foot (pound f)** by selecting it from the menu of choices.

❖ The units used in an FEA analysis <u>MUST</u> be consistent throughout the analysis.

8.5 Workplane Appearance

❖ The workplane is a construction tool; it is a coordinate system that can be moved in space. The size of the workplane display is only for our visual reference, since we can sketch on the entire plane, which extends to infinity.

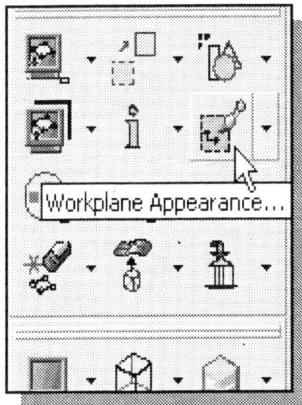

1. Choose **Workplane Appearance** in the icon panel. (The icon is located in the second row of the application icon panel.) The *Workplane Attributes* window appears.

2. Toggle **on** the *Display Border* switch as shown.

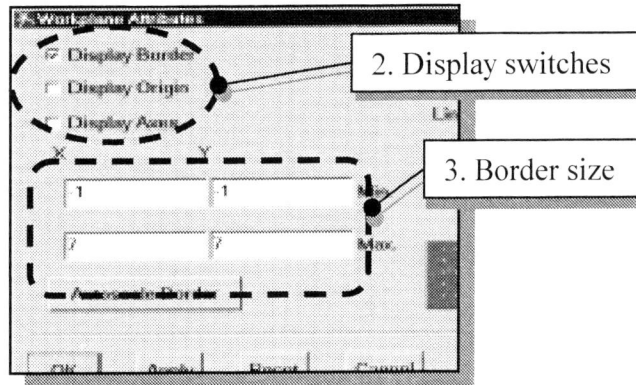

3. Adjust the **workplane border size** by entering the *Min.* & *Max.* values as shown in the figure above.

4. Click on the **OK** button to exit the *Workplane Attributes* window.

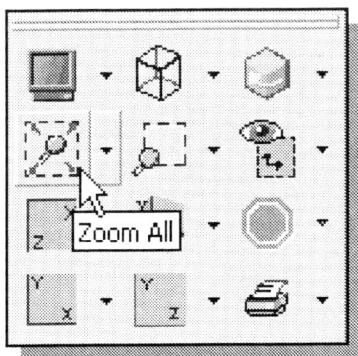

5. Choose **Zoom-All** in the display viewing icon panel

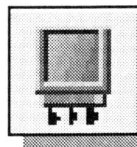

6. Choose **Refresh** in the icon panel. (The icon is located in the first row of the display icon panel.)

8.6 Using the Create Part option

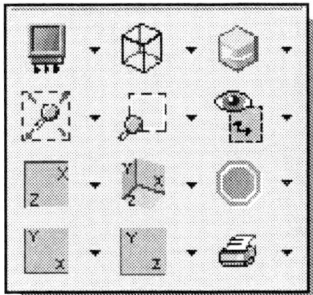

1. Choose *Isometric View* in the display viewing icon panel.

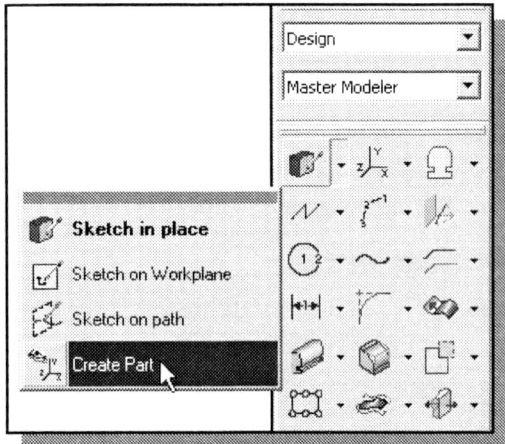

2. Choose **Create Part** in the icon panel.

➤ The icon is located in the first row of the task specific icon panel. The icon is located in the same stack as the *Sketch In Place* icon. Press and hold down the left-mouse-button on the icon stack to display the choice menu.

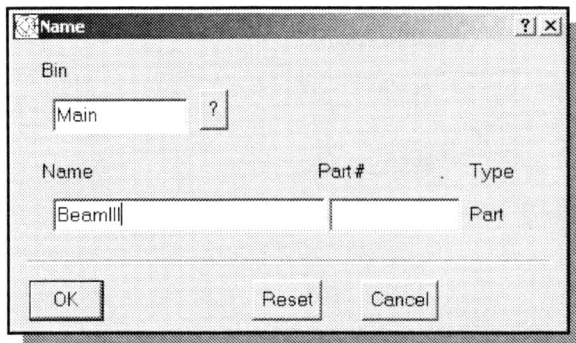

3. The *Name Part* window appears on the screen, enter **BeamIII** as the name of the part as shown.

4. Click on the **OK** button to proceed with the **Create Part** command.

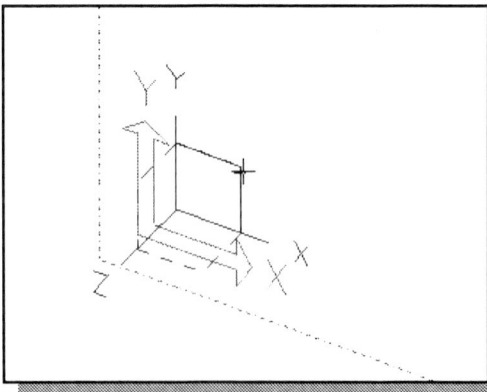

5. In the prompt window, the message "*Pick plane to sketch on*" is displayed. Pick the **XY** plane of the newly created coordinate system as shown. (Note that the default work plane, **blue** color, is still aligned to the XY plane of the world coordinate system. Aligning the sketch plane to the newly created coordinate system assures the proper association of the features to the part.)

6. Press the **ENTER** key once, or click once with the right-mouse-button to accept the placement of the workplane.

8.7 Creating a Wireframe Model of the System

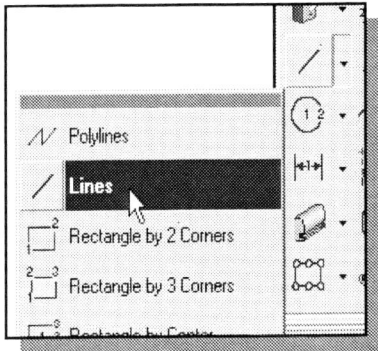

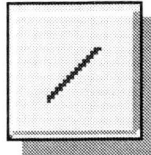

1. Pick **Polylines** in the icon panel. (The icon is located in the second row of the task specific icon panel.) The message "*Locate start*" is displayed in the prompt window.

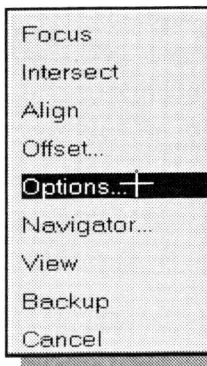

2. Move the cursor inside the graphics window. Press and hold down the right-mouse-button to display the popup option menu. Slide the cursor up and down in the popup menu and select **Options...**

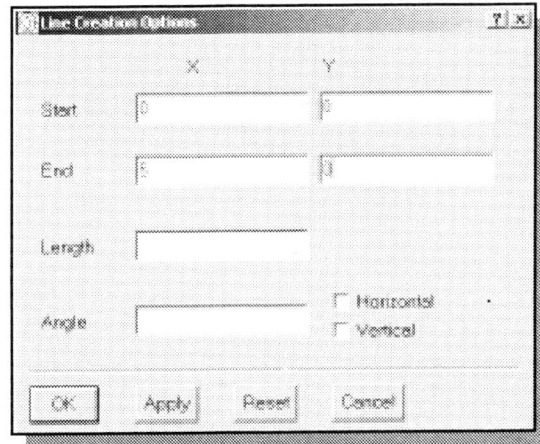

3. Fill in the **Start** and **End** point coordinates as shown.

4. Click the **OK** button to accept the settings and create a line.

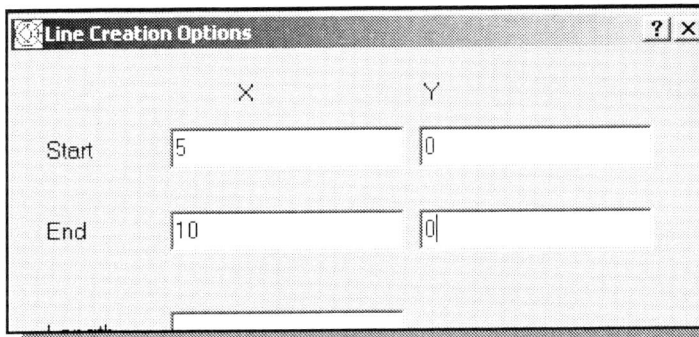

5. On your own, repeat the above steps and create another line at the coordinates shown.

6. Click the **OK** button to accept the settings and create the second line.

7. Press the **ENTER** key or the middle-mouse-button to end the *Lines* command.

➤ Note that we have intentionally created two separate line-segments for the example. The purpose of creating two separate line-segments is to illustrate (1) assuring a node point at a specific point, in this case, the midpoint, (2) creating a different type of mesh-length on the same model and (3) using the *Check Coincident Nodes* command.

8.8 Creating an Element Cross Section

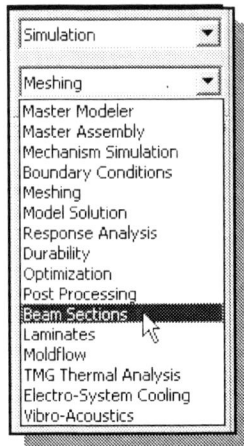

1. Switch to the **Beam Sections** task by selecting the task menu in the icon panel as shown.

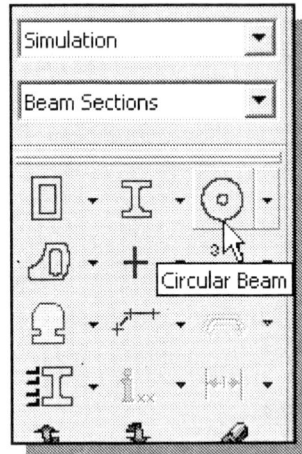

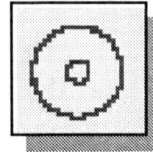

2. Choose **Circular Beam** in the icon panel. (The icon is located in the first row of the task icon panel. If the icon is not on top of the stack, press and hold down the left-mouse-button on the displayed icon to display all the choices.)

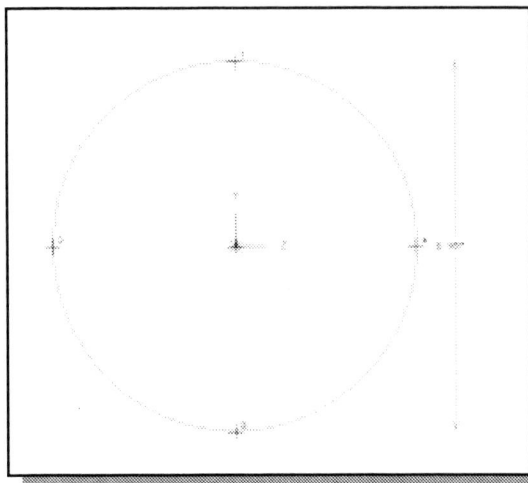

3. The message "*ENTER outside diameter*" is displayed. Enter **2/12** for the *outside diameter.*

4. Enter **0.0** for the *inside diameter*.

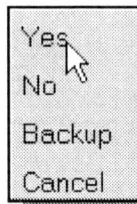

5. Pick **Yes** in the popup menu to accept the settings and complete creating the cross section.

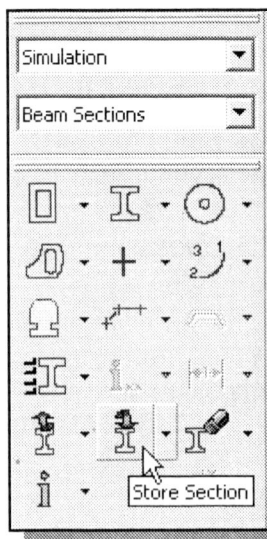

6. Choose **Store Section** in the icon panel. (The icon is located in the fifth row of the task icon panel.) The message "*Enter beam cross sect prop name or no (1-CIRCULAR 0.16667 x 0.0).*" is displayed in the prompt window.

7. Press the **ENTER** key to accept the default name *(1-CIRCULAR 0.16667 x 0.0).*

8.9 Setup of an FEA Model

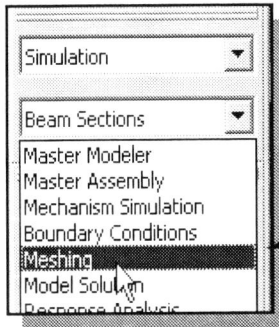

1. Switch back to the **Meshing** task by selecting the Meshing task in the task menu as shown.

Pick *Meshing*

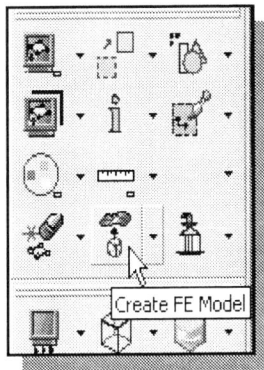

2. Choose **Create FE Model** in the icon panel. (The icon is located in the last row of the application specific icon panel.) The *FE Model Create* window appears.

3. In the *FE Model Create* window, enter **BeamIII** in the *FE Model Name* box.

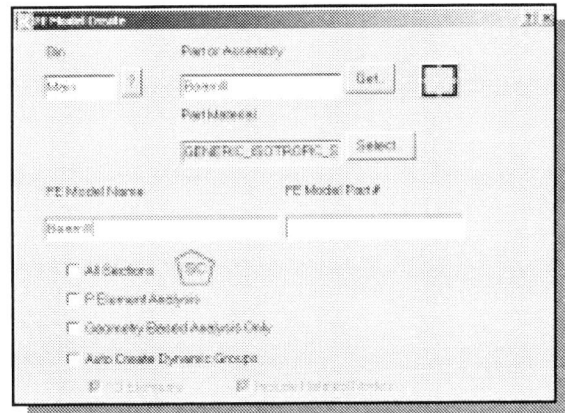

4. Click on the **OK** button to accept the settings.

❖ Note the *Part* name is displayed in the *FE Model Create* window.

8.10 Materials Property Table

1. Choose **Materials** in the icon panel. (The icon is located in the fourth row of the task icon panel.) The *Materials* window appears.

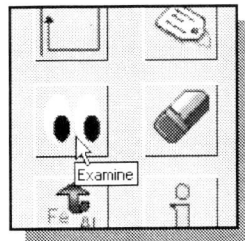

2. Choose the default material, **Generic_Isotropic_Steel**, in the list window.

3. Click on the **Examine** icon and review the material properties.

4. On your own, exit the *Examine* window and the *Materials* window.

8.11 Define Beam Auto-Mesh

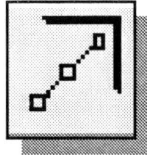

1. Choose **Define Beam Mesh** in the icon panel. (The icon is located in the first row of the task icon panel. Press and hold down the left-mouse-button on the displayed icon and select the *Define Beam Mesh* command by releasing the left-mouse-button when the option is highlighted.)

2. The message "*Pick edges/Curves*" is displayed in the prompt window. Pick the left line by left-clicking once on the line.

3. Press the **ENTER** key or the middle-mouse-button to accept the selection.

4. The *Define Mesh* window appears. Enter **1.0** in the *Element Length* box.

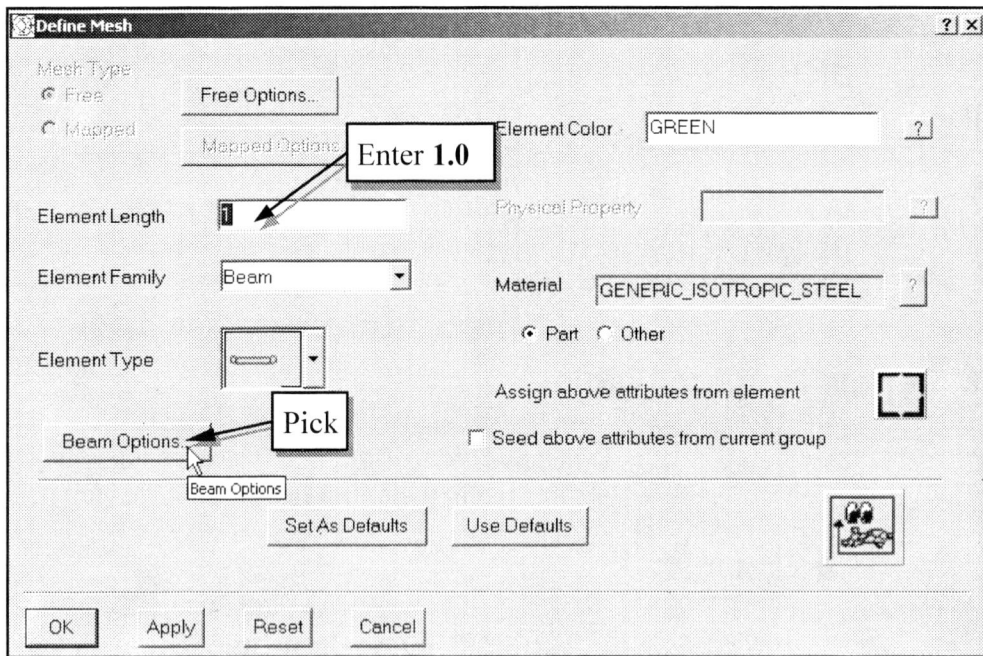

5. Pick **Beam Options** and confirm the cross section is set to *1- CIRCULAR 0.16667 x 0.0.*

6. Click on the **OK** button to exit the *Beam Options* window.

7. Click on the **OK** button to exit the *Define Beam Mesh* window.

8. Choose **Define Beam Mesh** in the icon panel. (The icon is located in the first row of the task icon panel. Press and hold down the left-mouse-button on the displayed icon and select the *Define Beam Mesh* command by releasing the left-mouse-button when the option is highlighted.)

9. The message "*Pick edges/Curves*" is displayed in the prompt window. Pick the right-line by left-clicking once on the line.

10. Press the **ENTER** key or the middle-mouse-button to accept the selection.

11. The *Define Mesh* window appears. Enter **0.5** in the *Element Length* box.

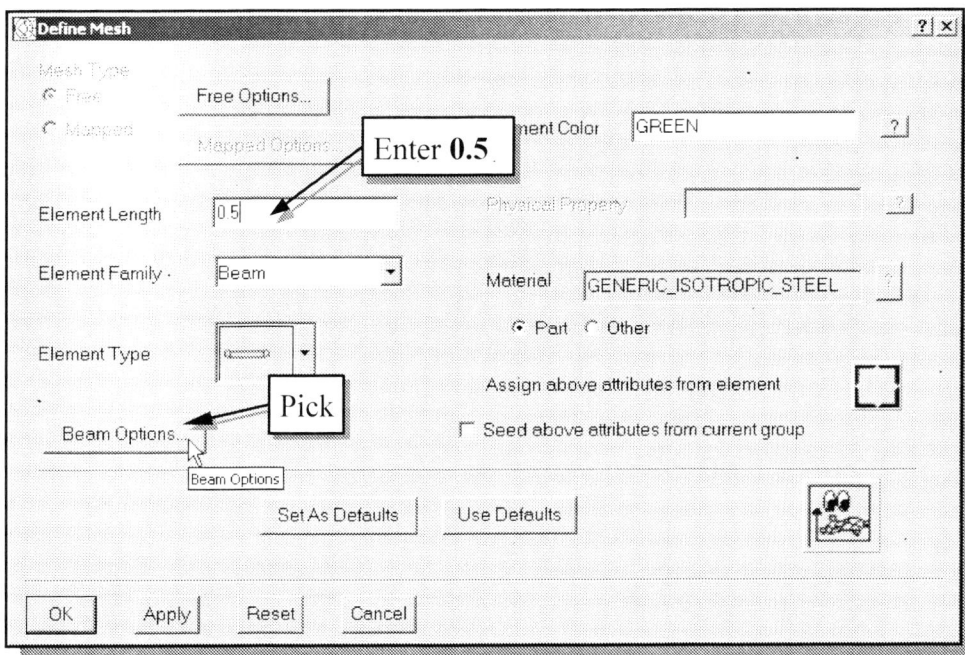

12. Pick **Beam Options** and confirm the cross section is set to *1- CIRCULAR 0.16667 x 0.0.*

13. Click on the **OK** button to exit the *Beam Options* window.

14. Click on the **OK** button to exit the *Define Beam Mesh* window.

8.12 Create Beam Auto-Mesh

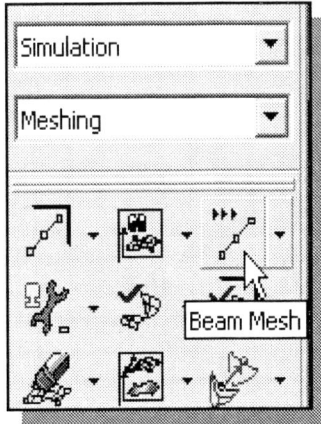

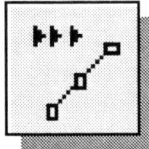

1. Choose **Create Beam Mesh** in the icon panel. (The icon is located in the first row of the task icon panel. Press and hold down the left-mouse-button on the displayed icon and select the *Create Beam Mesh* command by releasing the left-mouse-button when the option is highlighted.) The message "*Pick edges/Curves*" is displayed in the prompt window.

2. Pick both lines by holding down the **SHIFT** key and left-clicking each line.

3. Press the **ENTER** key or the middle-mouse-button to accept the selections. New nodes and elements are created. Note the different lengths of the two elements.

4. Press the **ENTER** key or the middle-mouse-button to accept the additions.

5. Pick **Isometric View** in the icon panel.

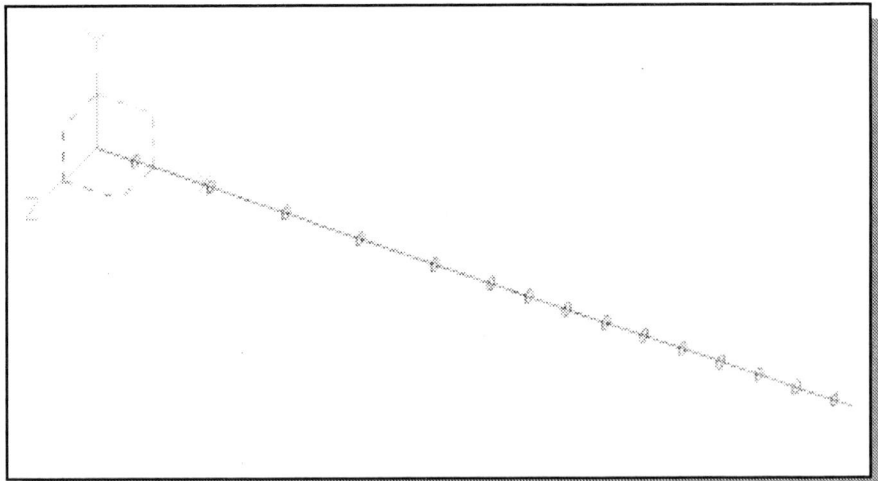

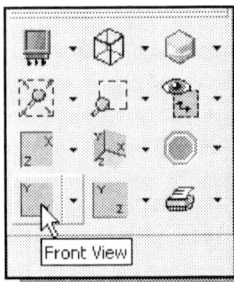

6. On your own, examine the FE model by using the dynamic *Rotate* command. Reset the display to **Front View** before proceeding to the next section.

8.13 Checking Coincident Nodes

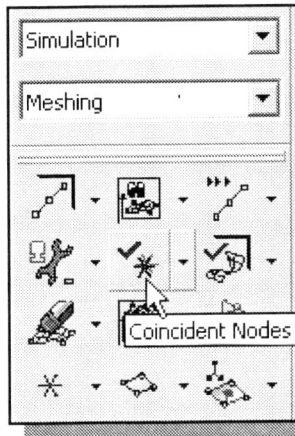

1. Choose **Coincident Nodes** in the icon panel. (The icon is located in the second row of the task icon panel.)

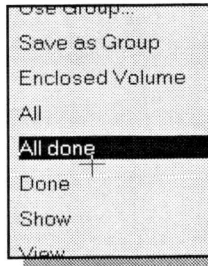

2. Move the cursor inside the graphics window. Press and hold down the right-mouse-button to display the popup option menu. Slide the cursor up and down in the popup menu and select **All done.**

3. The message "*Enter distance between nodes to be considered coincident (0.0000328084)*" is displayed in the prompt window. Press the **ENTER** key or the middle-mouse-button to accept the default value.

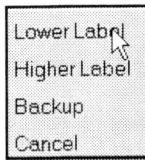

4. In the prompt window, the message "*Enter method to select coincident node (Lower_Label)*" is displayed. Select **Lower Label** in the popup menu.

5. The message "*OK to list element labels?*)" is displayed in the prompt window. Select **Yes** in the popup menu. *I-DEAS* now performs the coincident node check.

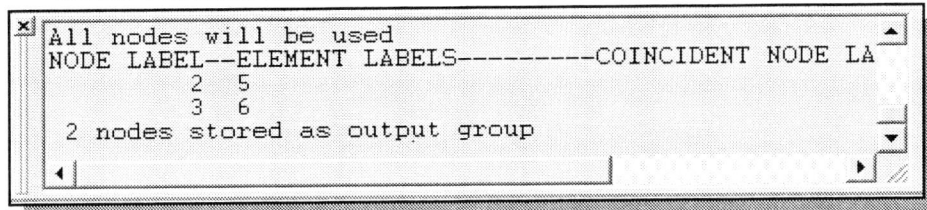

```
All nodes will be used
NODE LABEL--ELEMENT LABELS---------COINCIDENT NODE LA
            2    5
            3    6
   2 nodes stored as output group
```

6. The result of the check is displayed in the list window. *Node 2* and *Node 3*, which are located at the midpoint of the structure, coincide with each other. Note that this is due to the two separate line-segments we created.

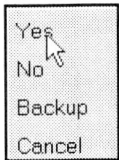

7. In the prompt window, the message "*OK to merge coincident nodes (No)*" is displayed. Select **Yes** in the popup menu.

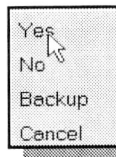

8. In the prompt window, the message "*OK to delete nodes that has been replaced (Yes)*" is displayed. Select **Yes** in the popup menu and click **Dismiss** in the warning window that appears.

➤ We have successfully removed coincident nodes in the FE model. Note that *I-DEAS* also provides other check options, such as the *Check Coincident Elements* option.

8.14 Apply Boundary Conditions

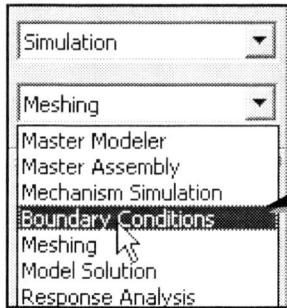

1. Switch to the **Boundary Conditions** task by selecting it in the task menu as shown.

   Pick *Boundary Conditions*

2. Choose ***Displacement Restraint*** in the icon panel. (The icon is located in the fourth row of the task icon panel.)

3. The message "*Pick Nodes/Centerpoints/Vertices*" is displayed in the prompt window. Pick ***Node 1*** (left endpoint of the member).

   Pick the node located at this end of the member.

4. Press the **ENTER** key or the middle-mouse-button to accept the selection. The *Displacement Restraint on Node* window appears.

5. Pick **Clamp** for the fixed-end support.

6. Click on the **OK** button to exit the *Displacement Restraint on Node* window.

7. Choose ***Displacement Restraint*** in the icon panel.

8. The message "*Pick Nodes/Centerpoints/Vertices*" is displayed in the prompt window. Pick the node at the right endpoint of the member.

   Pick the node located at this end of the member.

9. Press the **ENTER** key to continue. The *Displacement Restraint on Node* window appears.

10. Pick **Roller** and set *Axis* to **X Trans Z Rot**

11. Click on the **OK** button to exit the *Displacement Restraint on Node* window.

8.15 Apply external Loads

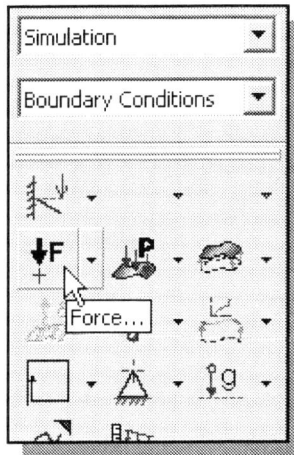

1. Choose **Force** in the icon panel. (The icon is located in the second row of the task icon panel.) The message "*Pick entities*" is displayed in the prompt window.

2. Pick the node (**N2**) that is at the center of the FE model.

3. Press the **ENTER** key or the middle-mouse-button to accept the selection. The *Force on Node* window appears.

4. The load is 1000lb. in the negative Y direction. Enter **-1000** in the *Y force* box.

5. Click on the **OK** button to exit the *Force on Node* window.

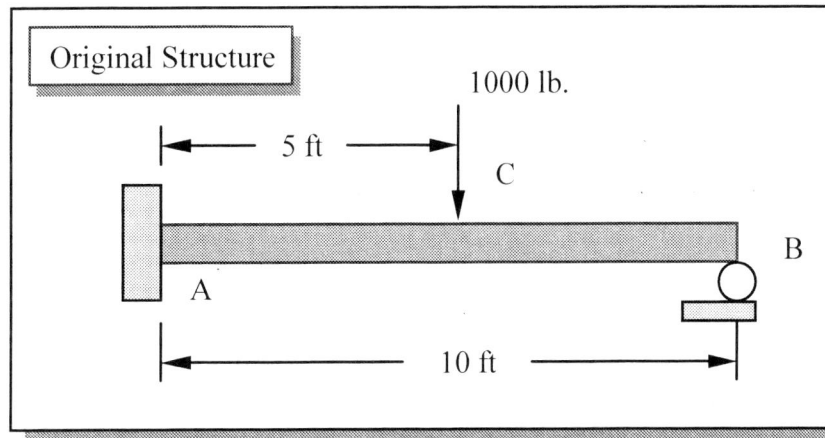

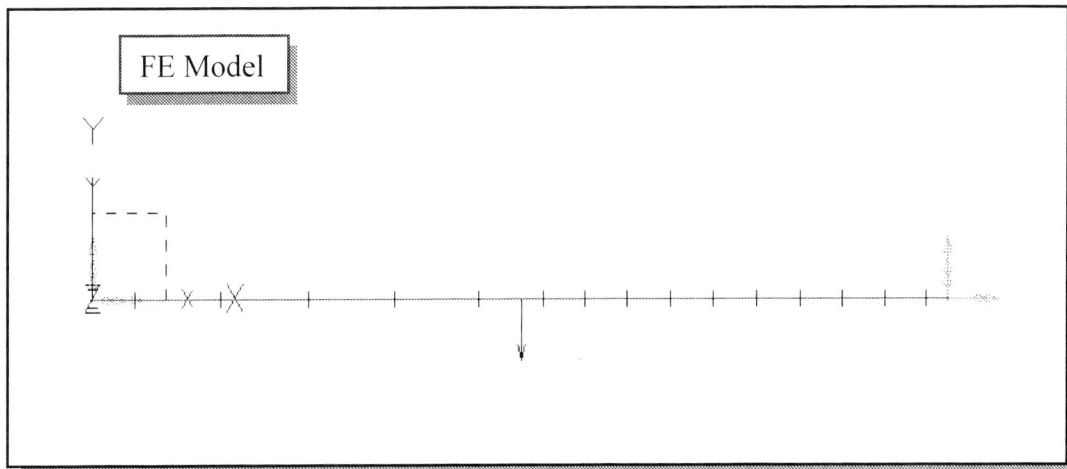

Original Structure

1000 lb.

5 ft

C

A

B

10 ft

FE Model

8.16 Build a Boundary Condition Set

❖ As boundary conditions are created, *I-DEAS* will automatically attach a boundary condition set to the FE model. The *Boundary Condition* option allows us to attach different boundary condition sets to the FE model.

1. Choose **Boundary Condition** in the icon panel. (The icon is located in the last row of the task icon panel.) The *Boundary Condition Set Management* window appears.

❖ Multiple *Restraint/Load Sets* can be setup in *I-DEAS*. The *Boundary Condition Set Management* option can be used to define different combination of the established *Restraint Sets* and *Load Sets*.

2. Confirm the **Restraint Set 1** and **Load Set 1** are preset as shown in the below figure.

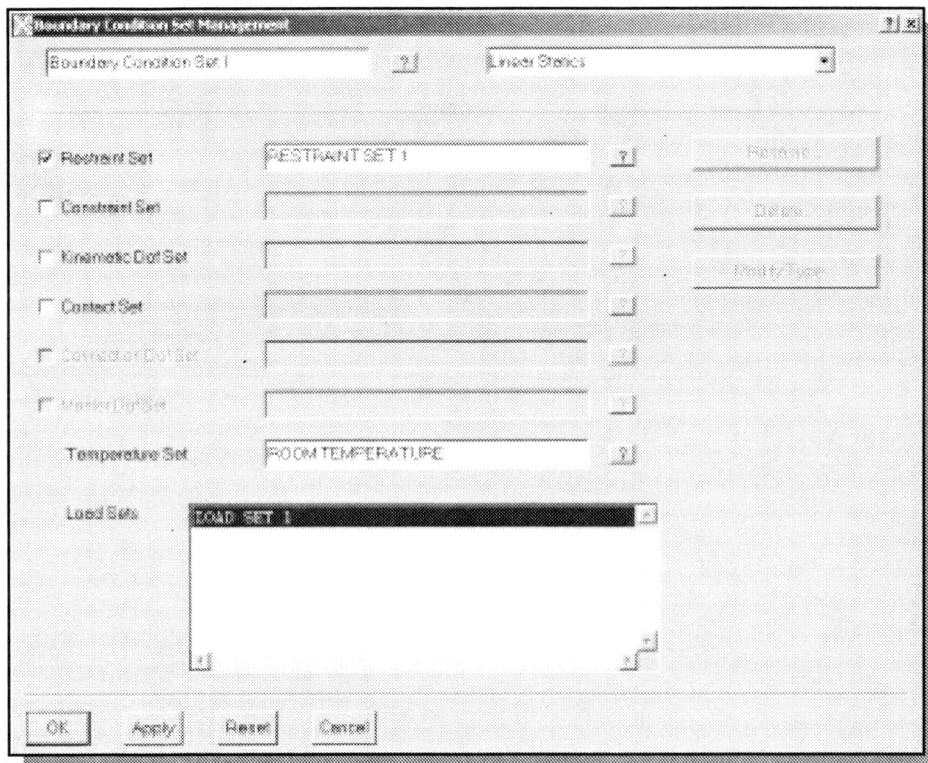

3. Click **OK** to close the *Boundary Condition Set Management* window.

8.17 Running the Solver

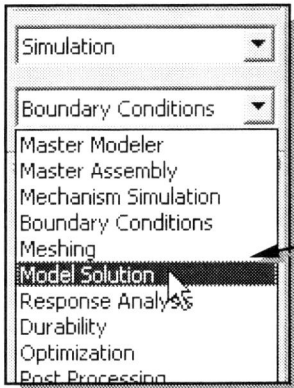

1. Switch to the **Model Solution** task by selecting the task in the task menu as shown.

Pick *Model Solution*

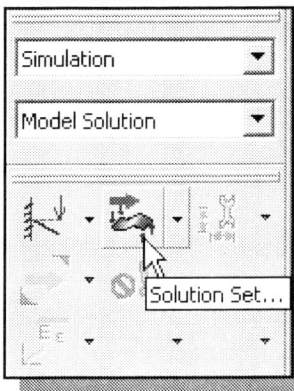

2. Choose **Solution Set** in the icon panel. (The icon is located in the first row of the task icon panel.) The *Manage Solution Sets* window appears.

3. Click on the **Create** button. The *Solution Set* window appears.

4. Click on the **Output Selection** button. The *Output Selection* window appears.

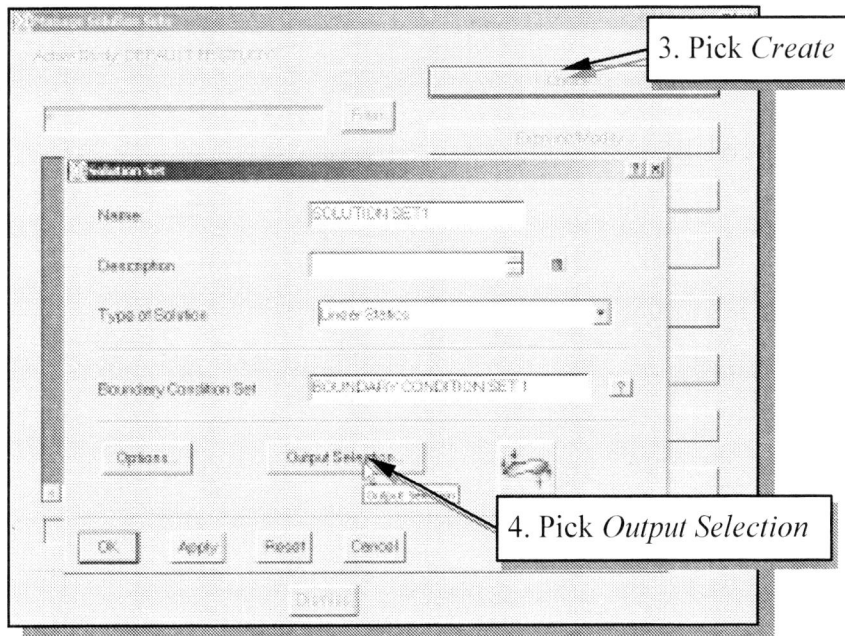

3. Pick *Create*

4. Pick *Output Selection*

5. On your own, set the **Element Forces** output option to **Store** in the *Output* list.

6. Click on the **OK** button to exit the *Output Selection* window.

7. Click on the **OK** button to exit the *Solution Set* window.

8. Pick **Dismiss** to exit the *Manage Solution Sets* window.

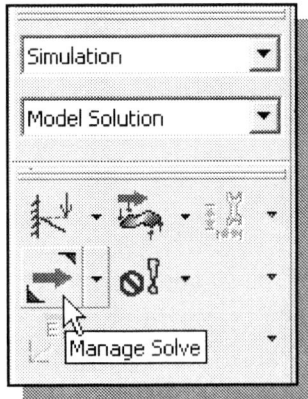

9. Choose **Manage Solve** in the icon panel. (The icon is located in the second row of the task icon panel.) The *Solve* window appears.

10. In the *Solve* window, pick the **Solve** icon to find the solutions.

➤ *I-DEAS* will begin the solving process. <u>DO NOT</u> close any windows. Any errors or warnings are displayed in the list window.

8.18 Viewing the results – Reaction Forces

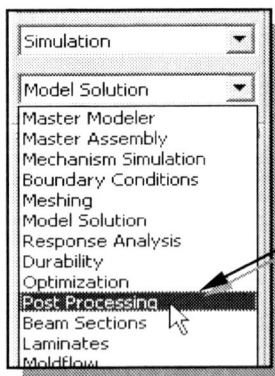

1. Switch to the **Post Processing** task by selecting in the task menu as shown.

Pick *Post Processing*

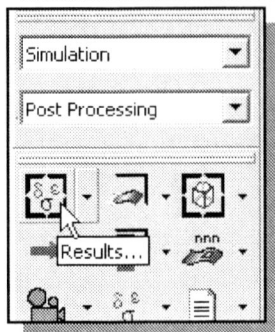

2. Choose **Results Selection** in the icon panel. (The icon is located in the first row of the task icon panel.)

3. The *Results Selection* window appears. Pick **Reaction Force** from the list.

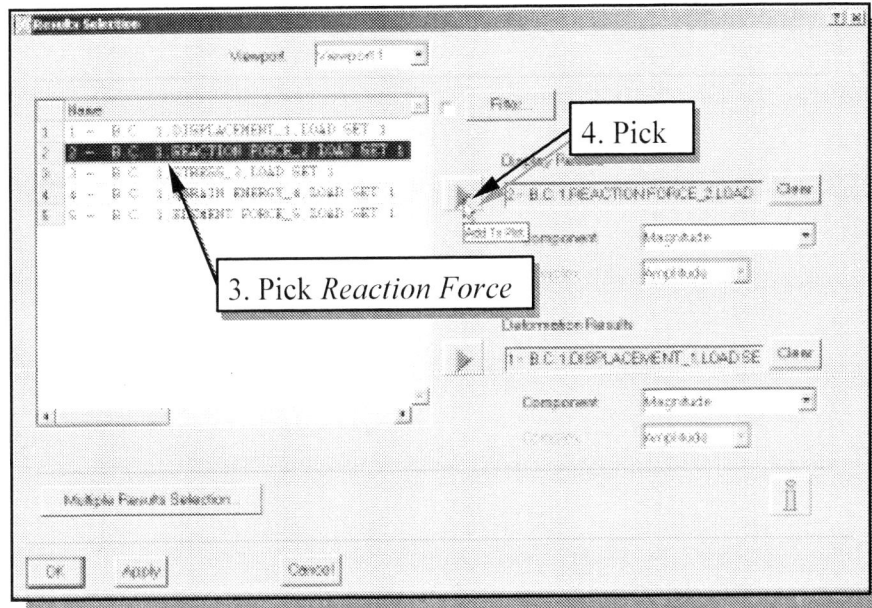

4. Click on the **Triangle** button to set the *Reaction Force* as the *Display Results*.

5. Click on the **OK** button to exit the *Results Selection* window.

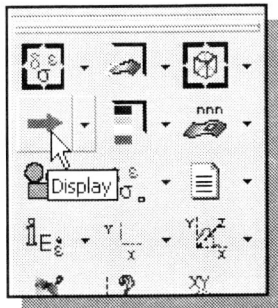

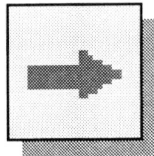

6. Choose **Results Display** in the icon panel. (The icon is located in the second row of the task icon panel.)

7. Pick the last element, **Element 15**, the element with one end resting on the roller support.

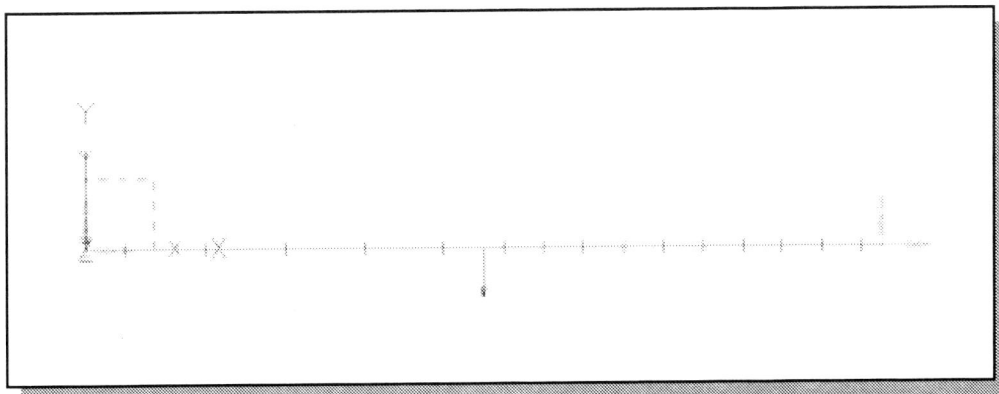

8. Press the **ENTER** key or the middle-mouse-button to accept the selection and display the reaction force associated with this element.

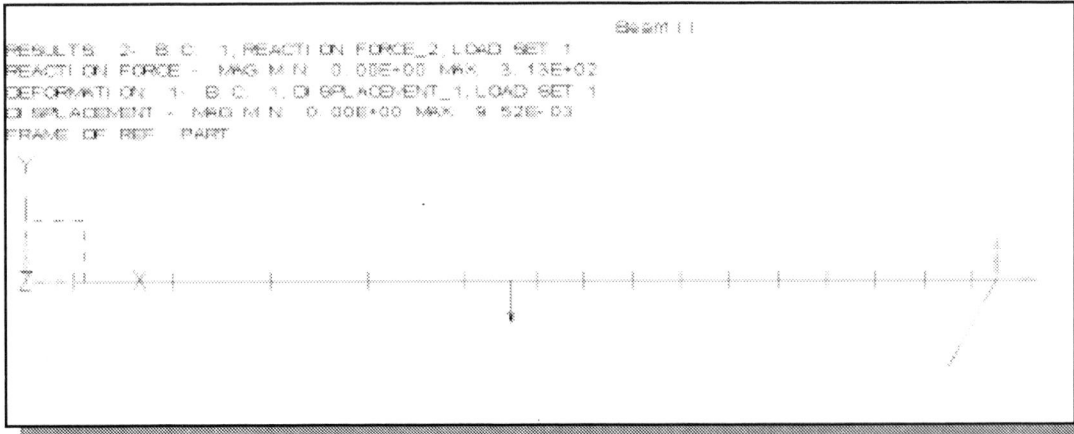

➢ The computer solution is **3.13E+02 lb.**, which matches with our hand calculation (see section 8.3 Preliminary Analysis).

9. Choose *Refresh* in the icon panel. (The icon is located in the first row of the display icon panel.)

8.19 View the Results – Reactions at Point A

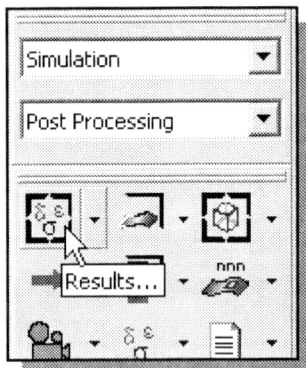

1. Choose *Results Selection* in the icon panel. (The icon is located in the first row of the task icon panel.) The *Results Selection* window appears.

2. Select **Y component** under the *Display Results Component*.

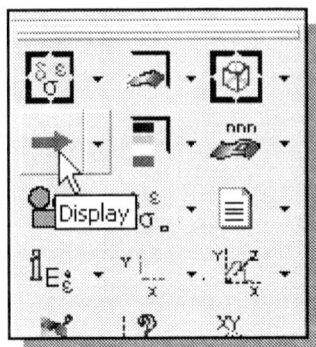

3. Click on the **OK** button to exit the *Results Selection* window.

4. Choose *Results Display* in the icon panel.

5. Pick the first element, **Element 1,** as shown.

Pick the first element

6. Press the **ENTER** key to display the reaction force associated with this element.

➤ The computer solution is **6.87 E+02 lb**, which matches our hand calculation of 687.5 lb (see section 8.3 Preliminary Analysis.)

7. Choose **Refresh** in the icon panel. (The icon is located in the first row of the display icon panel.)

8. Choose **Results Selection** in the icon panel. (The icon is located in the first row of the task icon panel.) The *Results Selection* window appears.

9. Select the **R$_Z$** component, which is the moment about the Z-axis, under the *Display Results Component*.

10. Click on the **OK** button to exit the *Results Selection* window.

11. Choose **Results Display** in the icon panel.

12. Pick the first element, **Element 1**.

13. Press the **ENTER** key to display the reaction associated with this element.

➤ The computer solution is **1.87 E+3 ft-lb**, which matches our hand calculation of 1875 ft-lb (see section 8.3 Preliminary Analysis.)

8.20 Shear Diagram

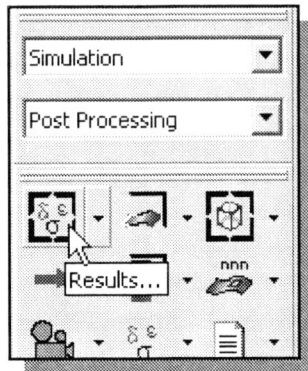

1. Choose **Results Selection** in the icon panel. (The icon is located in the first row of the task icon panel.)

2. The *Results Selection* window appears. Pick **Element Force** from the list.

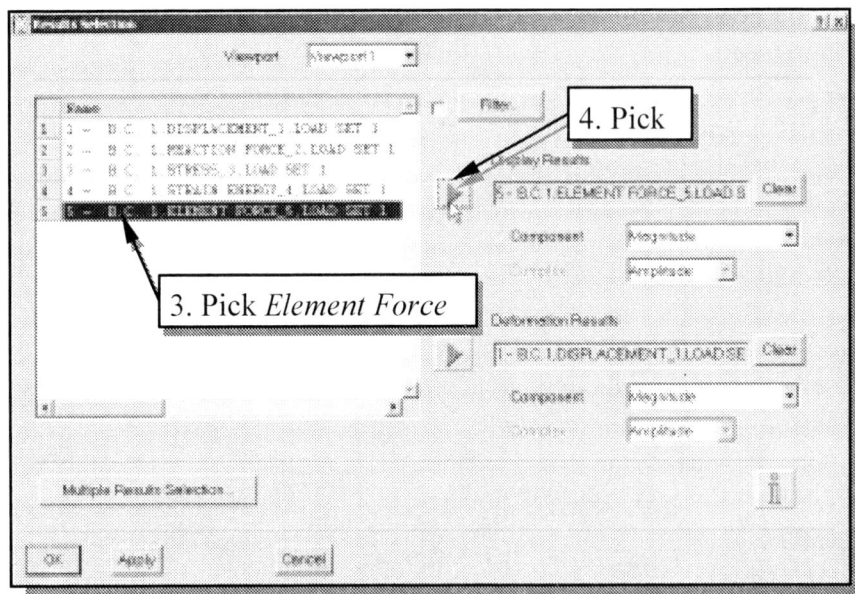

4. Pick

3. Pick *Element Force*

3. Click on the **Triangle** button to set the *Element Force* as the *Display Results*.

4. Click on the **OK** button to exit the *Results Selection* window.

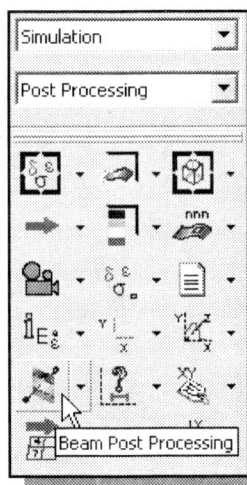

5. Choose **Beam Post-Processing** in the icon panel. (The icon is located in the fifth row of the task icon panel.) The cascading menu appears in the graphics window.

6. Select **Force & Stress** in the cascading menu.

7. Select **Data Component** in the popup menu.

8. Select **Force** in the popup menu.

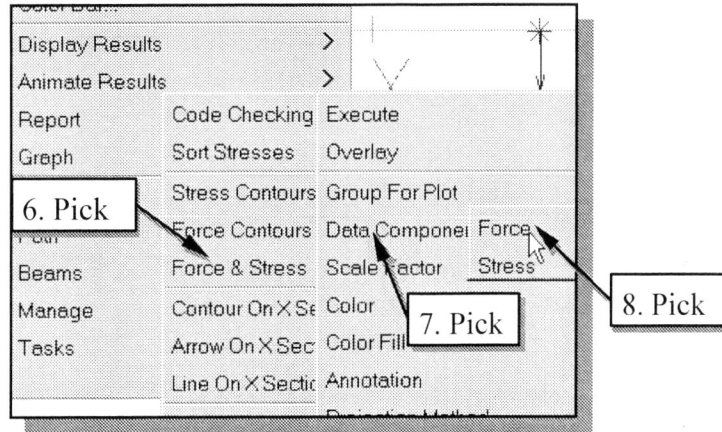

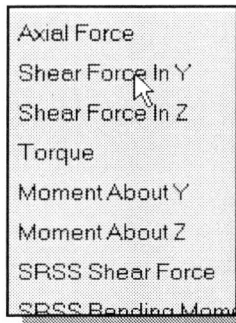

9. In the popup menu, select **Shear Force in Y**.

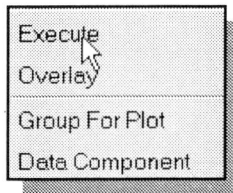

10. In the cascading menu, select **Execute**.

11. Press down the **CTRL** key and hit the **M** key to switch off the cascading menu.

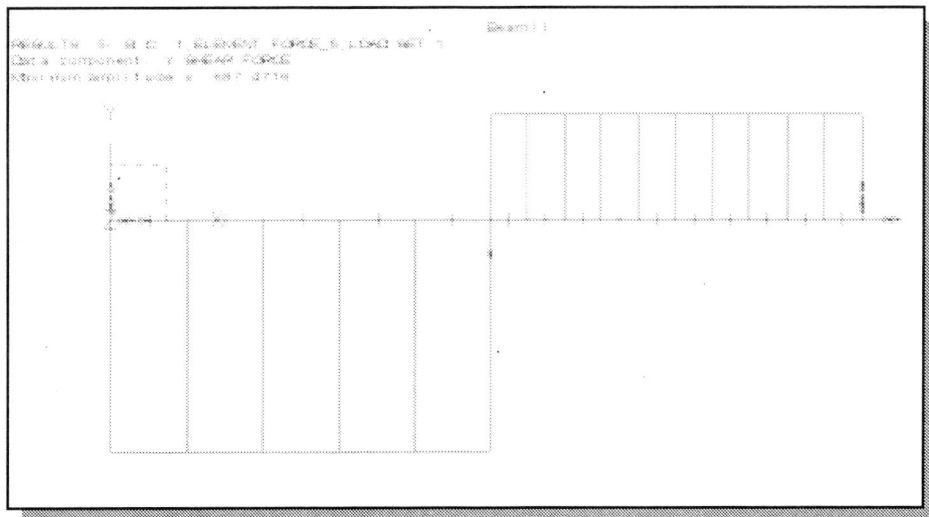

❖ The Y-shear diagram should appear as shown. Does the diagram match with the preliminary analysis we did at the beginning of the chapter?

8.21 Z Bending Moment Diagram

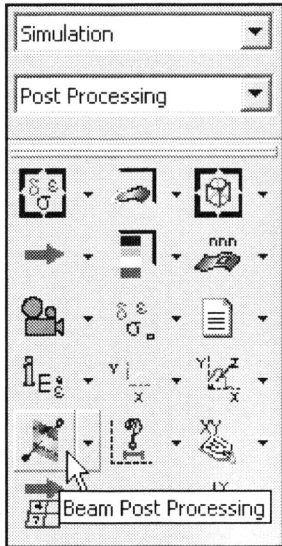

1. Choose **Beam Post-Processing** in the icon panel. (The icon is located in the fifth row of the task icon panel.) The cascading menu appears in the graphics window.

2. Select **Force & Stress** in the cascading menu.

3. Select **Data Component** in the popup menu.

4. Select **Force** in the popup menu.

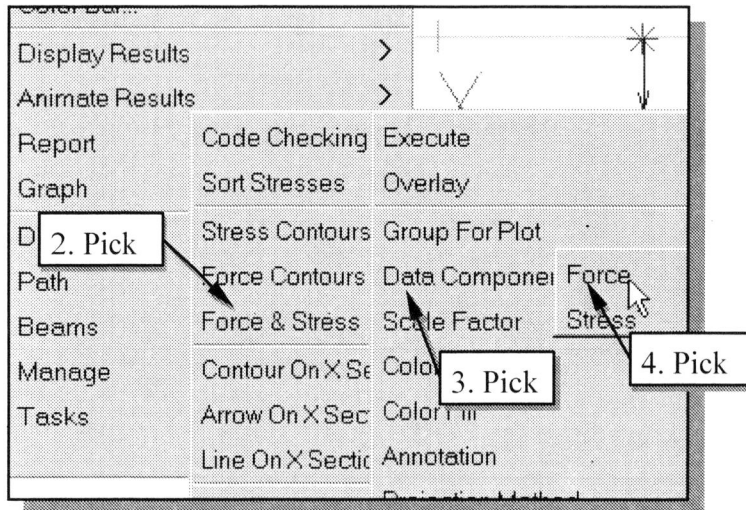

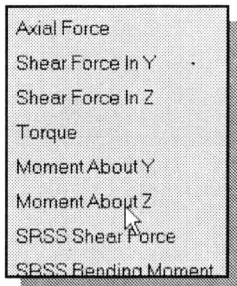

5. In the popup menu, select **Moment About Z.**

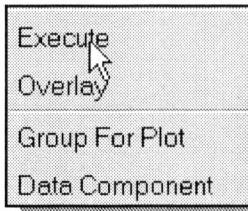

6. In the cascading menu, select **Execute**.

7. Press down the **CTRL** key and hit the **M** key to switch off the cascading menu.

❖ The bending moment diagram should appear as shown. Does the diagram match with the preliminary analysis we did at the beginning of the chapter? The moment diagram shows us that the maximum moment occurs at which location?

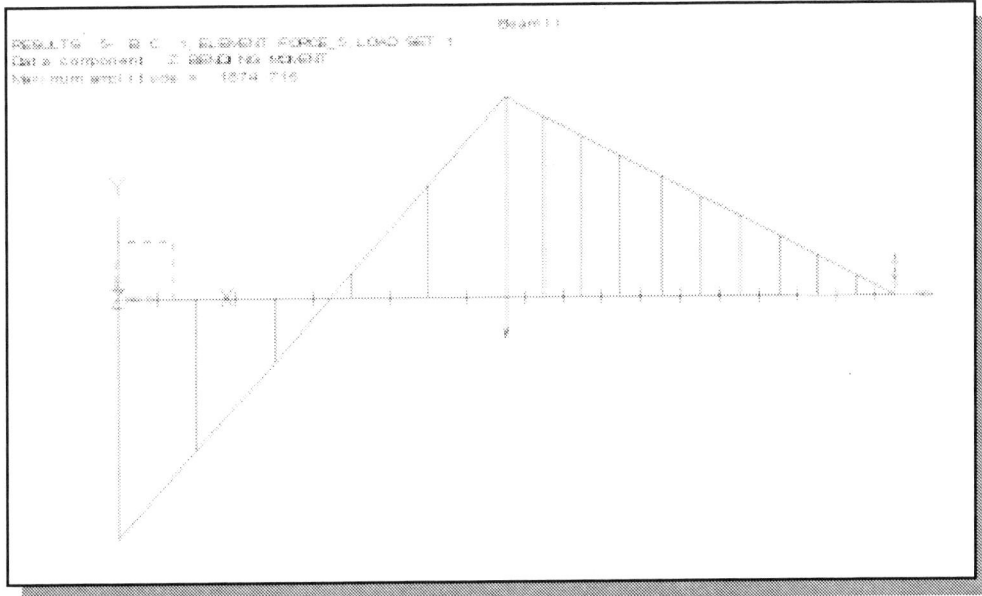

8.22 Maximum Bending Stress in the Member

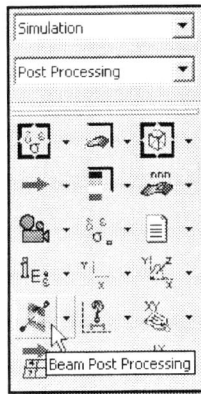

1. Choose **Beam Post-Processing** in the icon panel. (The icon is located in the fifth row of the task icon panel.) The cascading menu appears in the graphics window.

2. Select **Force & Stress** in the cascading menu.

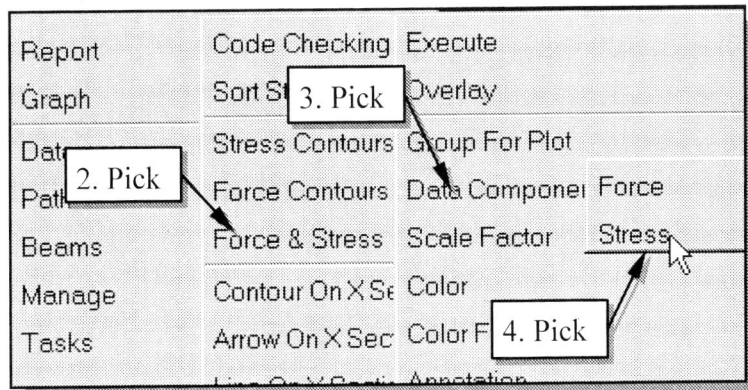

3. Select **Data Component** in the cascading menu.

4. Select **Stress** in the cascading menu.

Axial Stress
Shear Stress In Y
Shear Stress In Z
Torsional Stress
Bending Stress About Y
Bending Stress About Z
SRSS Shear Stress
SRSS Bending Stress
Von Mises Stress

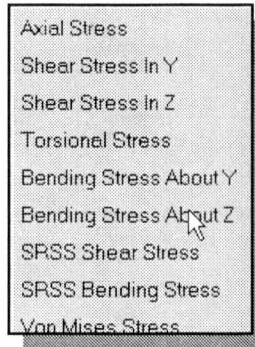

5. In the popup menu, select **Bending Stress About Z**.

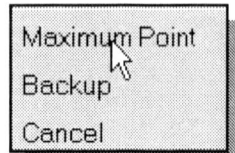

Maximum Point
Backup
Cancel

6. Pick **Maximum Point** in the popup menu.

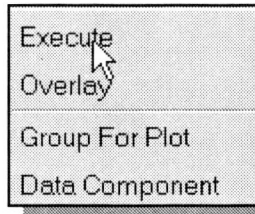

Execute
Overlay
Group For Plot
Data Component

7. In the cascading menu, select **Execute**.

8. Press down the **CTRL** key and hit the **M** key to switch off the cascading menu.

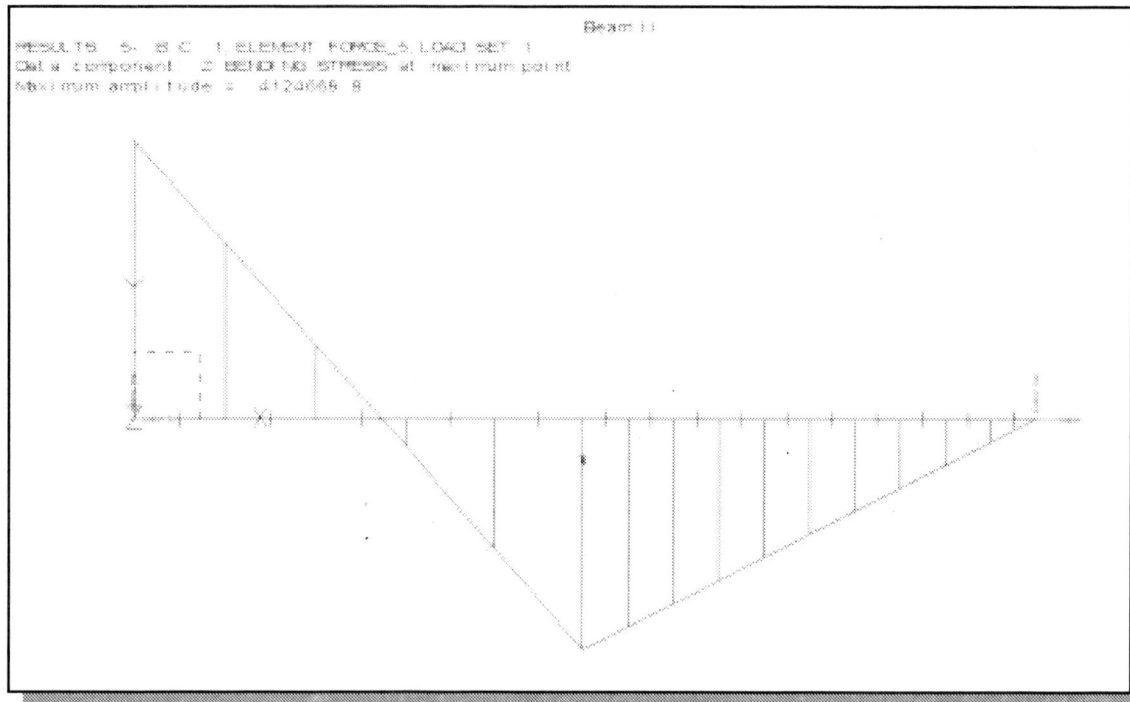

➤ The computer solution for the maximum stress is **4124669 lb/ft^2**, which occurs at the fixed-end of the member. Our preliminary calculation gives the value of 4.125E+6.

Questions:

1. What are the equations of *Statics* (Equations of Equilibrium)?

2. What is the type of structure where there are more supports than are necessary to maintain equilibrium?

3. How do we obtain more equations for *statically indeterminate* structures?

4. What is the key combination to toggle on/off the *cascading menu*?

5. Identify and describe the following commands:

(a)

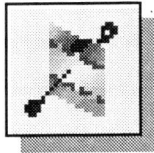

(b)

(c)

(d)

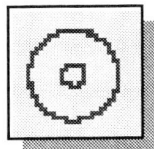

Exercises:

Determine the maximum stress produced by the loads and create the shear and moment diagrams.

1.   Material: Steel rod,
            Diamater 1.5 in.

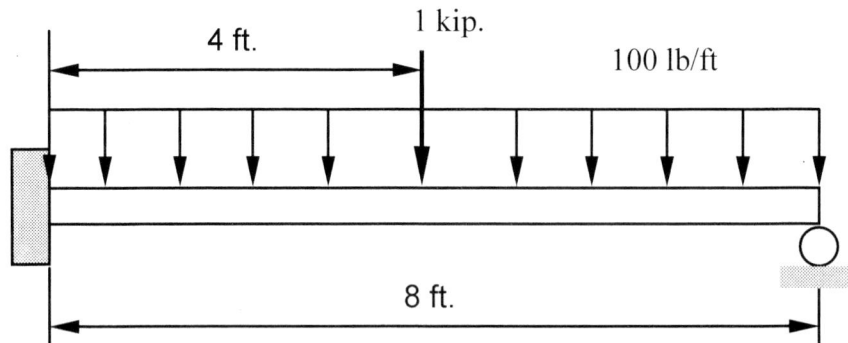

2.   Material: Aluminum Alloy 6061 T6

   1.5 in.

0.75 in.

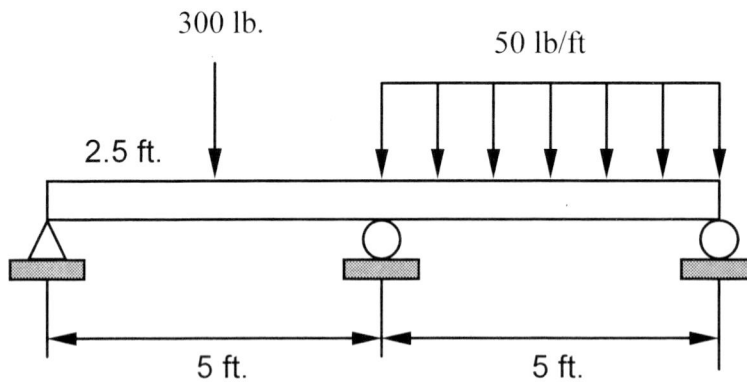

Chapter 9
Two Dimensional Solid Elements

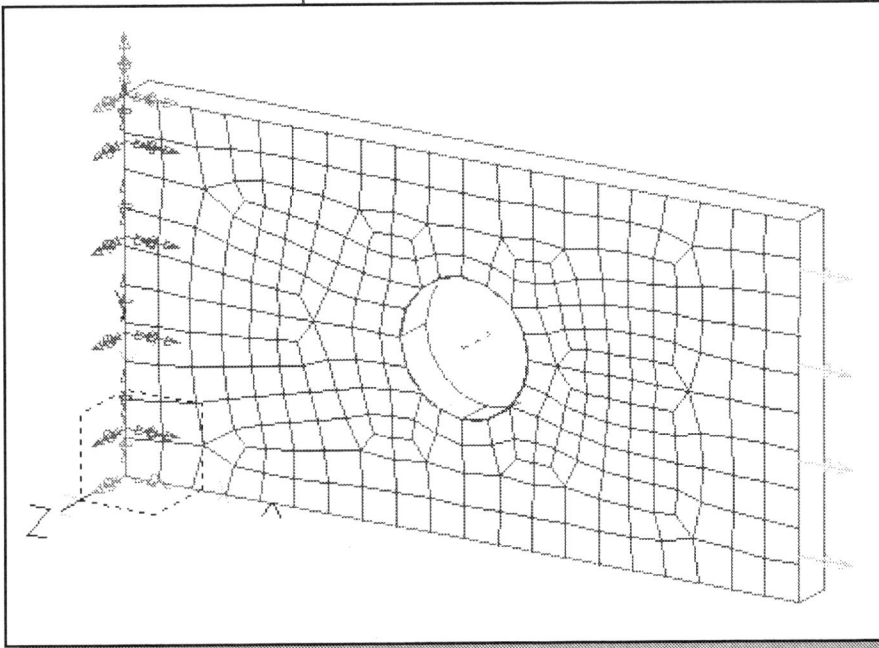

Learning Objectives

When you have completed this lesson, you will be able to:
◆ **Understand the basic assumptions for 2D elements.**
◆ **Use the Define Shell Mesh command.**
◆ **Create 2D solid elements using the Auto-Mesh.**
◆ **Understand Effects of Stress concentration.**
◆ **Perform Basic Plane Stress Analysis using I-DEAS.**

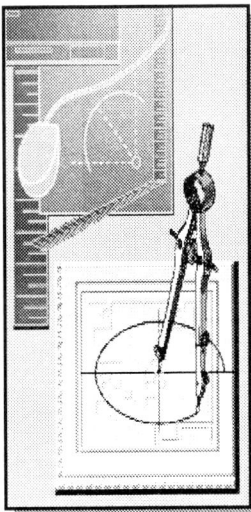

9.1 Introduction

In this chapter, we will examine the use of two-dimensional solid finite elements. Three-dimensional problems can be reduced to two dimensions if they satisfy the assumptions associated with two-dimensional solid elements. There are four basic types of two-dimensional solid elements available in *I-DEAS*:

1. **Plane Stress Elements**: Plane stress is defined to be a state of stress that the *normal stress* and the *shear stresses* directed perpendicular to the plane are assumed to be zero.

2. **Plane Strain Elements**: Plane strain is defined to be a state of strain that the *normal strain* and the *shear strains* directed perpendicular to the plane are assumed to be zero.

3. **Axisymmetric Elements**: Axisymmetric structures, such as storage tanks, nozzles and nuclear containment vessels, subjected to uniform internal pressures, can be analyzed as two-dimensional systems using *Axisymmetric Elements*.

4. **Thin-Shell Elements**: Thin-shell elements are flat planar surface elements with a thickness that is small compared to their other dimensions. In finite element modeling, we model the mid-surface that divides the thickness into equal halves.

The recent developments in computer technology have triggered tremendous advancements in the development and use of 2D and 3D solid elements in FEA. Many problems that once required sophisticated analytical procedures and the use of empirical equations can now be analyzed through the use of FEA.

At the same time, users of FEA software must be cautioned that it is very easy to fall into the trap of blind acceptance of the answers produced by the FEA software. Unlike the line elements (*Truss* and *Beam* elements), where the analytical solutions and the FEA solutions usually matched perfectly, more care must be taken with 2D and 3D solid elements since only very few analytical solutions can be easily obtained. On the other hand, the steps required to perform finite element analysis using 2D and 3D solid elements are in general less complicated than that of line elements.

This chapter demonstrates the use of 2D solid elements to solve the classical stress concentration problem – a plate with a hole in it. This project is simple enough to demonstrate the necessary steps for the finite element process, and the FEA results can be compared to the analytical solutions to assure the accuracy of the FEA procedure. We will demonstrate the use of *plane stress elements* for this illustration. Generally, members that are thin and whose loads act only in the plane can be considered to be under plane stress. For *plane stress/strain* analysis, *I-DEAS* requires the surface to be aligned with the **XY-plane**.

9.2 Problem Statement

Determine the maximum normal stress that loading produces in the steel plate.

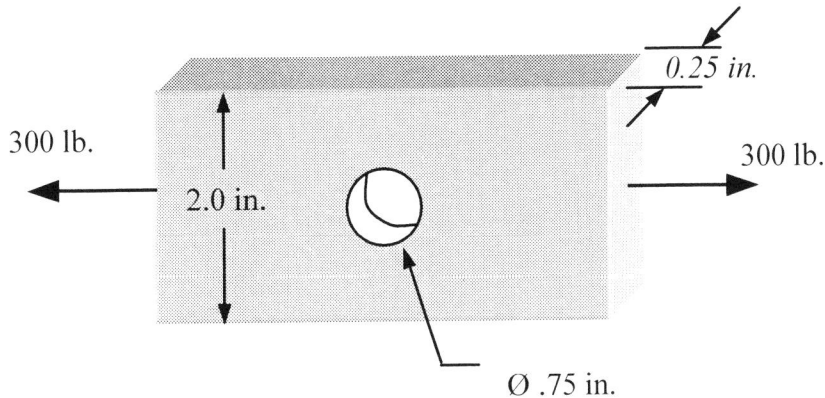

9.3 Preliminary Analysis

- **Maximum Normal Stress**

The nominal normal stress developed at the smallest cross section (through the center of the hole) in the plate is

$$\sigma_{nominal} = \frac{P}{A} = \frac{300}{(2 - 0.75) \times .25} = 960 \text{ psi.}$$

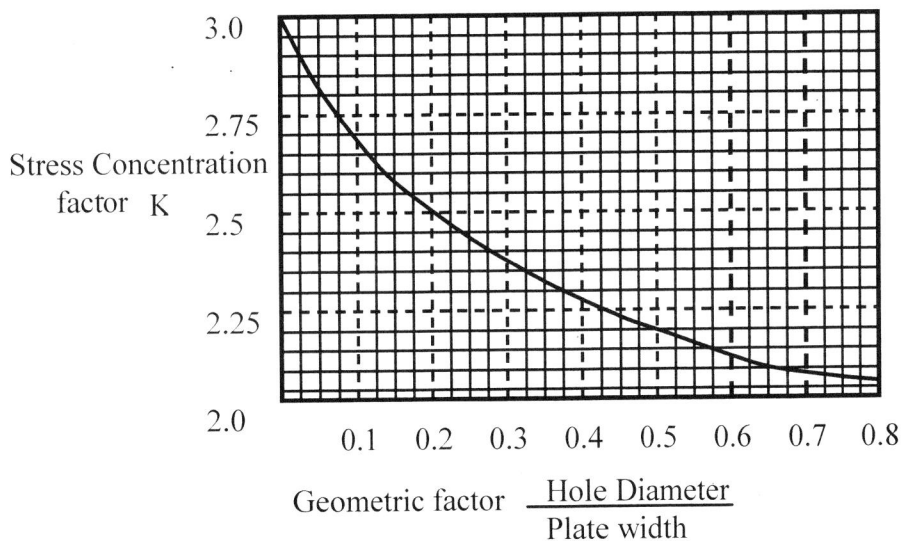

Geometric factor $\frac{\text{Hole Diameter}}{\text{Plate width}}$

Geometric factor = .75/2 = 0.375
Stress concentration factor K is obtained from the graph, **K = 2.27**

$$\sigma_{MAX} = K \, \sigma_{nominal} = 2.27 \times 960 = 2180 \text{ psi.}$$

- **Maximum Displacement**

We will also estimate the displacement under the loading condition. For a statically determinant system the stress results depend mainly on the geometry. The material properties can be in error and still the FEA analysis comes up with the same stresses. However, the displacements always depend on the material properties. Thus, it is necessary to always estimate both the stress and displacement prior to a computer FEA analysis.

The classic one-dimensional displacement can be used to estimate the displacement of the problem:

$$\delta = \frac{PL}{EA}$$

Where P=force, L=length, A=area, E= elastic modulus, and δ = deflection.

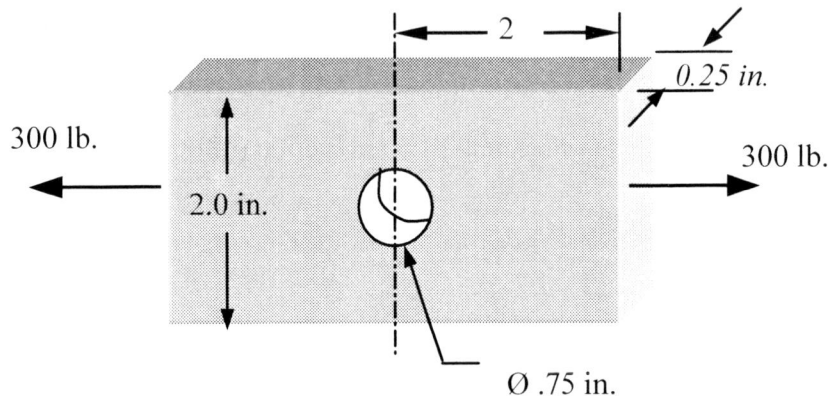

A lower bound of the displacement of the right-edge, measured from the center of the plate, is obtained by using the full area:

$$\delta_{lower} = \frac{PL}{EA} = \frac{300 \times 2}{30E6 \times (2 \times 0.25)} = 4.0E\text{-}5 \text{ in.}$$

and an upper bound of the displacement would come from the reduced section:

$$\delta_{upper} = \frac{PL}{EA} = \frac{300 \times 2}{30E6 \times (1.25 \times 0.25)} = 6.4E\text{-}5 \text{ in.}$$

but the best estimate is a sum from the two regions:

$$\delta_{average} = \frac{PL}{EA} = \frac{300 \times 0.375}{30E6 \times (1.25 \times 0.25)} + \frac{300 \times 1.625}{30E6 \times (2.0 \times 0.25)}$$

$$= 1.2E\text{-}5 + 3.25E\text{-}5 = 4.45E\text{-}5 \text{ in.}$$

9.4 Considerations of Finite Elements

➤ **Symmetry**

For *Linear Statics* analysis, designs with symmetrical features can often be reduced to expedite the analysis.

For the plate with a hole problem, there are two planes of symmetry. Thus, we only need to create an FE model that is one-fourth of the actual system. By taking advantage of symmetry, we can use a finer subdivision of elements that can provide more accurate and faster results.

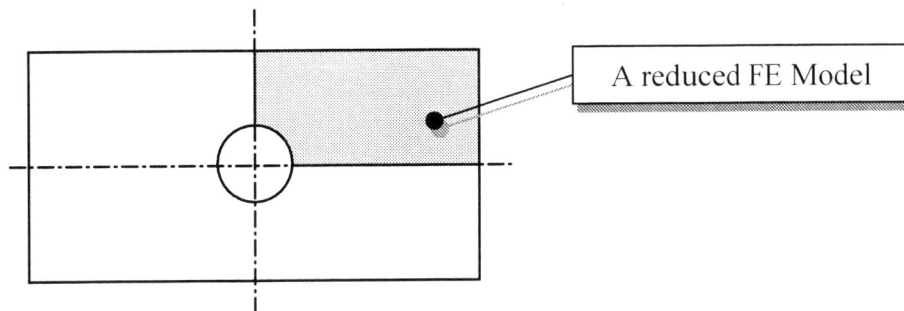

A reduced FE Model

One should also be cautious of using symmetrical characteristics in FEA. The symmetry characteristics of boundary conditions and loads should also be considered. The symmetry characteristic that is used in the *Linear Statics* analysis does not imply similar symmetrical results in vibration or buckling modes.

➤ **Aspect Ratio and Internal angles**

For plane stress analysis using two-dimensional quadrilateral elements, one should:

1. Keep the aspect ratio (the ratio between the element's longest to shortest dimensions) close to one.

2. Keep internal angles close to 90 degrees.

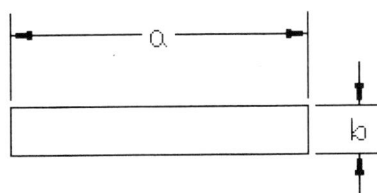

Large Aspect Ratio

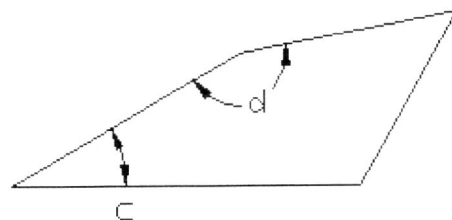

Very Large and small angles

Elements with less desirable shapes

9.5 Starting *I-DEAS*

1. Login to the computer and bring up *I-DEAS*. Start a new model file by filling in the items as shown below in the *I-DEAS Start* window:

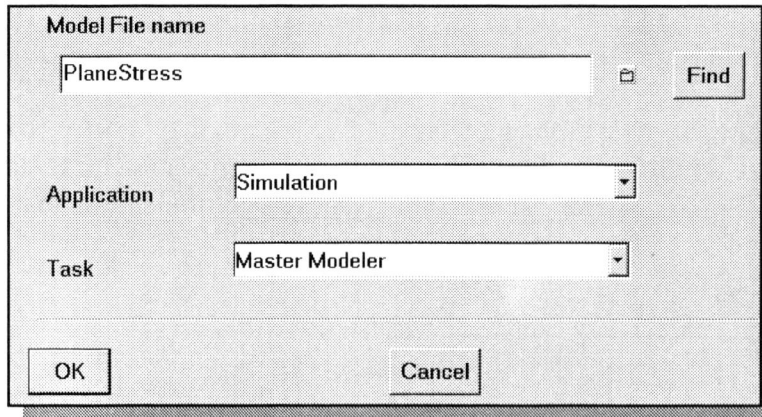

Model File name		
PlaneStress	🗀	Find
Application	Simulation	▾
Task	Master Modeler	▾
OK	Cancel	

2. After you click **OK**, a *warning window* will appear indicating a new model file will be created. Click **OK** to continue.

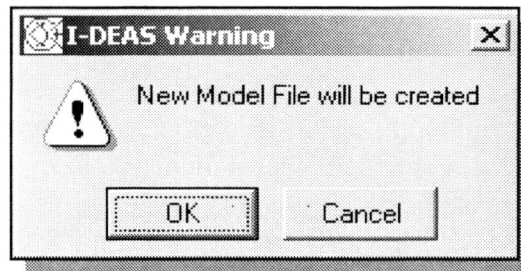

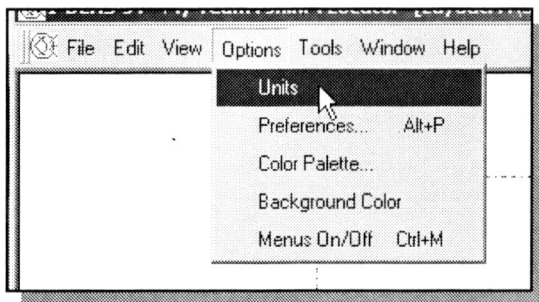

I-DEAS Warning   ✕

⚠ New Model File will be created

OK   Cancel

File Edit View Options Tools Window Help

Units
Preferences...   Alt+P
Color Palette...
Background Color
Menus On/Off   Ctrl+M

3. Use the left-mouse-button and select the **Options** menu in the icon panel.

4. Select the **Units** option.

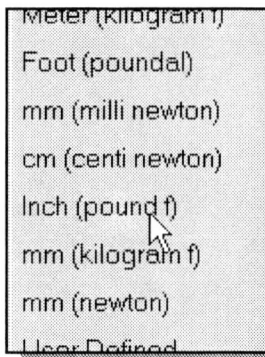

Meter (kilogram f)
Foot (poundal)
mm (milli newton)
cm (centi newton)
Inch (pound f)
mm (kilogram f)
mm (newton)
User Defined

5. Set the units to **Inch (pound f)** by selecting it from the menu of choices.

❖ The units used in an FE analysis MUST be consistent throughout the analysis.

9.6 Workplane Appearance

❖ The workplane is a construction tool; it is a coordinate system that can be moved in space. The size of the workplane display is only for our visual reference, since we can sketch on the entire plane, which extends to infinity.

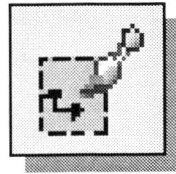

1. Choose **Workplane Appearance** in the icon panel. (The icon is located in the second row of the application icon panel.) The *Workplane Attributes* window appears.

2. Toggle **on** the *Display Border* switch as shown.

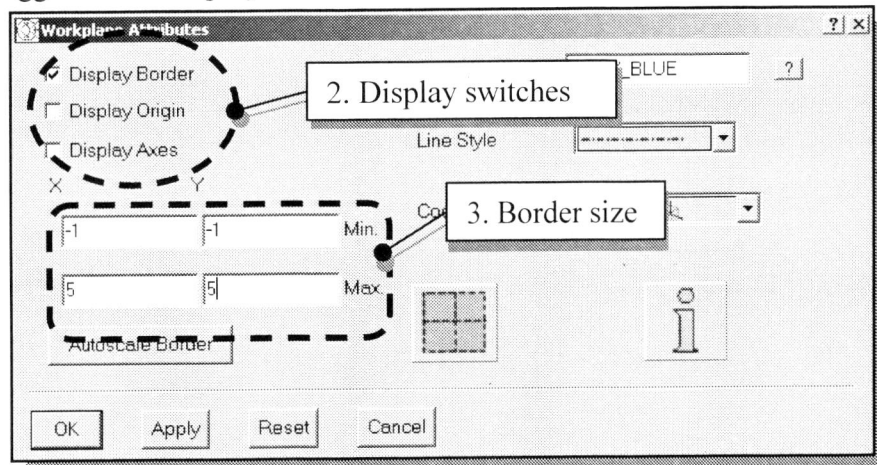

2. Display switches

3. Border size

3. Adjust the workplane border size by entering the *Min.* & *Max.* values as shown in the figure above.

4. Click on the **OK** button to exit the *Workplane Attributes* window.

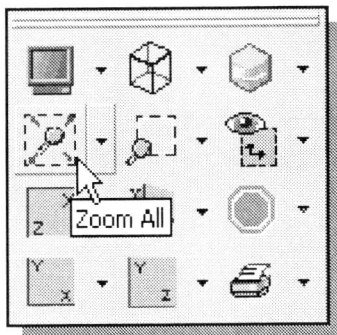

5. Choose **Zoom-All** in the display viewing icon panel

6. Choose **Refresh** in the icon panel. (The icon is located in the first row of the display icon panel.)

9.7 using the Create Part option

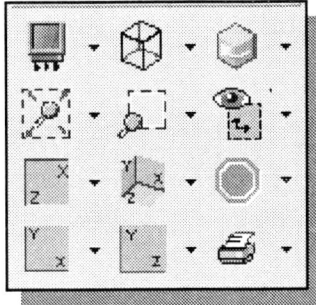

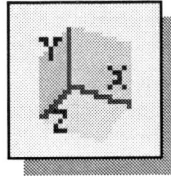

1. Choose **Isometric View** in the display viewing icon panel.

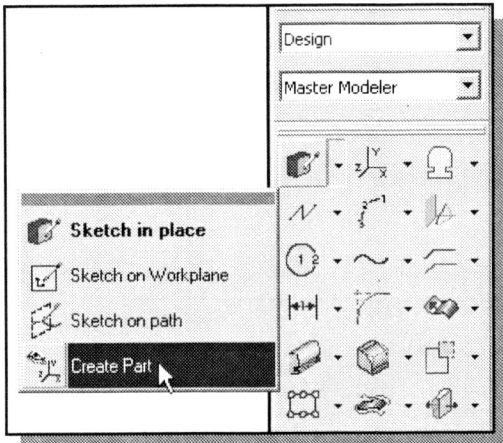

2. Choose **Create Part** in the icon panel.

➤ The icon is located in the first row of the task specific icon panel. The icon is located in the same stack as the *Sketch In Place* icon. Press and hold down the left-mouse-button on the icon stack to display the choice menu.

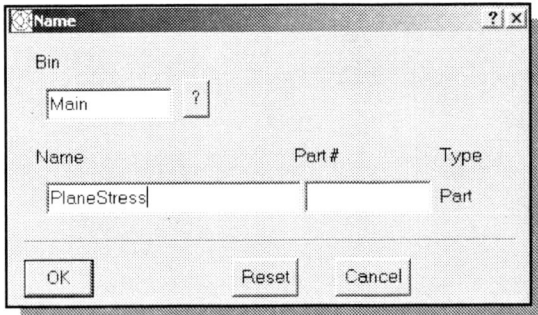

3. The *Name Part* window appears on the screen, enter **PlaneStress** as the name of the part as shown.

4. Click on the **OK** button to proceed with the **Create Part** command.

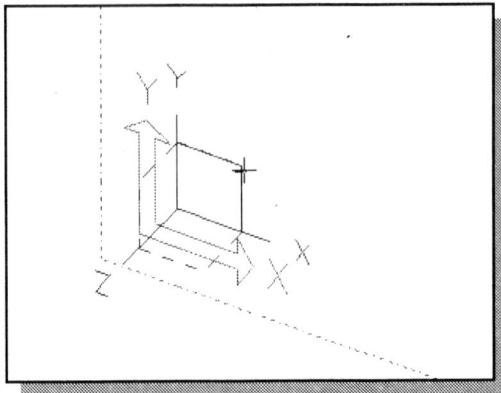

5. In the prompt window, the message "*Pick plane to sketch on*" is displayed. Pick the **XY** plane of the newly created coordinate system as shown. (Note that the default work plane, **blue** color, is still aligned to the XY plane of the world coordinate system. Aligning the sketch plane to the newly created coordinate system assures the proper association of the features to the part.)

6. Press the **ENTER** key once, or click once with the right-mouse-button to accept the placement of the workplane.

9.8 Creating a CAD Model of the System

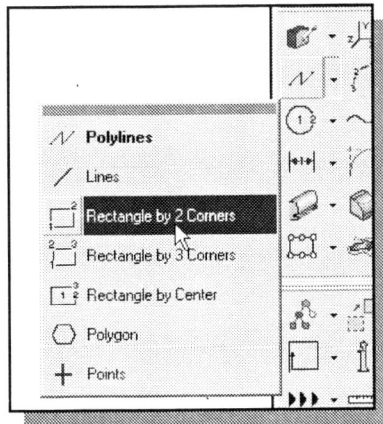

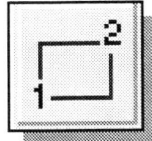

1. Choose **Rectangle by 2 Corners** in the icon panel. The message "*Locate first corner*" is displayed in the prompt window.

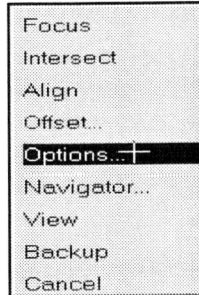

2. Move the cursor inside the graphics window. Press and hold down the right-mouse-button to select **Options...** from the option menu. The *Rectangle by Two Points Options* window appears.

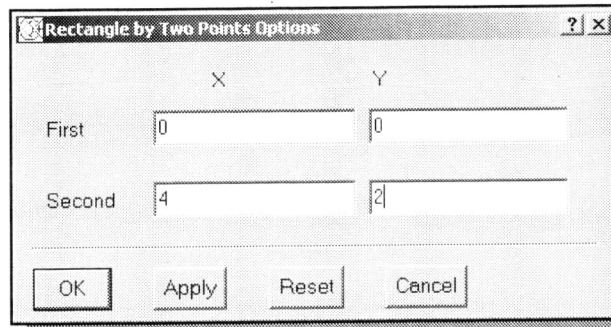

3. Enter (**0,0**) and (**4,2**) for the first and second corners of the rectangle.

4. Click on the **OK** button to proceed with creating the rectangle.

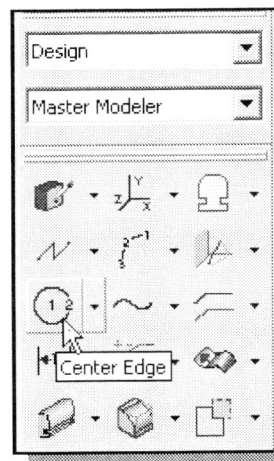

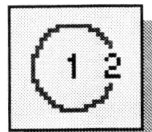

5. Choose **Circle - Center Edge** in the icon panel.

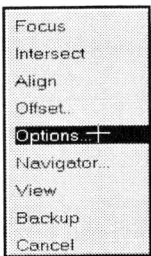

6. Move the cursor inside the graphics window. Press and hold down the right-mouse-button to select **Options...**

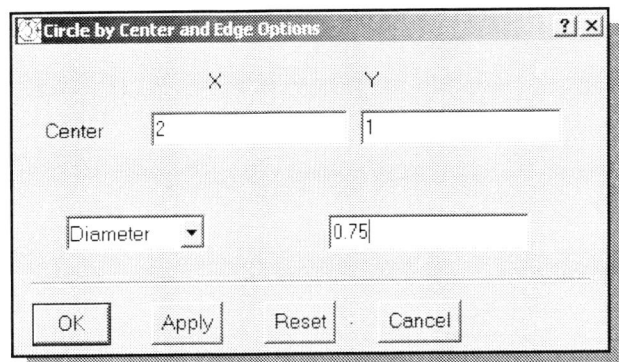

7. In the *Circle by Center and Edge Options* window, enter (**2,1**) to position the circle at the center of the rectangle and **.75** as the diameter of the circle.

8. Click the **OK** button to accept the settings and create the circle.

9.9 Add dimensions

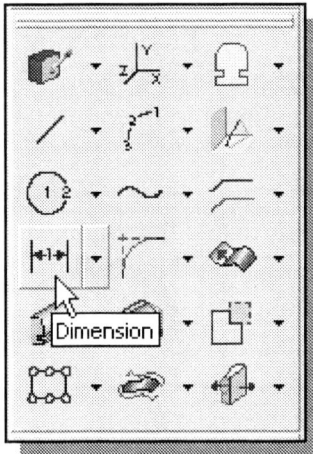

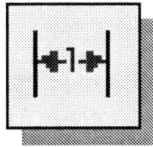

1. To fully control the location of the circle, we will add two location dimensions. Choose **Dimension** in the icon panel.

2. Pick the center of the circle.

3. Pick the left side of the rectangle.

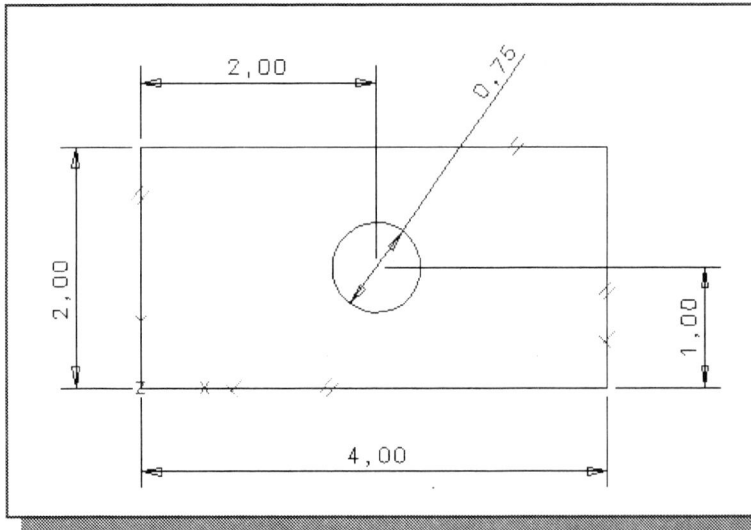

4. Place the text to the top or bottom of the rectangle.

5. Pick the center of the circle.

6. Pick the bottom side of the rectangle.

7. Place the text to the right or left of the rectangle.

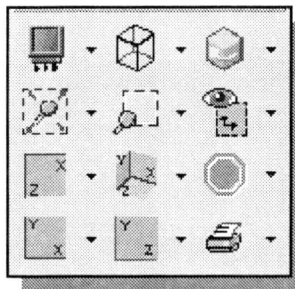

8. Press the **ENTER** key or click the middle-mouse-button to end the *Dimension* command.

9. Choose **Isometric View** in the display viewing icon panel.

10. Choose **Zoom-All** in the display viewing icon panel

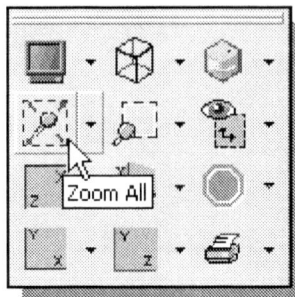

11. Choose **Refresh** in the icon panel. (The icon is located in the first row of the display icon panel.)

9.10 Extrusion

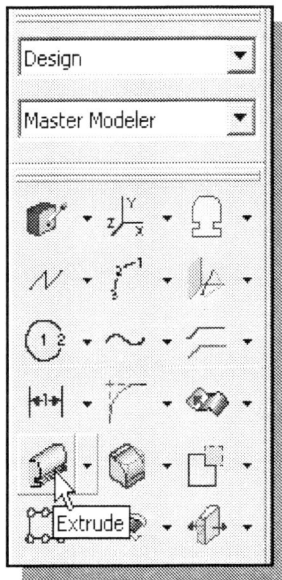

1. Choose **Extrude** in the icon panel. (The **Extrude** icon is located in the fifth row of the task specific icon panel. To select a different icon, press and hold down the left-mouse-button on the icon to display all the choices and slide the mouse up and down to switch between different commands.)

2. In the prompt window, the message "*Pick curve or section*" is displayed. Pick any edge of the **rectangle**. By default, the *Extrude* command will automatically select all segments of the shape that form a closed region shape.

3. Notice the *I-DEAS* prompt "*Pick curve to add or remove.*" We can select more geometry entities or deselect any entity that has been selected. Picking the same geometry entity will toggle the selection of the entity "*on*" or "*off*" with each left-mouse-button click. Pick the **circle**.

4. Press the **ENTER** key or click the middle-mouse-button to accept the selected entities.

5. The *Extrude Section* window will appear on the screen. Click on the **Arrows** button to flip the extrusion direction so that it points in the negative Z-direction.

6. In the *Distance* box, enter **0.25**.

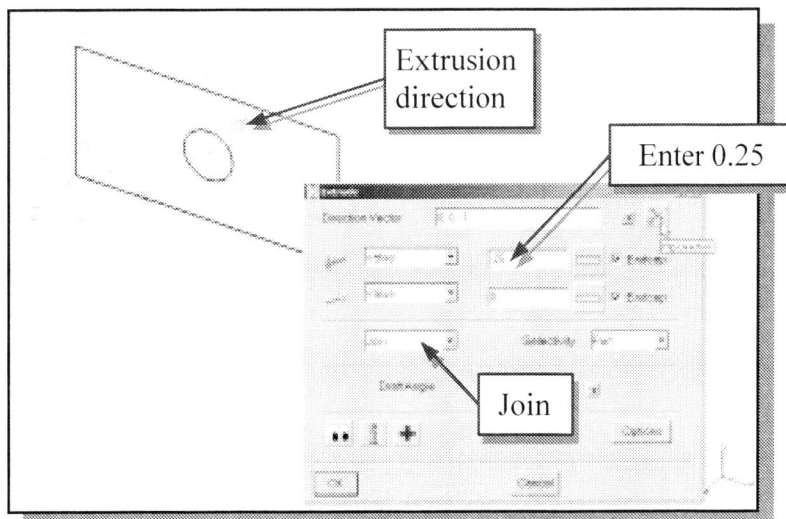

7. Confirm the **Join** option is switched on and click on the **OK** button.

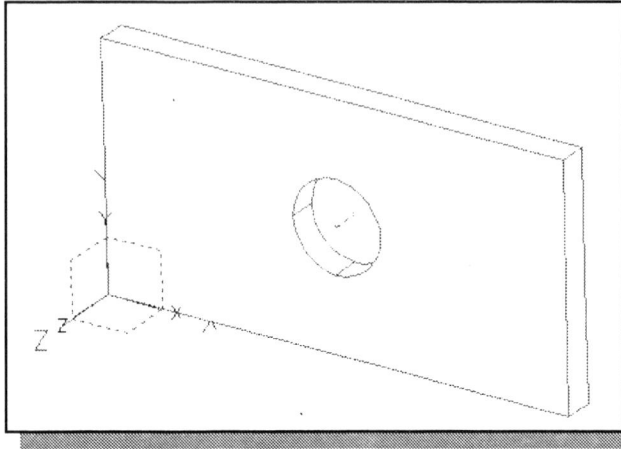

❖ The 2D sketch is extruded into a 3D solid model. Note the front face of the solid model is aligned to the XY-Plane, which is the plane we will create the FE model.

9.11 Create a Cut Feature

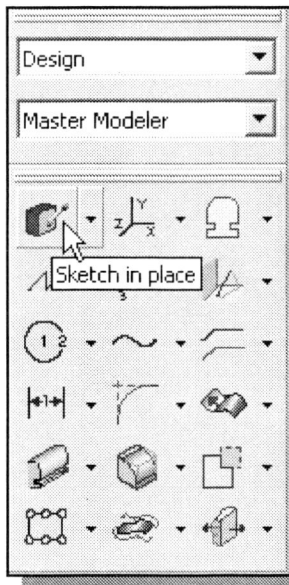

1. Choose **Sketch in Place** in the icon panel. In the prompt window, the message "*Pick plane to sketch on*" is displayed.

2. Pick the front face of the horizontal portion of the 3D object by left-clicking the surface, when it is highlighted as shown in the figure below.

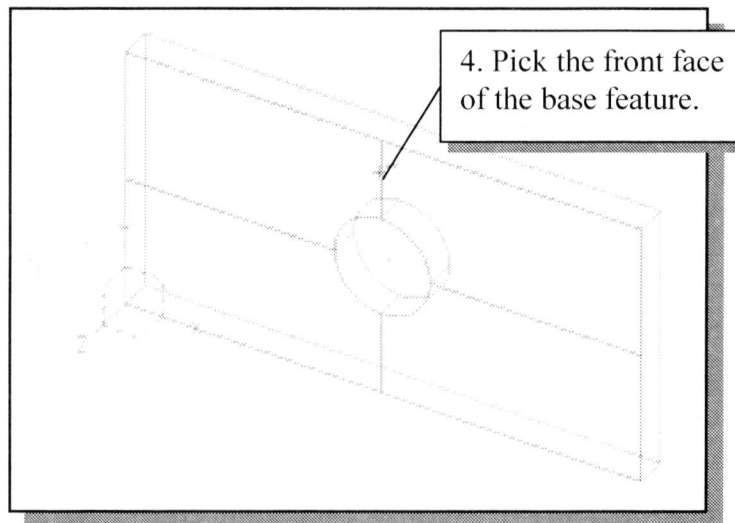

4. Pick the front face of the base feature.

❖ Notice that, as soon as the front face of the model is picked, *I-DEAS* automatically orients the workplane to the selected surface. The surface selected is highlighted with a different color to indicate the attachment of the workplane.

- Next, we will create another 2D sketch, which will be used to create a cut feature to remove three-quarters of the existing solid object.

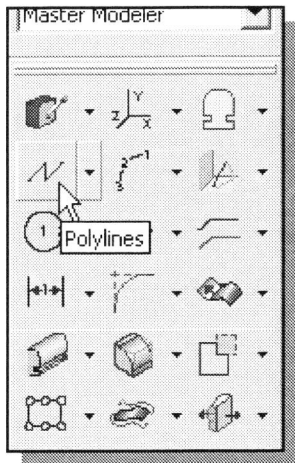

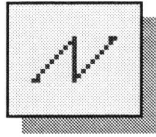

3. Pick **Polylines** in the icon panel. (The icon is located in the second row of the task specific icon panel. If the icon is not on top of the stack, press and hold down the left-mouse-button on the displayed icon to display all the choices. Select the desired icon by clicking with the left-mouse-button when the icon is highlighted.)

4. Create a sketch, starting at the upper-left-corner of the existing solid (Point 1), with segments perpendicular/parallel and passing through midpoints to the existing edges of the solid model as shown below.

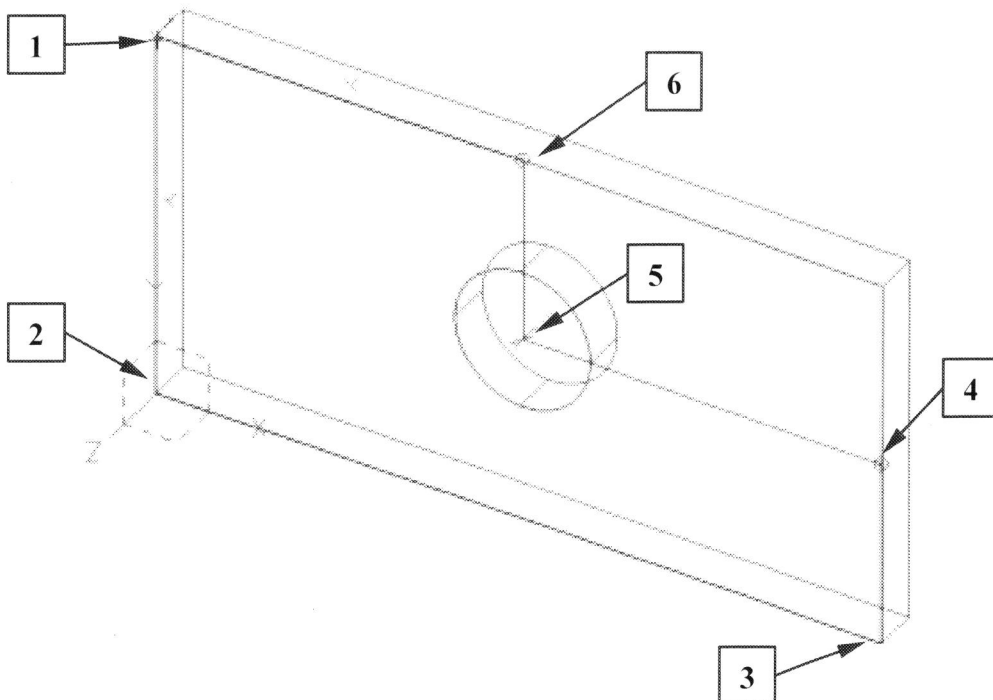

- Note that the edges of the new sketch are either perpendicular or parallel to the existing edges of the solid model. Also note that none of the edges are aligned to the mid-points or corners of the existing solid model.

Design

Master Modeler

Extrude

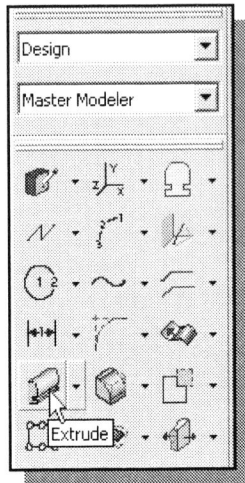

5. Choose **Extrude** in the icon panel. The *Extrude* icon is located in the fifth row of the task specific icon panel.

6. In the prompt window, the message "*Pick curve or section*" is displayed. Pick any edge of the 2D shape. By default, the *Extrude* command will automatically select all neighboring segments of the selected segment to form a closed region. Notice the different color signifying the selected segments.

7. Pick the segment in between the displayed two small circles so that the highlighted entities form a closed region as shown.

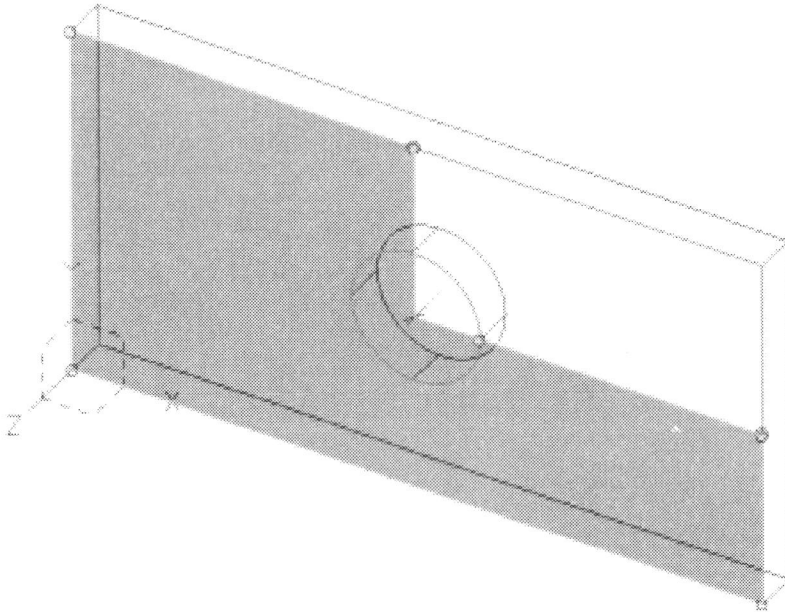

8. Press the **ENTER** key once, or click once with the middle-mouse-button, to accept the selected entity.

❖ Attempting to select a line where two entities lie on top of one another (i.e. coincide) causes confusion as indicated by the double line cursor ╫ symbol and the prompt window message "*Pick curve to add or remove (Accept)***". This message indicates *I-DEAS* needs you to confirm the selected item. If the correct entity is selected, you can continue to select additional entities. To reject an erroneously selected entity, press the [**F8**] key to select a neighboring entity or press the right-mouse-button and highlight **Deselect All** from the popup menu.

9. Press the **ENTER** key once, or click once with the middle-mouse-button, to proceed with the *Extrude* command.

10. The *Extrude Section* window appears. Set the **extrude** option to **Cut**. Note the extrusion direction displayed in the graphics window.

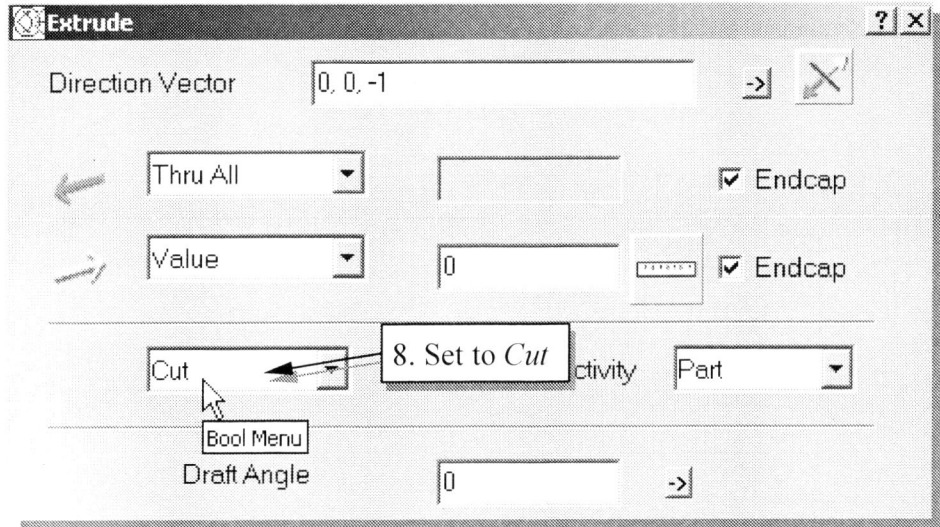

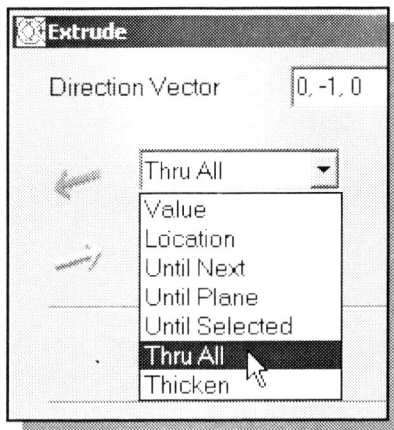

11. Click and hold down the left-mouse-button on the **depth** menu and select the **Thru All** option. *I-DEAS* will calculate the distance necessary to cut through the part.

12. Click on the **OK** button to accept the settings. The 2D sketch is extruded and the volume removed from the 3D model.

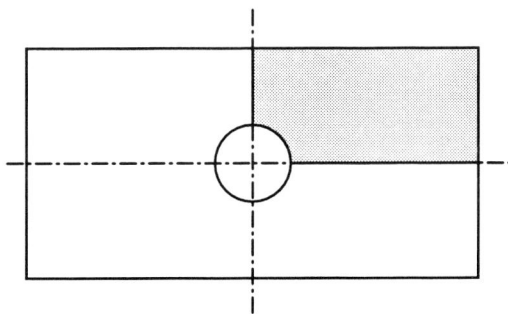

9.12 Geometric Based Analysis

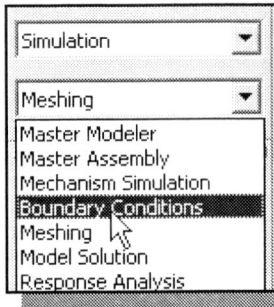

1. Switch to the **Boundary Conditions** task by selecting it in the task menu as shown.

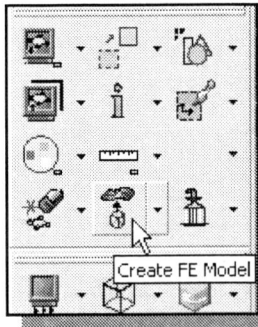

2. Choose **Create FE Model** in the icon panel. (The icon is located in the last row of the application specific icon panel.) The *FE Model Create* window appears.

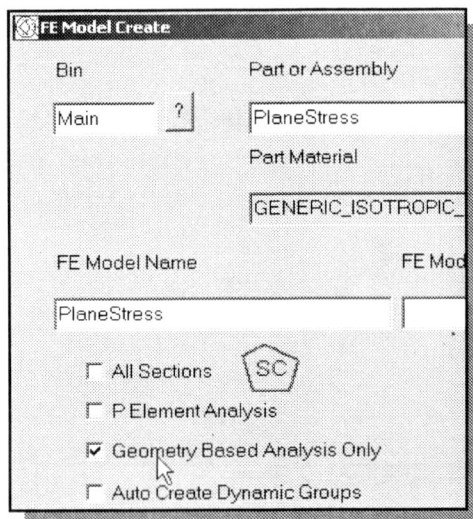

3. In the *FE Model Create* window, enter **PlaneStress** in the *FE Model Name* box.

4. Switch *on* the *Geometric Based Analysis Only* button.

5. Click on the **OK** button to accept the settings.

➢ *Geometric Based Analysis* allows us to apply *boundary conditions* to the constructed geometries so that the boundary conditions remain independent of the finite element mesh. The *Geometric Based Analysis* option eliminates the need to reapply the boundary conditions when different meshes are applied to the model. Adjusting the mesh is a common practice, and sometimes necessary, in using 2D and 3D solid elements. This is especially the case when using the **h-method** of FEA, which is what we are using in this example. The *Geometric Based Analysis* option allows the boundary conditions be applied to the FE model first. *I-DEAS* will automatically set up the proper boundary conditions to the corresponding nodes and elements generated by the *Auto-Mesh* operation. Next, we will apply the boundary conditions to the FE model before the nodes and elements are created.

9.13 Apply Boundary Conditions

- In plane stress analysis, only the x- and y-displacements are active. The constraints in all other directions can be set to *free* as a reminder of which ones are the true unknowns.

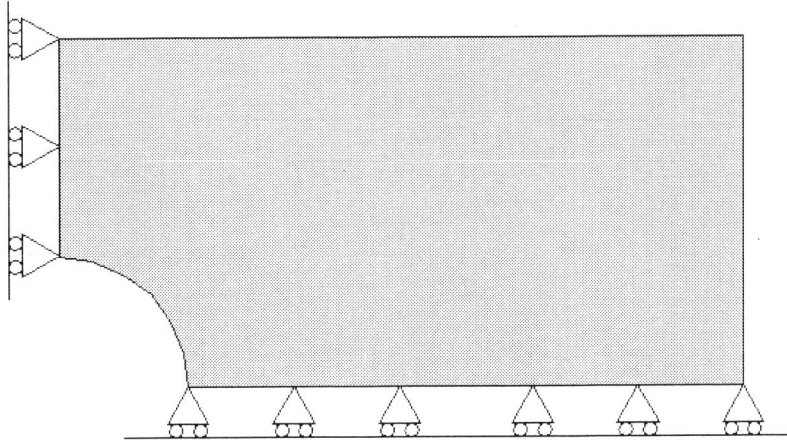

1. Choose **Displacement Restraint** in the icon panel. (The icon is located in the fourth row of the task icon panel.)

2. The message "*Pick Nodes/Centerpoints/Vertices*" is displayed in the prompt window. Pick the front left edge of the plate as shown. Note the displayed **Exx** symbol as the cursor is moved on top of the edge.

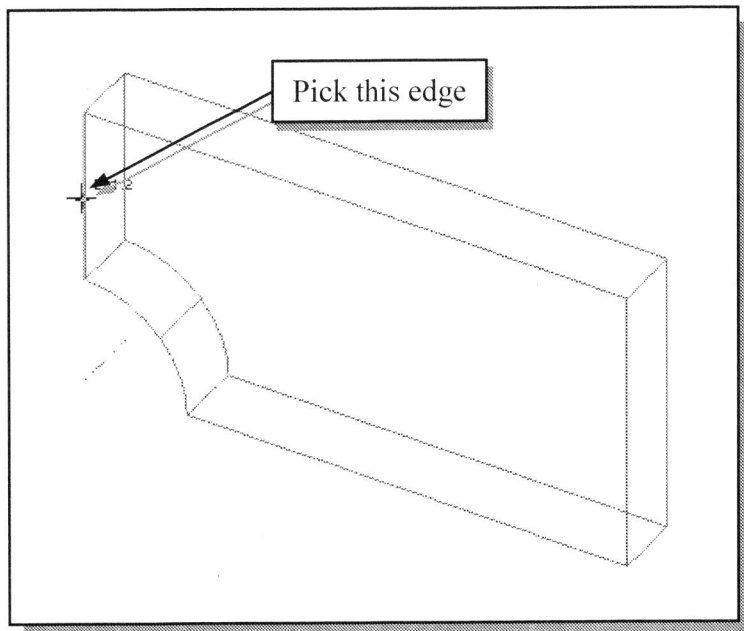

Pick this edge

3. Press the **ENTER** key to accept the selection and continue with the constraint command. The *Displacement Restraint on Edge* window appears.

4. We will restrict the displacement in the X-direction, and free all other restraints; of the left edge of the plate.

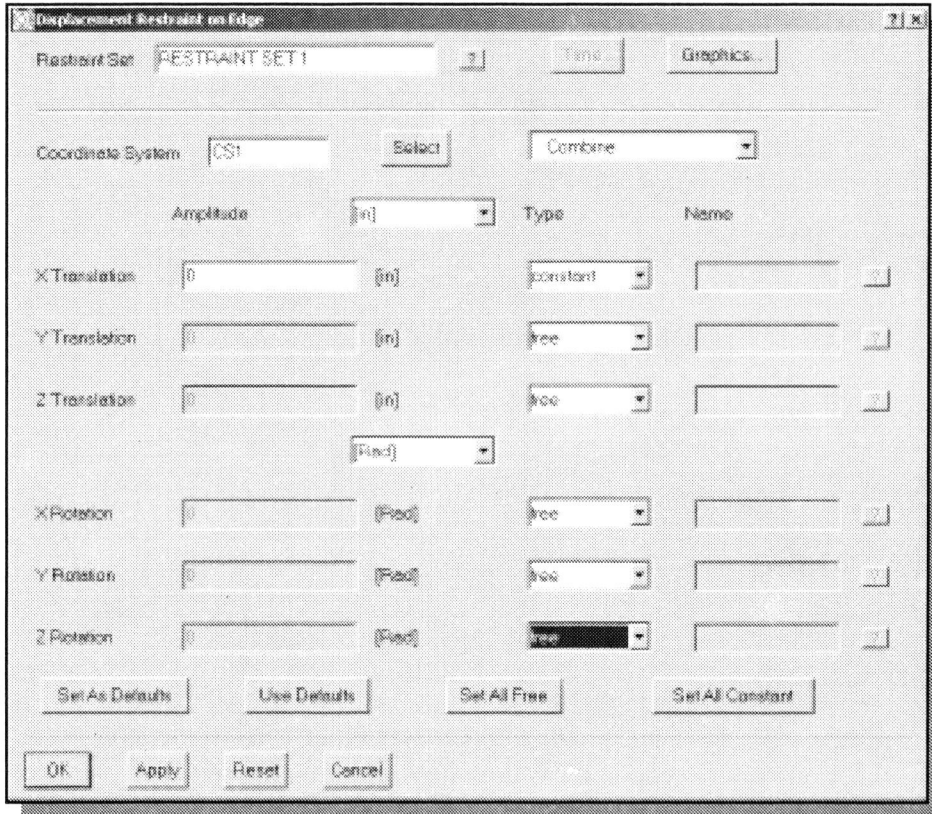

5. Click on the **OK** button to accept the settings.

6. On your own, repeat the above steps to restrict the displacement in the Y-direction, and all other restraints are set to free, of the bottom edge of the model.

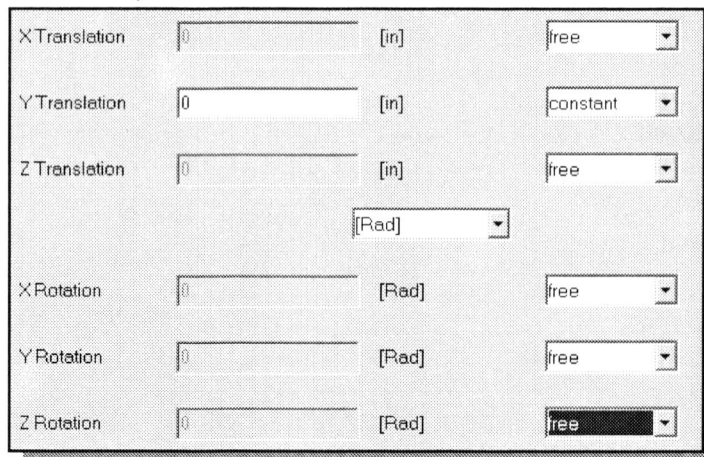

9.14 Apply the External Loads

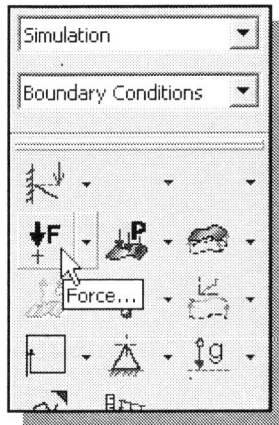

1. Choose **Force...** in the icon panel. (The icon is located in the second row of the task icon panel.)

2. The message "*Pick entities*" is displayed in the prompt window. Pick the right edge of the front surface.

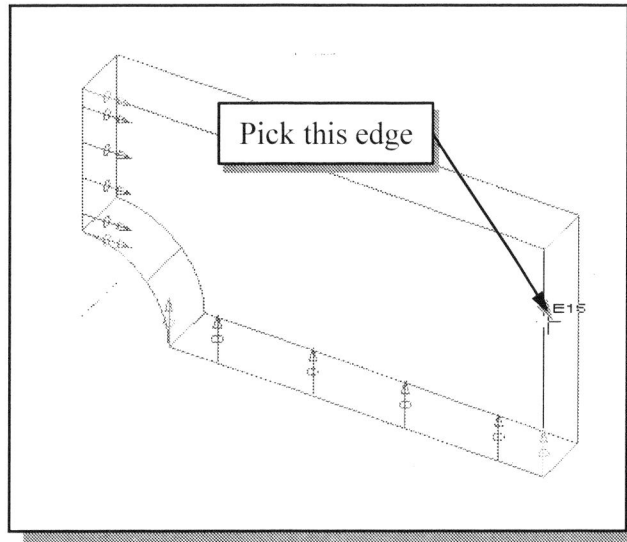

Pick this edge

3. Press the **ENTER** key to accept the selection. Two surfaces are highlighted in the graphics window.

4. The message "*Pick surface to set direction of forces on edge 2*" is displayed in the prompt window. Pick the front surface of the plate.

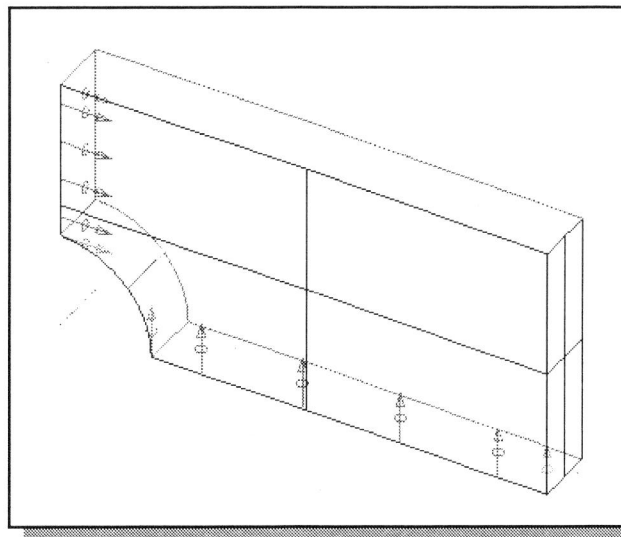

5. Set to *Total Force*.

4. Pick the front face to set the force direction.

6. Enter *–150.*

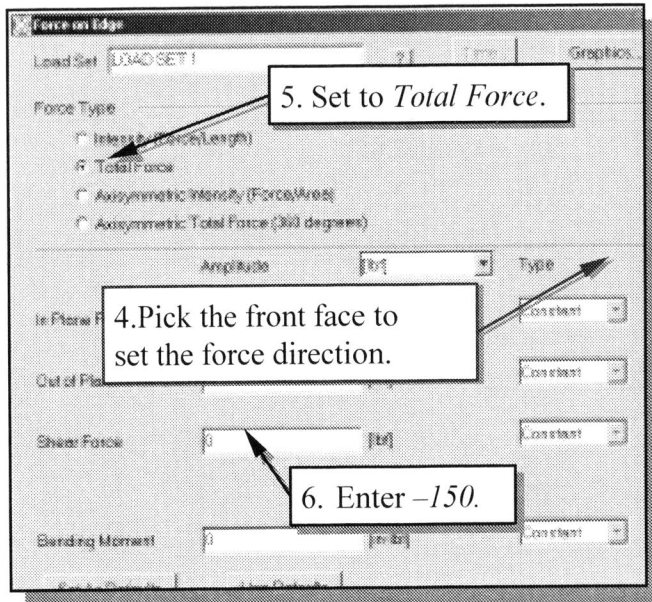

5. In the *Force on Edge* window, switch **on** the *Total Force* option.

6. The load is 300lb. Enter **-150** in the *In Plane Force* box.

7. Click on the **OK** button to exit the *Force on Edge* window.

9.15 Boundary Condition Set

1. Choose **Boundary Condition Set Management** in the icon panel. (The icon is located in the sixth row of the task icon panel.)

2. Switch **on** the *Restraint Set.*

3. Select **Load Set 1** in the *Load Sets* list.

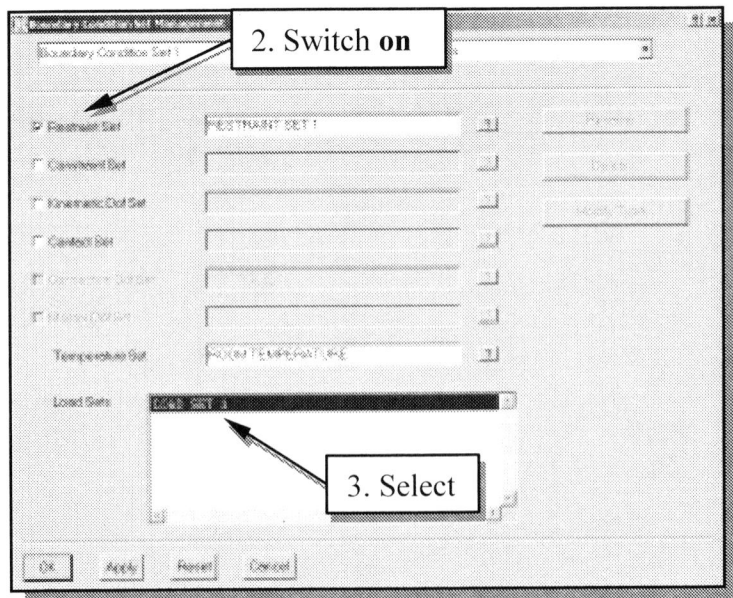

4. Click on the **OK** button to accept the settings and exit the *Boundary Condition Set Management* window.

2. Switch **on**

3. Select

9.16 Material Property Table

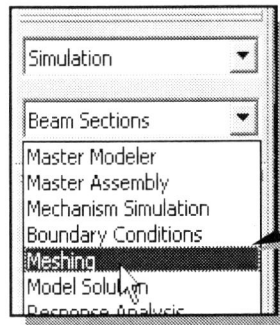

Pick *Meshing*

1. Switch back to the **Meshing** task by selecting the Meshing task in the task menu as shown.

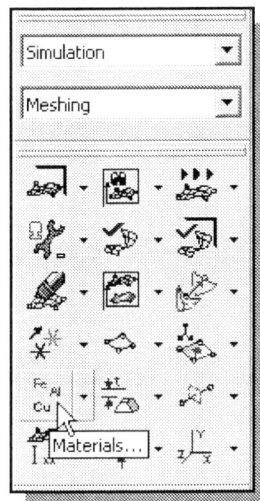

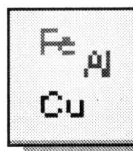

2. Choose **Materials** in the icon panel. (The icon is located in the fourth row of the task icon panel.) The *Materials* window appears.

❖ In this example, we will use the default material property set, *Generic_Isotropic_Steel*, for our analysis.

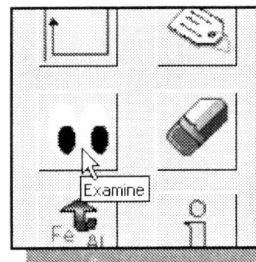

3. On your own, use the **Examine** option to review the material properties.

4. Click on the **OK** button to exit the *Examine* window.

5. Click on the **OK** button to exit the *Materials* window.

9.17 Define Auto-Mesh

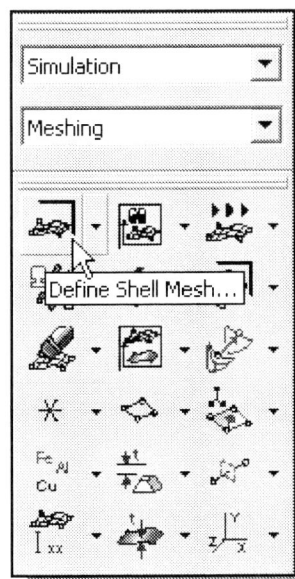

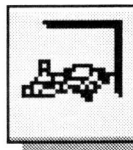

1. Choose **Define Shell Mesh** in the icon panel. (The icon is located in the first row of the task icon panel.) The message "*Pick Surfaces*" is displayed in the prompt window.

2. Pick the front surface of the plate.

3. Press the **ENTER** key or click the middle-mouse-button to accept the selection. The *Define Mesh* window appears.

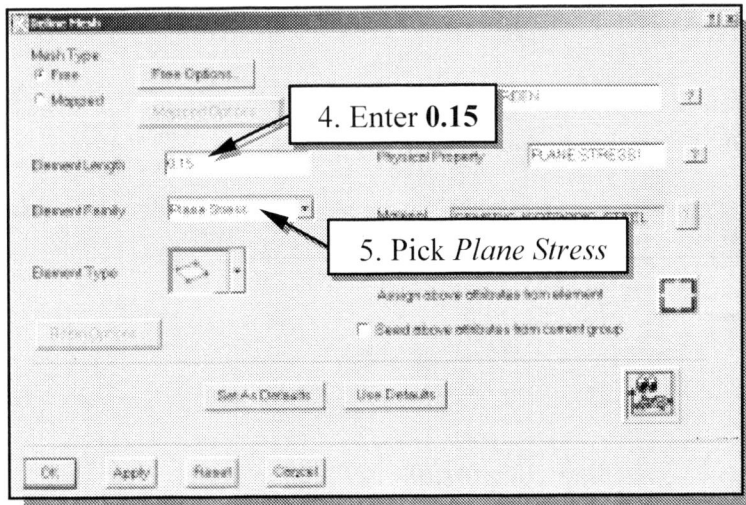

4. Enter **0.15** in the
    Element Length box.

5. Choose **Plane Stress**
    in the *Element Family*
    list.

6. Click on the **OK** button
    to exit the *Define Mesh*
    window.

9.18 Create the Mesh

1. Choose **Create Shell Mesh** in the icon
    panel. (The icon is located in the first row of
    the task icon panel.)

2. The message "*Pick surfaces*" is displayed in the prompt
    window. Pick the front surface of the plate.

3. Press the **ENTER** key or the middle-mouse-button to accept
    the selection. New nodes and elements are created. Record
    the number of elements and nodes generated by *Auto-Mesh*.

4. Press the **ENTER** key or the **middle-mouse-button** to accept the
    additions.

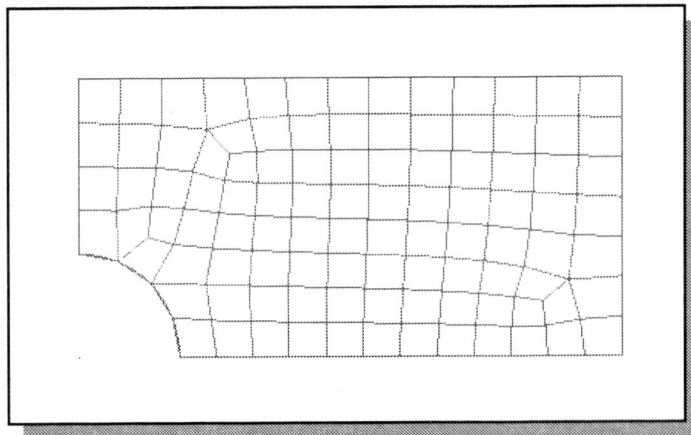

➤ For the initial analysis, the elements generated by *I-DEAS* appear to be quite adequate.
We will use the results from the initial analysis to decide which portion of the model
needs further refinement.

9.19 Define the Thickness of the Elements

1. Choose **Physical Property Table** in the icon panel. (The icon is located in the fifth row of the task icon panel.)

2. In the *Physical Property Tables* window, select **2D** and **Plane Stress** as shown.

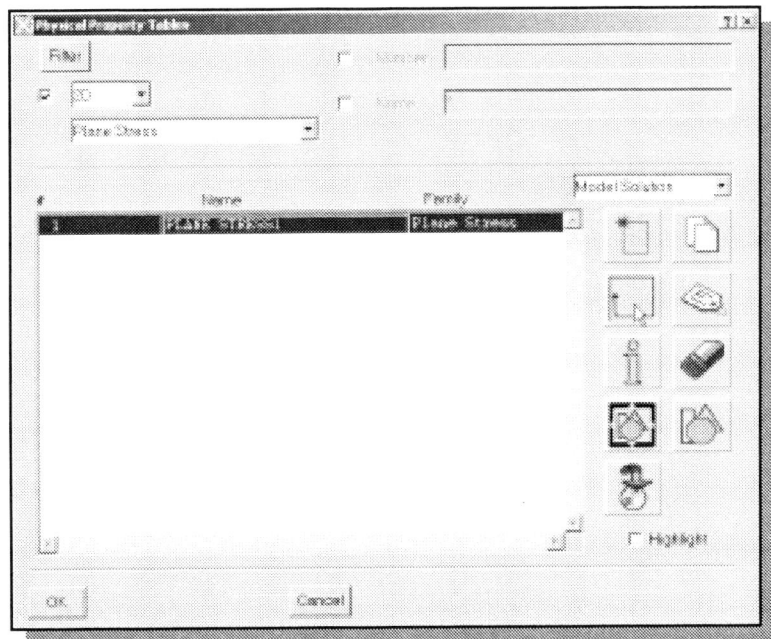

3. Select the Plane Stress 1 set from the list area.

4. Pick **Modify Table** in the *Physical Property Tables* window.

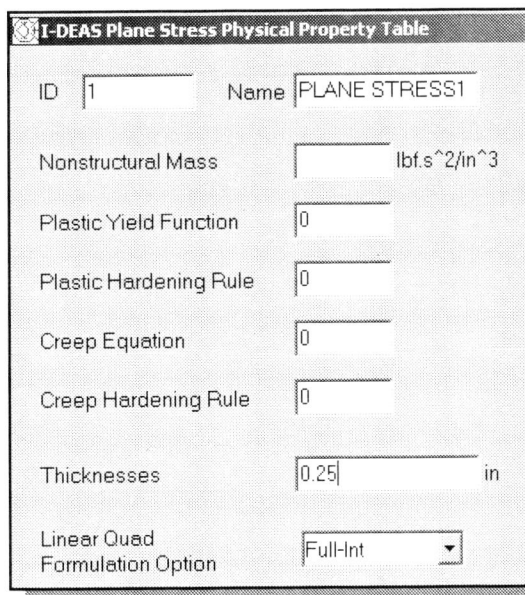

5. Enter **0.25** as the thickness of the plane stress FE model we have created.

❖ The FE model uses the geometric information of the CAD model. For the plane stress analysis, the thickness of the CAD model is not recognized.

6. Pick **OK** accept the settings.

7. Pick **OK** to end the *Physical Property Table* command.

9.20 Run the Solver

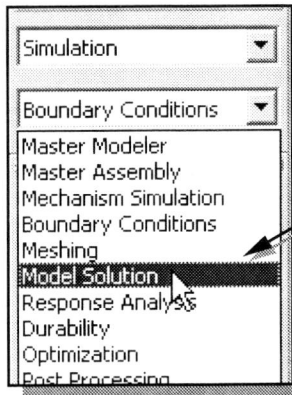

1. Switch to the **Model Solution** task by selecting the task in the task menu as shown.

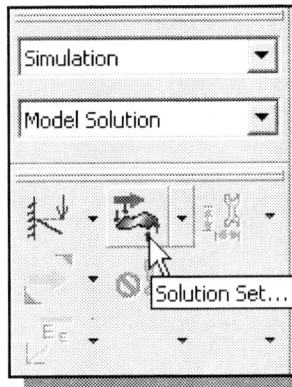

   Pick *Model Solution*

2. Choose ***Solution Set*** in the icon panel. (The icon is located in the first row of the task icon panel.)

3. The *Manage Solution Sets* window appears.

4. Click on the **Create** button. The *Solution Set* window appears.

5. We will use the default settings; click the **OK** button to exit the *Solution Set* window.

6. Pick **Dismiss** to exit the *Manage Solution Sets* window.

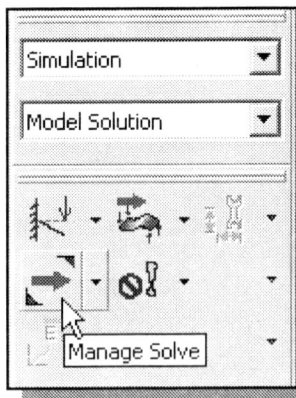

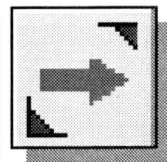

7. Choose ***Manage Solve*** in the icon panel. (The icon is located in the second row of the task icon panel.) The *Solve* window appears.

8. In the *Solve* window, pick the **Solve** icon to find the solutions.

➢ *I-DEAS* will begin the solving process. <u>DO NOT</u> close any windows. Any errors or warnings are displayed in the list window.

9.21 View the Results

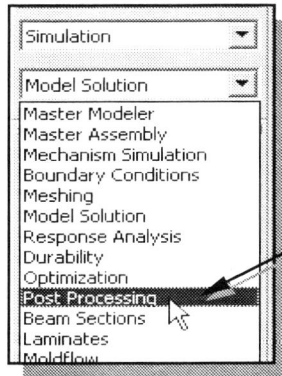

1. Switch to the **Post Processing** task by selecting in the task menu as shown.

Pick *Post Processing*

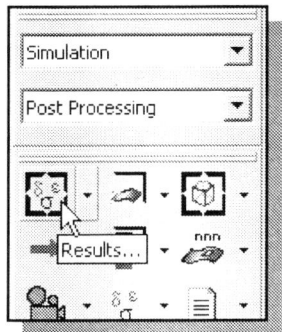

2. Choose **Results Selection** in the icon panel. (The icon is located in the first row of the task icon panel.) The *Results Selection* window appears.

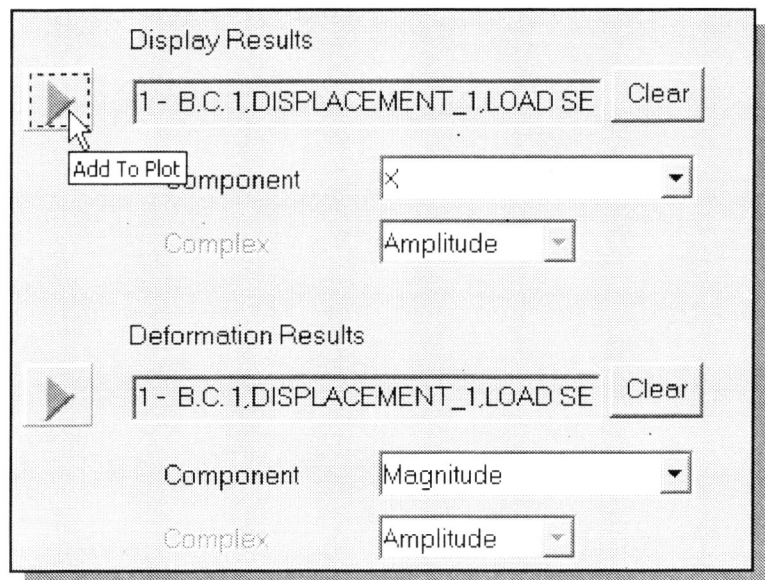

Display Results

1 - B.C. 1,DISPLACEMENT_1,LOAD SE   Clear

Add To Plot Component   X

Complex   Amplitude

3. Set the *Display Results* to **Displacement**.

4. Set the *Display Component* to **X** direction.

Deformation Results

1 - B.C. 1,DISPLACEMENT_1,LOAD SE   Clear

5. Click on the **OK** button to exit the *Results Selection* window.

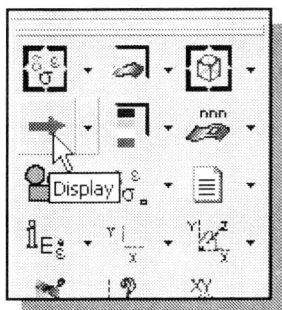

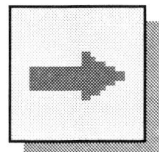

Component   Magnitude

Complex   Amplitude

6. Choose **Results Display** in the icon panel. (The icon is located in the second row of the task icon panel.)

7. The message "*Pick elements*" is displayed in the prompt window. Press the **ENTER** key to continue. (Displacements on all elements will be processed.)

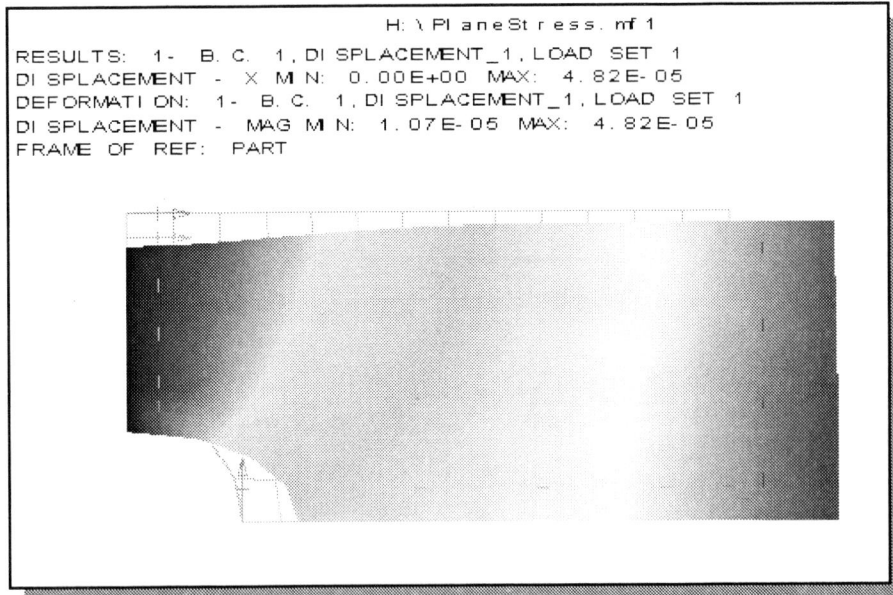

> From the *I-DEAS* solution, the *Maximum Displacement* is 4.82E-5 in., which is very close to the value, 4.45E-5 in., calculated in the preliminary analysis (Page 9-4).

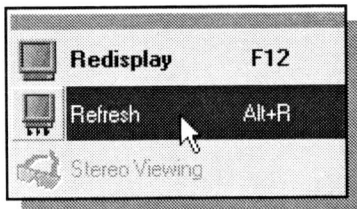

8. Choose **Refresh** in the icon panel. (The icon is located in the first row of the display icon panel.)

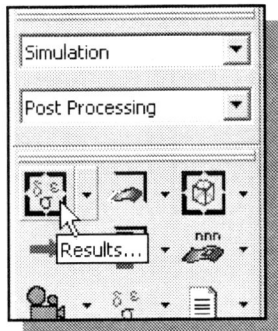

9. Choose **Results Selection** in the icon panel. (The icon is located in the first row of the task icon panel.) The *Results Selection* window appears.

10. Set the *Display Results* to **Stress**.

11. Set the *Display Component* to **Maximum Principal**.

12. Click on the **OK** button to exit the *Results Selection* window.

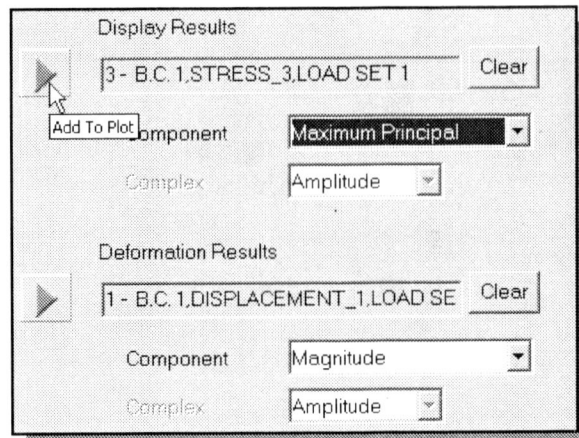

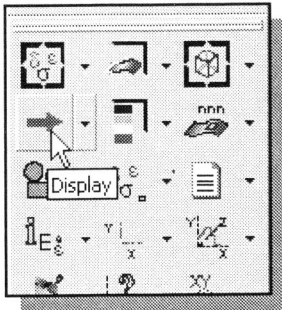

13. Choose **Results Display** in the icon panel. (The icon is located in the second row of the task icon panel.)

14. The message *"Pick elements"* is displayed in the prompt window. Press the **ENTER** key to continue. (Stresses on all elements will be processed.)

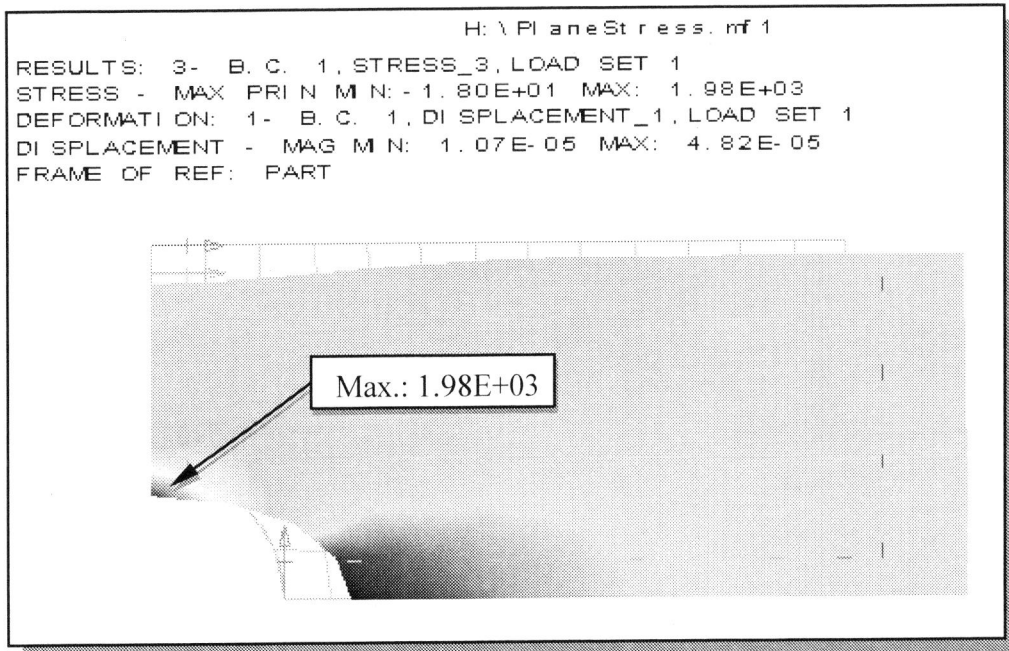

```
                          H:\PlaneStress.mf 1
RESULTS:  3-  B.C.  1,STRESS_3,LOAD SET 1
STRESS -  MAX PRIN MIN:-1.80E+01 MAX:  1.98E+03
DEFORMATION:  1-  B.C.  1,DISPLACEMENT_1,LOAD SET 1
DISPLACEMENT -  MAG MIN: 1.07E-05 MAX:  4.82E-05
FRAME OF REF:  PART
```

Max.: 1.98E+03

➤ The computer solution of the maximum displacement and the stress distribution matched our expectations, but the maximum stress value is lower than our preliminary solution. A more refined mesh in the higher stresses area is required to accurately obtain the stresses than the displacements. As a rule, start with a relatively small number of elements and progressively move to more refined models. In most cases, use of a complex and very refined model is not justifiable since it most likely provides computational accuracy at the expense of unnecessarily increased processing time. The following sections demonstrate two procedures to refine the mesh of the current FE model.

9.22 Delete the Solution Set and the Mesh

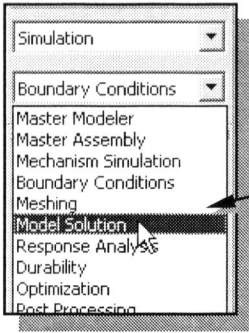

1. Switch to the **Model Solution** task by selecting the task in the task menu as shown.

Pick *Model Solution*

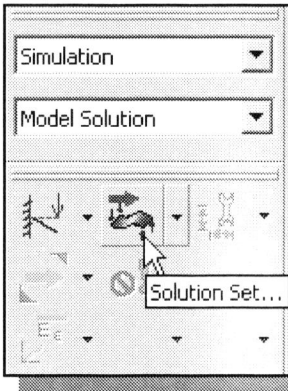

2. Choose **Solution Set** in the icon panel. (The icon is located in the first row of the task icon panel.)

3. The *Manage Solution Sets* window appears. Select **Solution SET1**.

4. Click on the **Delete** button to remove the solution set**.**

5. Click on the **OK** button to confirm deletion of **Solution Set1**.

6. Click on the **Dismiss** button to exit the *Manage Solution Sets* window.

❖ The solution set needs to be removed in order to modify the FE model,

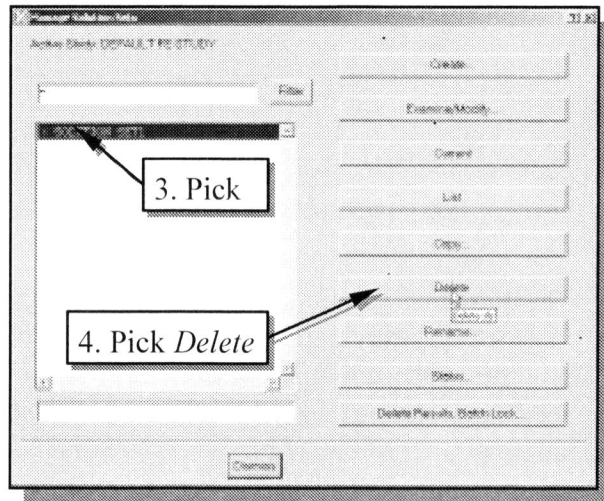

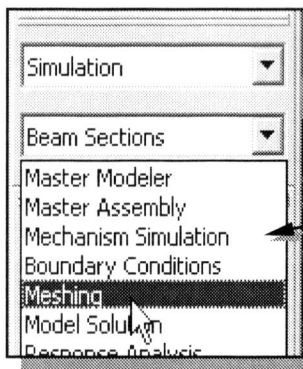

3. Pick

4. Pick *Delete*

7. Switch back to the **Meshing** task by selecting the Meshing task in the task menu as shown.

Pick *Meshing*

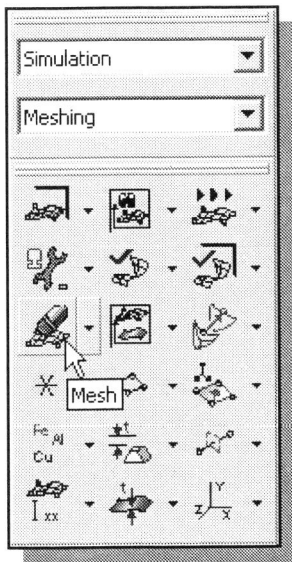

8. Choose **Delete Mesh** in the icon panel. (The icon is located in the third row of the task icon panel.)

9. The message "*Pick surfaces*" is displayed in the prompt window. Pick the front surface of the plate.

10. Press the **ENTER** key or the middle-mouse-button to accept the selection. The mesh is removed.

9.23 Refining the Mesh – Method 1: Free Options

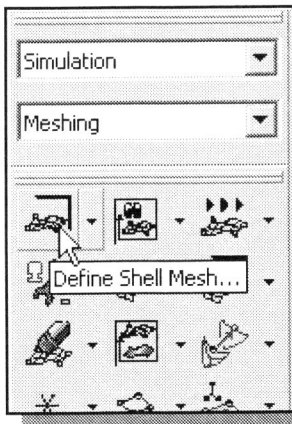

1. Choose **Define Shell Mesh** in the icon panel. (The icon is located in the first row of the task icon panel.)

2. The message "*Pick Surfaces*" is displayed in the prompt window. Pick the front surface of the plate.

3. Press the **ENTER** key or the middle-mouse-button to accept the selection. The *Define Mesh* window appears.

4. Click on the **Free Options** button to set the meshing options.

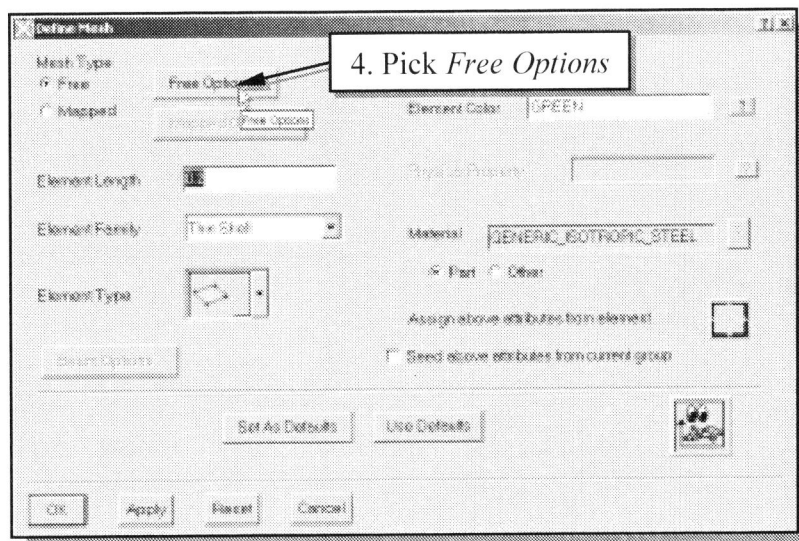

5. In the *Define Free Meshing Options* window, select *Percent Deviation*.

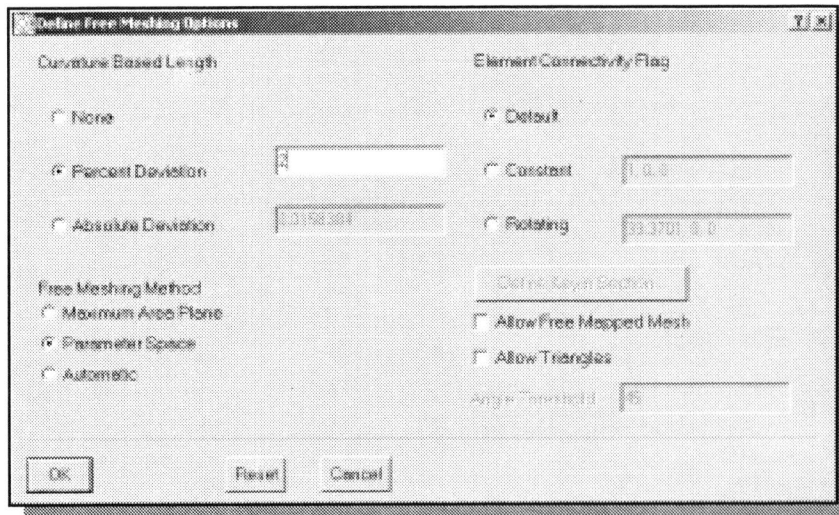

6. Enter **2** in the *Percent Deviation* box. This option sets the element length to be based on the percent deviation from a curve expressed as: chord height/chord length.

❖ The **Curvature Based Length** option is one of the easiest ways to refine an FEA mesh. In general, curves in FE models are approximated with line-segments. By adjusting the number of line-segments, we adjust the number of elements used along the curves. Note that *I-DEAS* also allows us to enter an **Absolute Derivation** value to adjust the line-segments along curves.

7. Click on the **OK** button to exit the *Define Free Meshing Options* window.

8. Click on the **OK** button to exit the *Define Mesh* window.

9.24 Create the Mesh

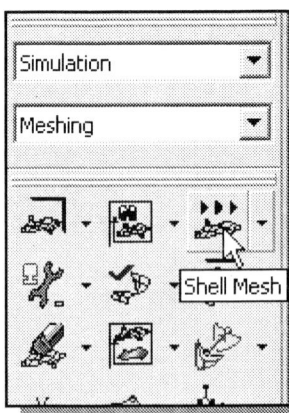

1. Choose **Create Shell Mesh** in the icon panel. (The icon is located in the first row of the task icon panel.) The message "*Pick surfaces*" is displayed in the prompt window.

2. Pick the front surface of the plate.

3. Press the **ENTER** key or the middle-mouse-button to accept the selection. New nodes and elements are created.

4. Press the **ENTER** key or the middle-mouse-button to accept the additions. Compare the number of elements and nodes to the first analysis we did.

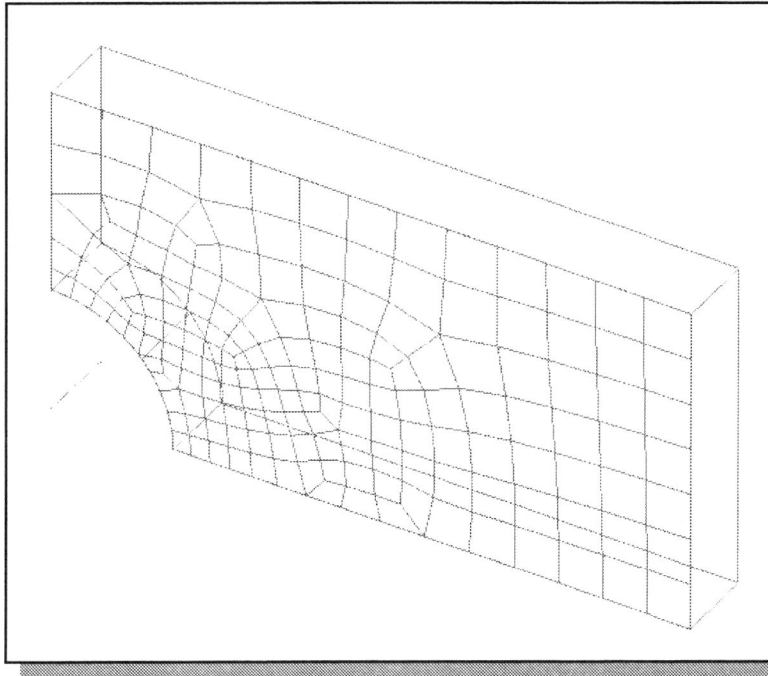

➤ On your own, complete the FE model and run the solver. Compare the results to the first analysis. What kind of percent difference is achieved by this refinement? Is the answer closer to the hand-calculation performed in 9.3?

9.25 Delete the Solution Set and the Mesh

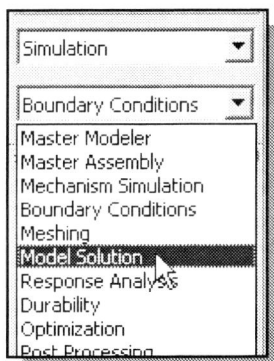

1. Switch to the **Model Solution** task by selecting the task in the task menu as shown.

2. Choose **Solution Set** in the icon panel and delete the solution set.

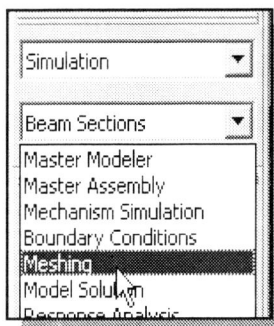

3. Switch back to the **Meshing** task by selecting the Meshing task in the task menu as shown.

4. Choose **Delete Mesh** in the icon panel and delete the FE mesh.

9.26 Refining the Mesh – Method 2: Define Free Local

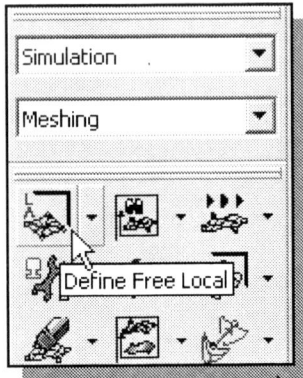

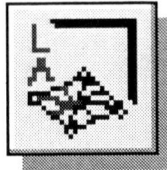

1. Choose **Define Free Local** in the icon panel. (The icon is located in the first row of the task icon panel.)

2. The message *"Pick Vertices/Edges/Surfaces/Curves/Curve Points"* is displayed in the prompt window. Pick the circle that is in the front surface of the plate.

3. Press the **ENTER** key or the middle-mouse-button to accept the selection.

4. The message *"Enter number of elements on edge (3)"* is displayed in the prompt window. The *Define Free Local* option allows us to enter the number of elements along the select entities. In our first mesh, we had 4 elements along the arc (as shown in the below figure.) We will increase that number to 12. Enter **12** as the number of elements along the circle.

First mesh, p. 9-22

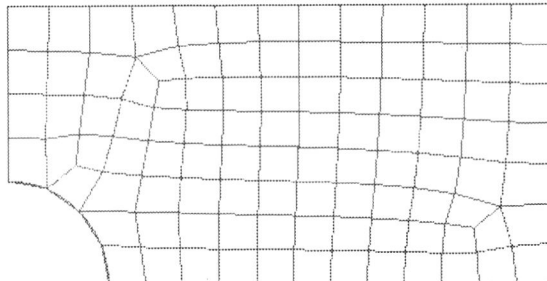

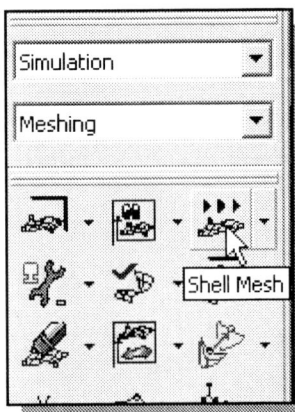

5. Press the **ENTER** key or the middle-mouse-button to end the *Define Free Local* command.

6. Choose **Create Shell Mesh** in the icon panel.

7. The message *"Pick surfaces"* is displayed in the prompt window. Pick the front surface of the plate.

8. Press the **ENTER** key or the middle-mouse-button to accept the selection. New nodes and elements are created.

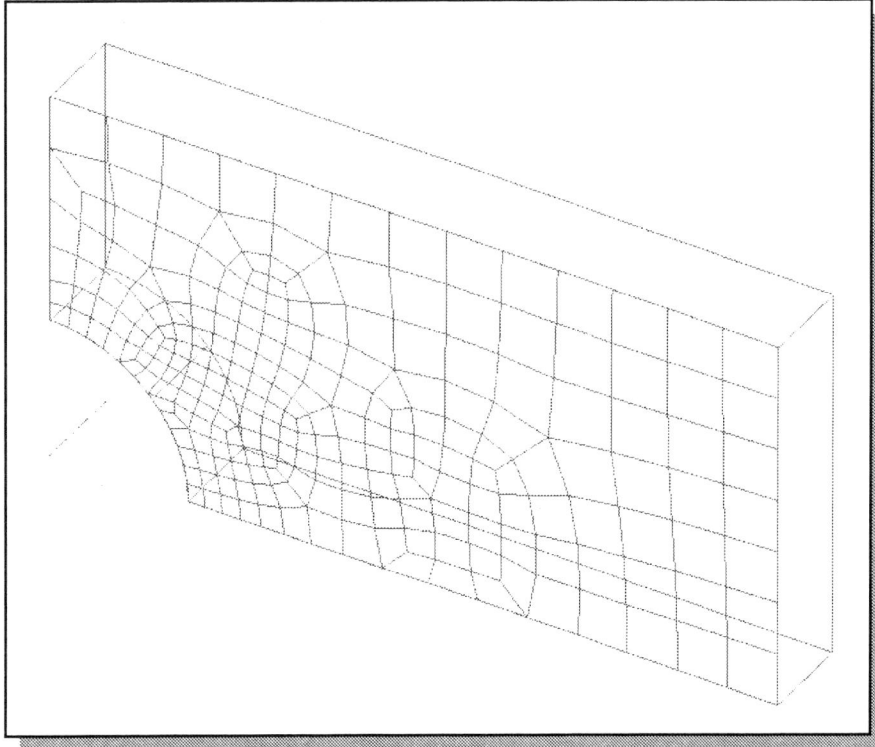

➤ On your own, complete the FE model and run the solver. Compare the results to the previous two analyses. What kind of percent difference is achieved by this refinement? Does the answers appear to be converging as refinements are made? Is the answer closer to the hand-calculation performed in 9.3? You are also encouraged to experiment with the different *Free Options* and *Free Local* options. Also, as a comparison, compare the number of elements generated with each refinement and the time it took to run the solver.

Questions:

1. What are the four basic types of two-dimensional solid elements?

2. What type of situation is more suitable to perform a *Plane Stress analysis* than a *Plane Strain analysis*?

3. Which type of two-dimensional solid elements is most suitable for members that are thin and whose loads act only in the plane?

4. For plane stress analysis, what *element aspect ratio* and *element angle* are more desirable?

5. Identify and describe the following commands:

   (a)

   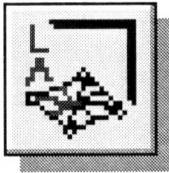

   (b)

   (c)

   (d)

Exercises:

Determine the maximum stress produced by the loads.

1.

Material: Steel Plate
Thickness: 25 mm

Ø 150 mm

0.15 m

0.3 m

0.15 m

5 KN

0.25 m   0.15 m   0.15 m

2.

Material: Steel
Thickness: 8 mm

R 34

R 64   R 22   13

R 3

9

32

15°

38

800 N   R 25   11

250

Notes:

Chapter 10
Three-Dimensional Solid Elements

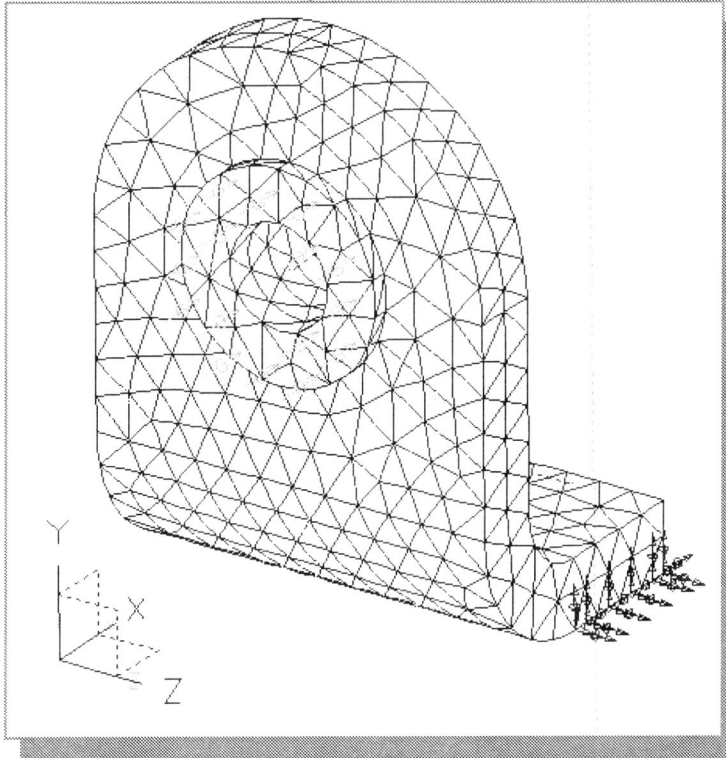

Learning Objectives

When you have completed this lesson, you will be able to:
◆ **Perform 3D Finite Element Aanalysis.**
◆ **Understand the concepts of Failure Criteria.**
◆ **Create Tetrehedral solid elements using Auto-Mesh.**
◆ **Create 3D Solid models and FE models.**
◆ **Use the I-DEAS Geometric Based Analysis.**
◆ **Use the different display options to display Stress Results.**

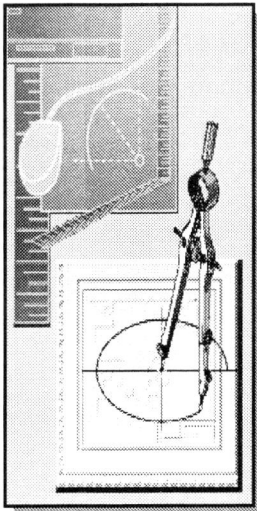

10.1 Introduction

In this final chapter, the general FEA procedure to using three-dimensional solid elements is illustrated. A finite element model using three-dimensional solid elements may look the most realistic as compared to the other types of FE elements. However, this type of analysis also requires more elements, which implies more mathematical equations and therefore more computational resources and time.

The main objective of finite element analysis is to calculate the stresses and displacements for specified loading conditions. Another important objective is to determine if *failure* will occur under the effect of the applied loading. It should be pointed out that the word *failure* as used for *failure criteria* is somewhat misleading. In the elastic region of the material, the system's deformation is recoverable. Once the system is stressed beyond the elastic limit, even in a small region of the system, deformation is no longer recoverable. This does not necessarily imply that the system has failed and cannot carry any further load.

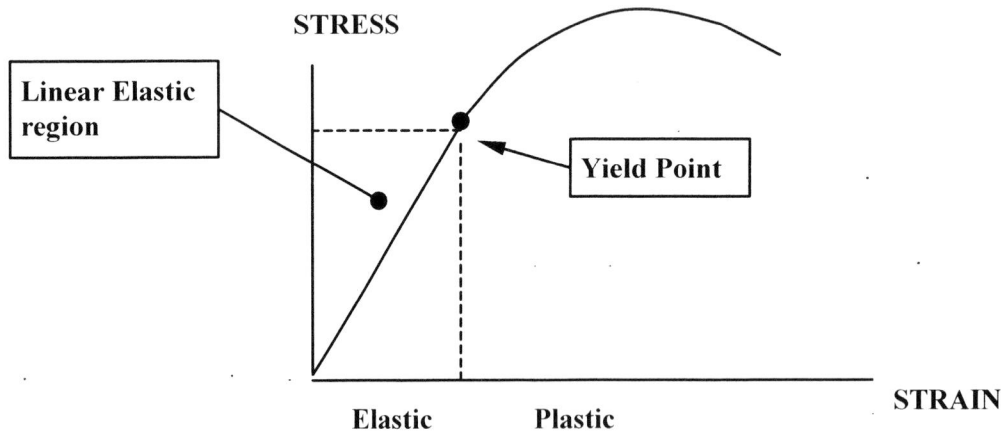

Stress-Strain diagram of typical ductile material

Several theories propose different failure criteria. In general, all these theories provide fairly similar results. The most widely used failure criteria are the **Von Mises yield criterion** and the **Tresca yield criterion**. Both the *Von Mises* and *Tresca* stresses are scalar quantities, and when compared with the yield stress of the material they indicate whether a portion of the system has exceeded the elastic state.

Von Mises Criterion: (Note that σ_1, σ_2, σ_3 are the three principal stresses.)

$$2\,\sigma_{yp}{}^2 = (\sigma_1 - \sigma_2)^2 + (\sigma_2 - \sigma_3)^2 + (\sigma_3 - \sigma_1)^2$$

Tresca Yield Criterion:

$$\frac{1}{2}\,\sigma_{yp} = \frac{1}{2}\,(\sigma_1 - \sigma_3)$$

This chapter illustrates the general FEA procedure of using three-dimensional solid elements. The creation of a solid model is first illustrated and *tetrahedral solid elements* are generated using the *I-DEAS Auto-Mesh* command. The procedure involved in performing a three-dimensional solid FEA analysis is very similar to that of a two-dimensional solid FEA analysis, as was demonstrated in Chapter 9. As one might expect, the number of node-points involved in a typical three-dimensional solid FEA analysis is usually much greater than that of a two-dimensional solid FEA analysis.

10.2 Starting *I-DEAS*

1. Login to the computer and bring up *I-DEAS*. Start a new model file by filling in the items as shown below in the *I-DEAS Start* window:

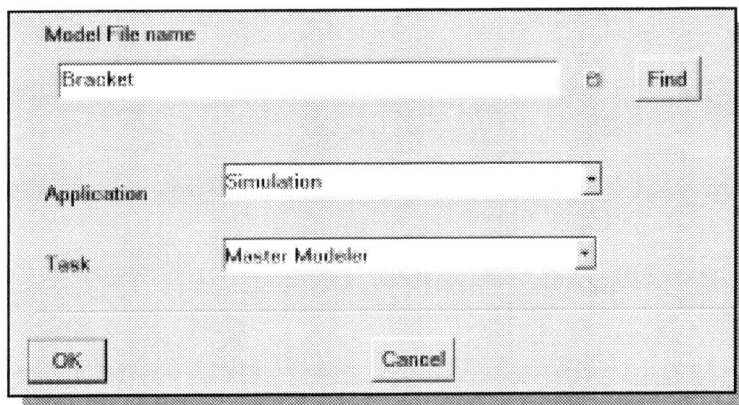

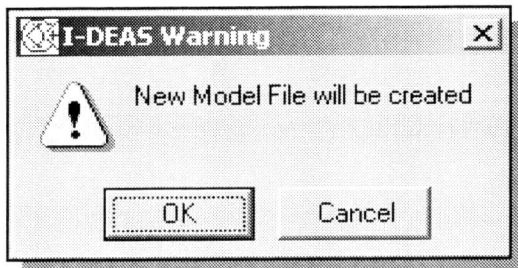

2. After you click **OK**, a *warning window* will appear indicating a new model file will be created. Click **OK** to continue.

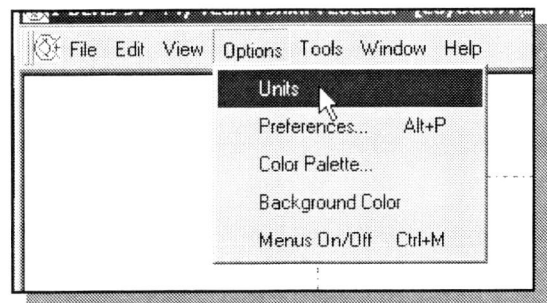

3. Use the left-mouse-button and select the **Options** menu in the icon panel.

4. Select the **Units** option.

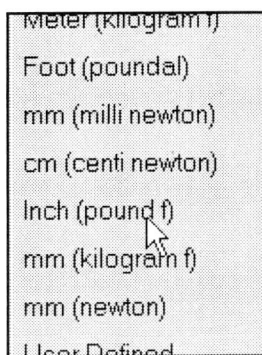

5. Set the units to **Inch (pound f)** by selecting it from the menu of choices.

❖ The units you used in an FE analysis <u>MUST</u> be consistent throughout the analysis.

10.3 Workplane Appearance

❖ The workplane is a construction tool; it is a coordinate system that can be moved in space. The size of the workplane display is only for our visual reference, since we can sketch on the entire plane, which extends to infinity.

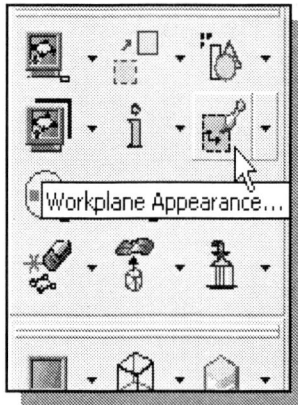

1. Choose **Workplane Appearance** in the icon panel. (The icon is located in the second row of the application icon panel.) The *Workplane Attributes* window appears.

2. Toggle *on* the *Display Border* switch as shown.

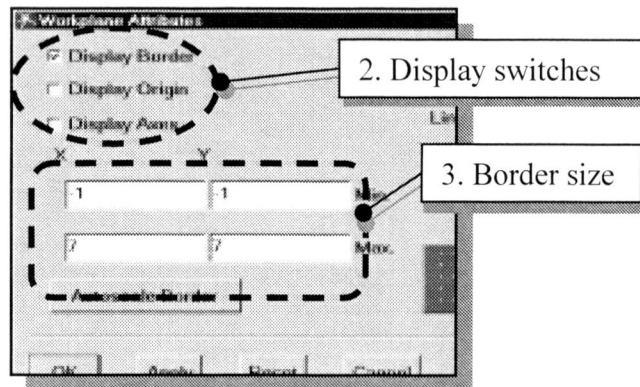

3. Adjust the **workplane border size** by entering the *Min.* & *Max.* values as shown in the figure above.

4. Click on the **OK** button to exit the *Workplane Attributes* window.

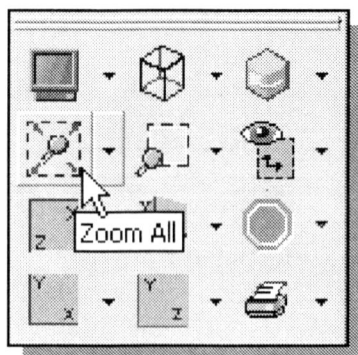

5. Choose **Zoom-All** in the display viewing icon panel

6. Choose **Refresh** in the icon panel. (The icon is located in the first row of the display icon panel.)

10.4 Using the Create Part option

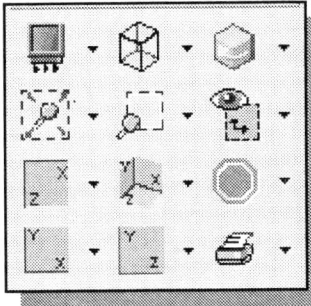

1. Choose **Isometric View** in the display viewing icon panel.

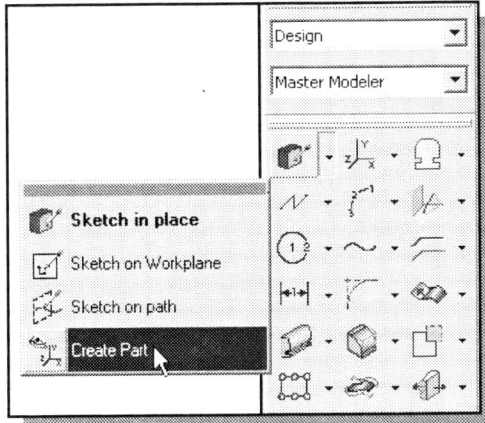

2. Choose **Create Part** in the icon panel.

➤ The icon is located in the first row of the task specific icon panel. The icon is located in the same stack as the *Sketch In Place* icon. Press and hold down the left-mouse-button on the icon stack to display the choice menu.

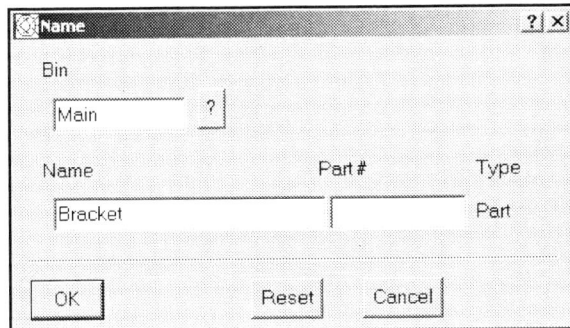

3. The *Name Part* window appears on the screen, enter **Bracket** as the name of the part as shown.

4. Click on the **OK** button to proceed with the **Create Part** command.

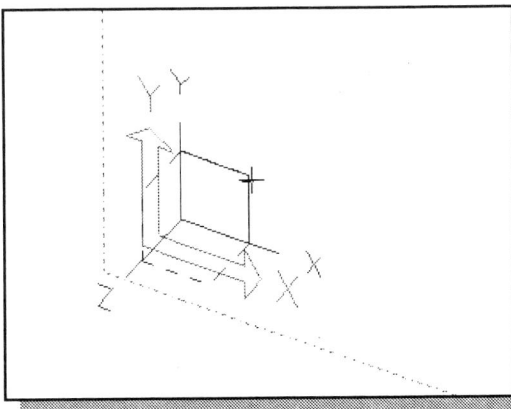

5. In the prompt window, the message "*Pick plane to sketch on*" is displayed. Pick the **XY** plane of the newly created coordinate system as shown. (Note that the default work plane, **blue** color, is still aligned to the XY plane of the world coordinate system. Aligning the sketch plane to the newly created coordinate system assures the proper association of the features to the part.)

6. Press the **ENTER** key once, or click once with the right-mouse-button to accept the placement of the workplane.

10.5 Creating a CAD Model of the System

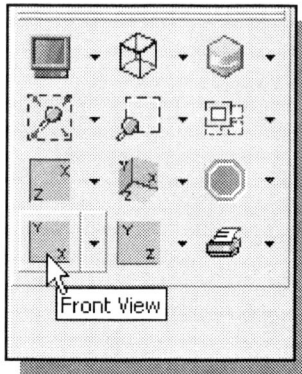

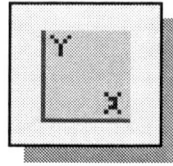

1. Click on the **Front View** icon in the display icon panel to switch the display to front view as shown.

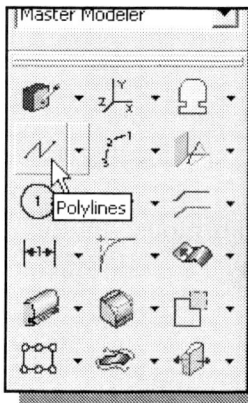

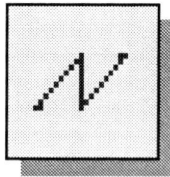

1. Choose **Polylines** in the icon panel. (The icon is located in the second row of the task specific icon panel.)

2. The message *"Locate start"* is displayed in the prompt window.

3. Left-click a starting point for the shape, roughly at the center of the graphics window.

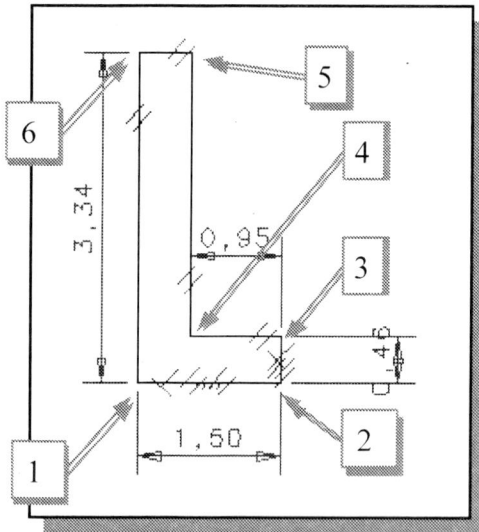

4. Move the graphics cursor horizontally to the right, *Point 2*, and create a horizontal line.

5. Complete the sketch as shown. Do not be overly concerned that the dimensions displayed on your screen might be different than what is shown. For the last location of the sketch, move the graphics cursor near the starting point of the sketch. Notice the *Dynamic Navigator* will snap to the end points of entities.

6. In the prompt window, the message *"Locate start"* is displayed. By default, *I-DEAS* remains in the *Polylines* command and expects you to start a new sequence of lines.

7. Press the **ENTER** key or the middle-mouse-button to end the *Polylines* command.

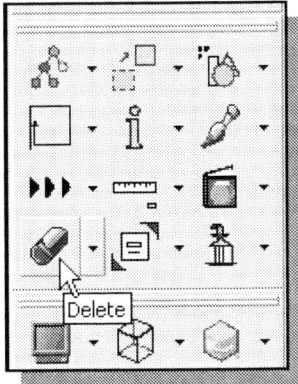

8. Choose **Delete** in the icon panel.

9. Pick the dimension as shown in the figure.

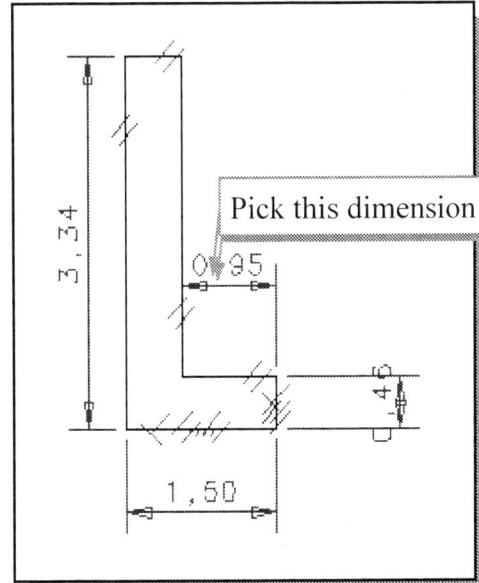

Pick this dimension

10. Press the **ENTER** key or the middle-mouse-button to accept the selection and continue with the delete command.

11. In the prompt window, the message *"OK to delete 1 dimension? (Yes)"* is displayed. Press the **ENTER** key or the middle-mouse-button to remove the selected dimension.

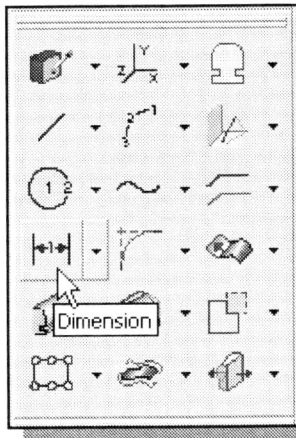

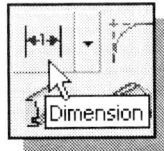

12. Choose **Dimension** in the icon panel.

13. The message *"Pick the first entry to dimension"* is displayed in the prompt window. Pick the vertical line as shown below.

14. Pick this line as the 2nd entity to dimension.

13. Pick this line as the 1st entity to dimension.

14. Pick the left vertical line as the second entity to dimension

15. Place the text to the top of the model by left-clicking once at the desired location.

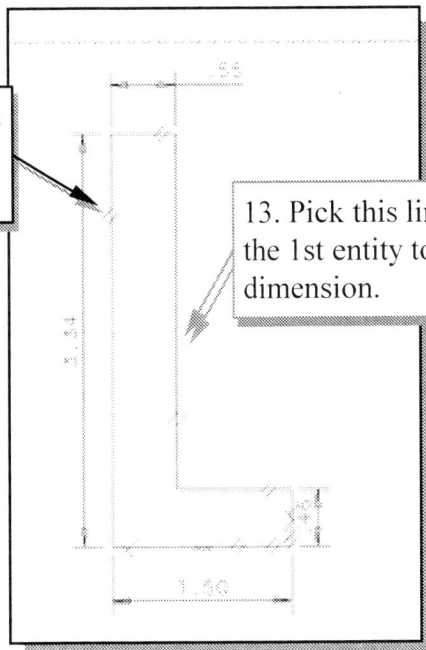

16. Press the **ENTER** key or the **middle-mouse-button** to end the current command.

17. Pre-select all of the dimensions by holding down the **SHIFT** key and left-clicking on each dimension.

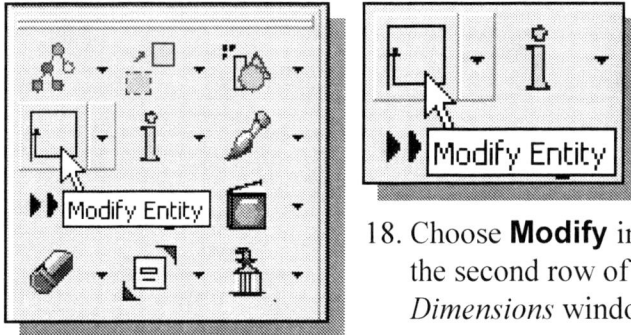

18. Choose **Modify** in the icon panel. The icon is located in the second row of the application icon panel. The *Dimensions* window appears.

19. Click on one of the dimensions in the pop-up window. The selected dimension will be highlighted in the graphics window. To see the selected dimension, move the pop-up window around by "dragging" the window title area with the left-mouse-button. You can also use the *Dynamic Viewing* functions to adjust the scale and location of the entities displayed in the graphics window (**F1** and the mouse, **F2** and the mouse).

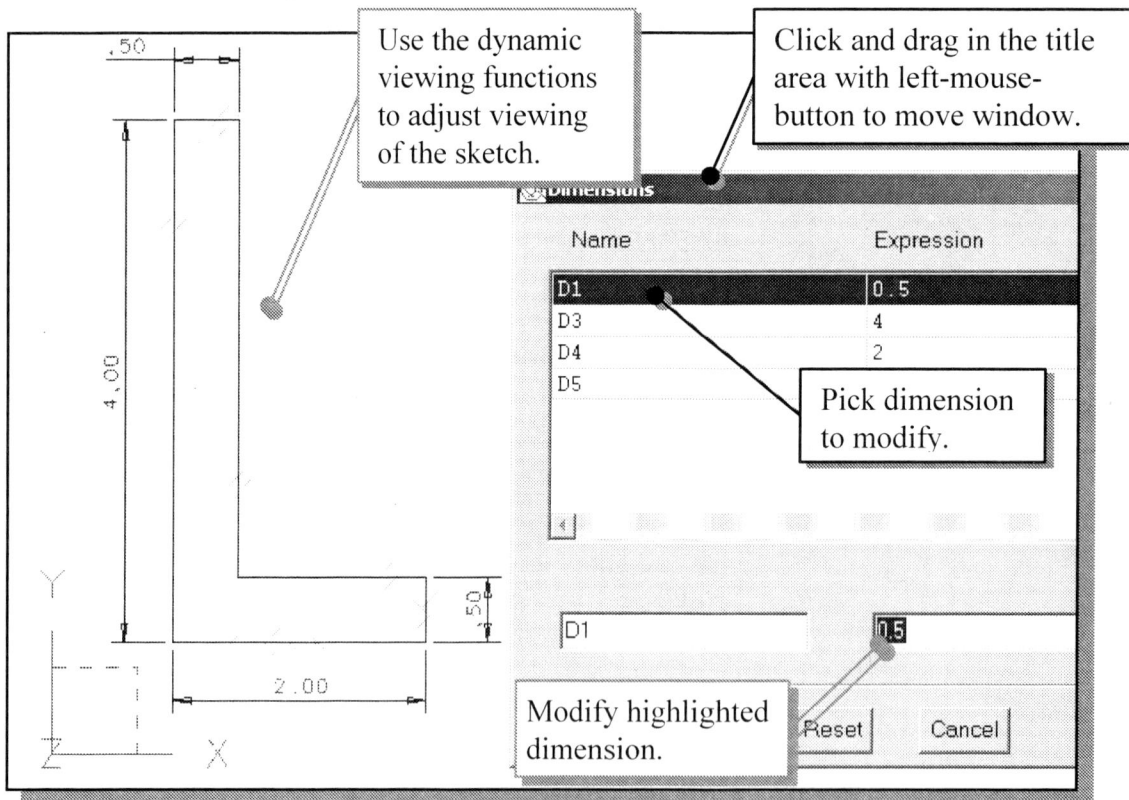

20. Modify the dimensions so that they appear as shown in the above figure.

10.6 Rounded Corners – 2D Fillets

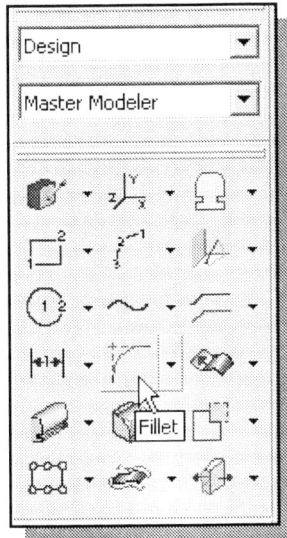

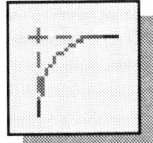

1. Choose **Fillet** in the icon panel. There are two fillet commands available in *I-DEAS Master Modeler*. Select the 2D **Fillet** icon. The message "*Pick section, curve or corner to fillet*" is displayed in the prompt window.

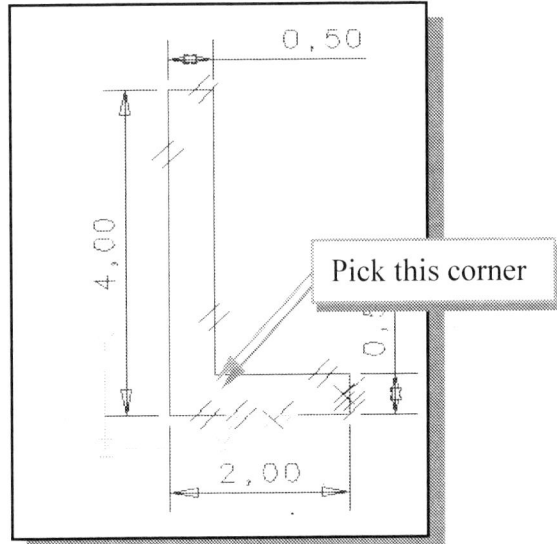

Pick this corner

2. Pick the corner as shown. The *Fillet* window appears.

3. In the *Fillet* window, enter **0.5** as the radius of the fillet.

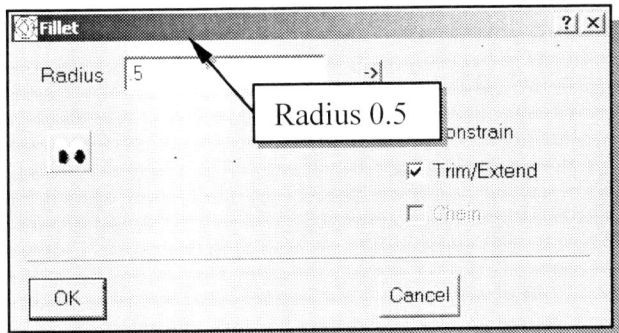

Radius 0.5

4. Confirm the *Constrain* and *Trim/Extend* options are toggled *on* and click on the **OK** button to create the fillet.

5. On your own, repeat the above steps and create the other rounded corner as shown below.

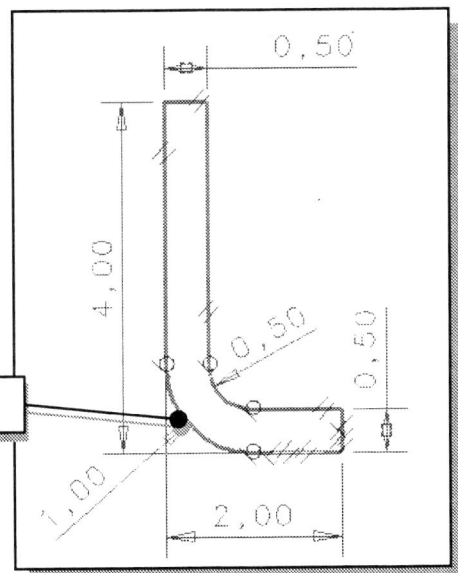

Radius of 1.0

10.7 Complete the Extrusion

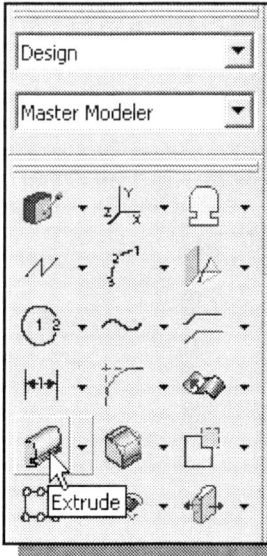

1. Choose **Extrude** in the icon panel. In the prompt window, the message "*Pick curve or section*" is displayed.

2. Pick any edge of the 2D shape. By default, the **Extrude** command will automatically select all segments of the shape that form a closed region shape.

3. At the *I-DEAS* prompt "*Pick curve to add or remove,*" press the **ENTER** key to accept the selected entities.

4. The *Extrude Section* window will appear on the screen. Fill in the items as shown below.

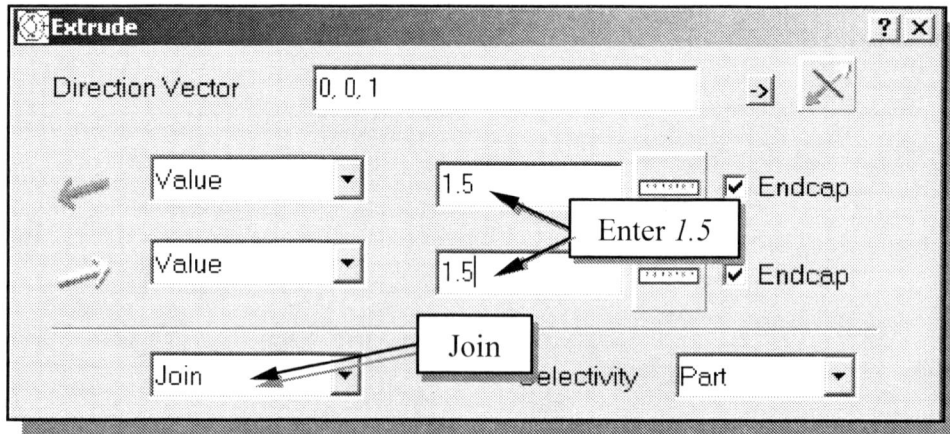

5. Click on the **OK** button to create the solid feature.

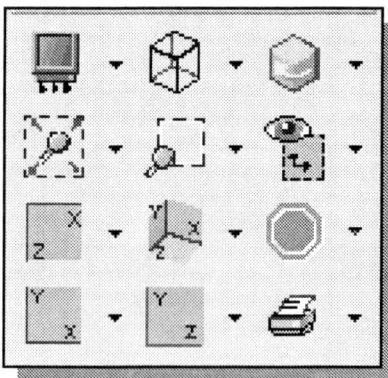

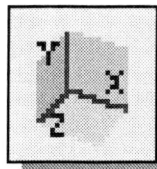

6. Choose **Isometric View** in the display viewing icon panel.

7. Choose **Zoom-All** in the display viewing icon panel.

10.8 Create Another Protrude Feature

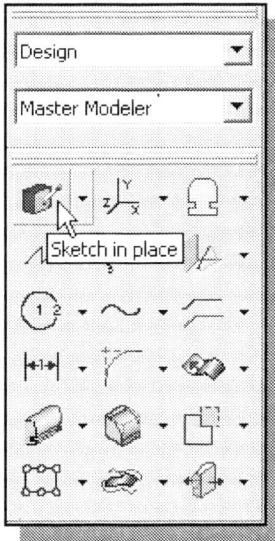

1. Choose **Sketch in Place** in the icon panel. In the prompt window, the message "*Pick plane or surface*" is displayed.

2. Pick the vertical face of the 3D object as shown below.

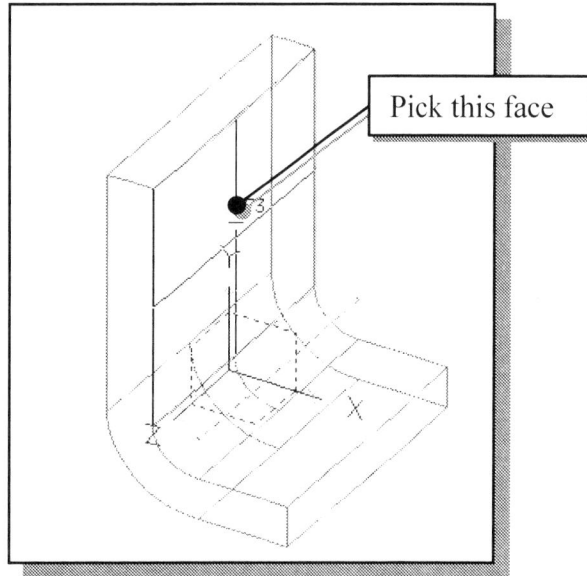

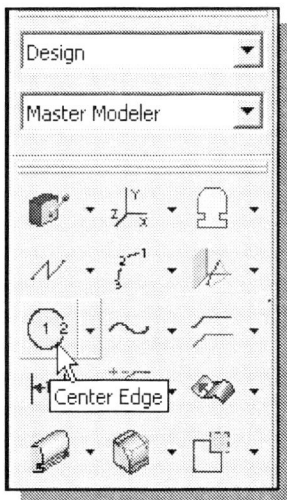

Pick this face

3. Choose **Circle - Center Edge** in the icon panel.

4. Position the center of the circle to align with the center of the highlighted surface. (Watch for the two dashed lines showing the alignments to the midpoints of the two edges of the face.)

5. Move the cursor inside the graphics window. Press and hold down the right-mouse-button and pick **Options** in the pop-up menu. The *Circle by Center-Edge* window appears.

6. Enter the radius of **0.75** in the *Radius* box.

7. Select the **OK** icon to accept the settings. A circle is drawn.

8. Press the **ENTER** key or click the center-mouse-button to exit the *Circle By Center-Edge* command.

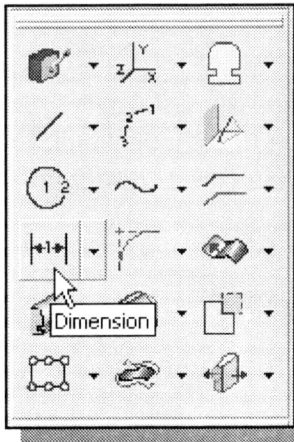

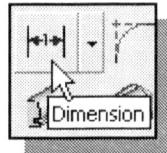

9. Choose **Dimension** in the icon panel. The message "*Pick the first entry to dimension*" is displayed in the prompt window.

10. Pick the center of the circle as the first entity and the left edge of the surface as the second entity.

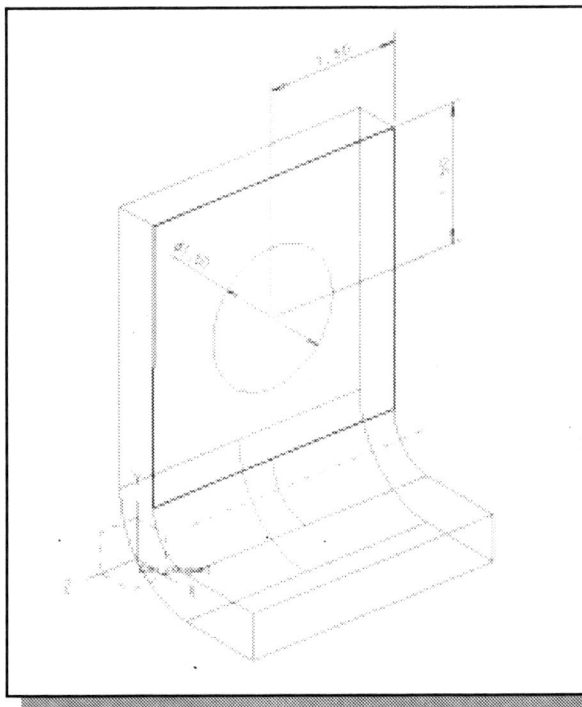

11. Place the dimension text above the surface that the sketch plane is aligned to.

12. Repeat the above steps and create the vertical dimension for the location of the hole as shown.

13. If necessary, make modifications to the dimensions so that the values are as shown.

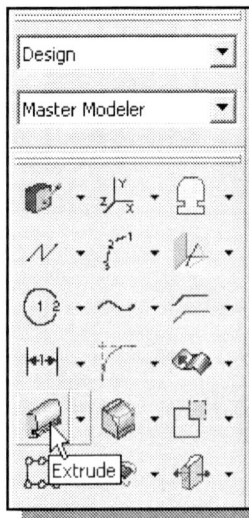

14. Choose **Extrude** in the icon panel. In the prompt window, the message "*Pick curve or section*" is displayed. Pick the circle we just created.

15. Press the **ENTER** key to accept the selected entity. The *Extrude Section* window appears

16. **Toggle the extrusion direction**

Extrude

Direction Vector   `-1, 9.10616E-017, 0`

Value   `625`   ☑ Endcap

17. **Enter** *0.625*

Value   `0`

Join   Selectivity   `Part`

18. **Join option**

Draft Angle   `0`

Options

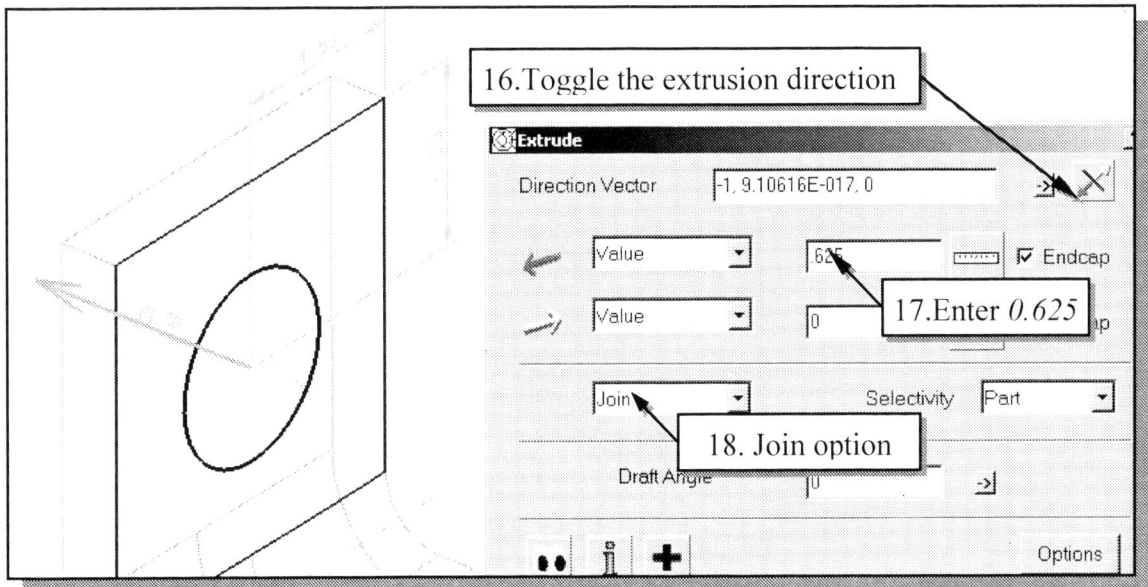

16. Click on the **Arrows** button to flip the extrude direction (into the vertical section of the solid model).

17. Enter **0.625** in the *Distance* box.

18. Confirm the *Join* option is set and click on the **OK** button to accept the settings.

10.9 Create a Cutout Feature

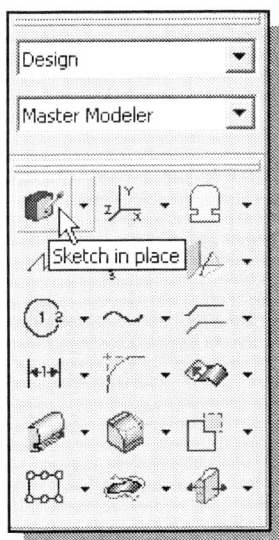

Design

Master Modeler

Sketch in place

1. Choose **Sketch in Place** in the icon panel. In the prompt window, the message *"Pick plane or surface"* is displayed.

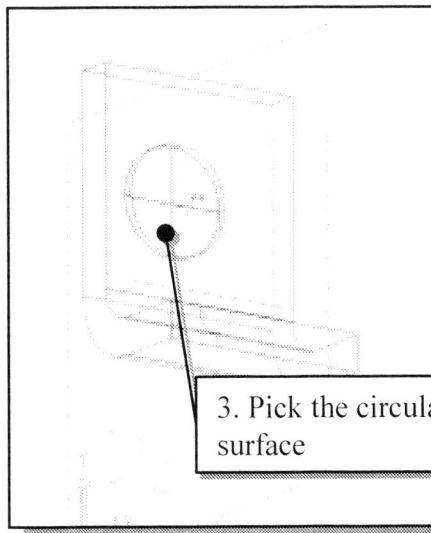

2. Use the *Dynamic Rotate* command to view the left vertical face of the model as shown.

3. Pick the circular surface of the 3D object as shown.

3. Pick the circular surface

4. Choose **Circle - Center Edge** in the icon panel.

5. Pick the center of the circle to be used as the center of the new circle.

6. Move the cursor inside the graphics window. Press and hold down the right-mouse-button and pick **Options** in the option menu. The *Circle by Center-Edge* window appears.

7. Enter the radius of **0.375** in the *Radius* box.

8. Select the **OK** icon to accept the settings. A circle is drawn.

9. Press the **ENTER** key or click the middle-mouse-button to create the circle and exit the *Circle by Center-Edge* command.

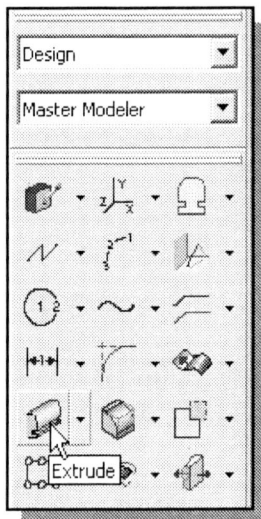

10. Choose **Extrude** in the icon panel. In the prompt window, the message "*Pick curve or section*" is displayed.

11. Pick the circle we just created.

12. Press the **ENTER** key or click the middle-mouse-button to accept the selected entity. The *Extrude Section* window appears.

13. Select **Cut** and **Thru All**.

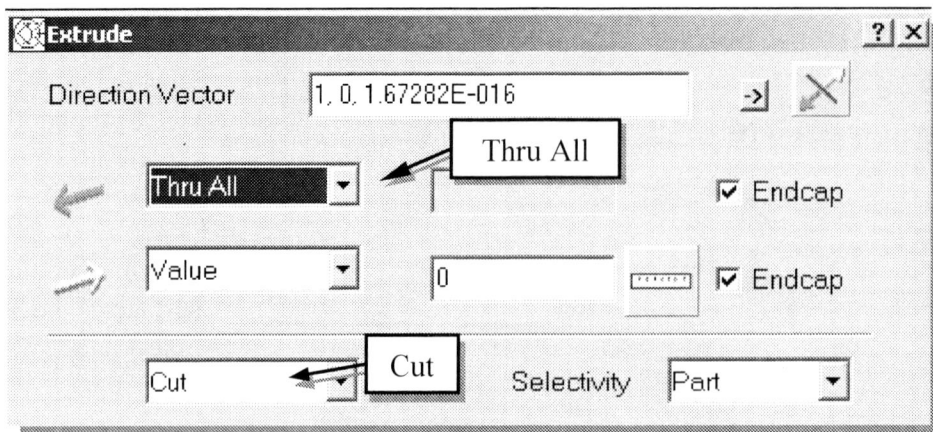

14. Click on the **OK** button to accept the settings and create the cutout.

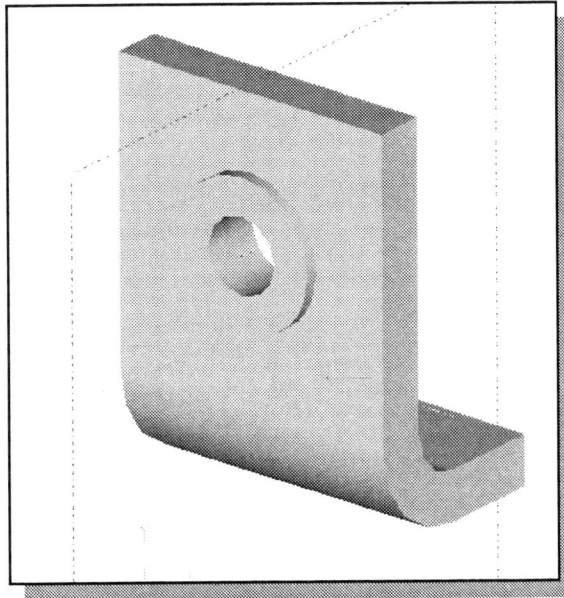

10.10 Rounded Corners – 3D Fillets

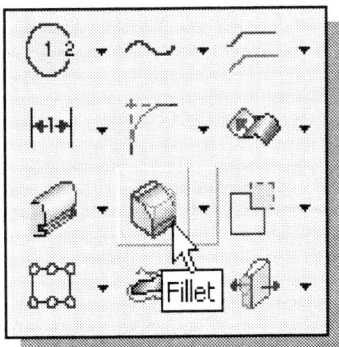

1. Choose **3D Fillet** in the icon panel.

2. The message "*Pick edges, vertices or surfaces to fillet/round*" is displayed in the prompt window. Pick the two edges by holding down the **SHIFT** key while selecting.

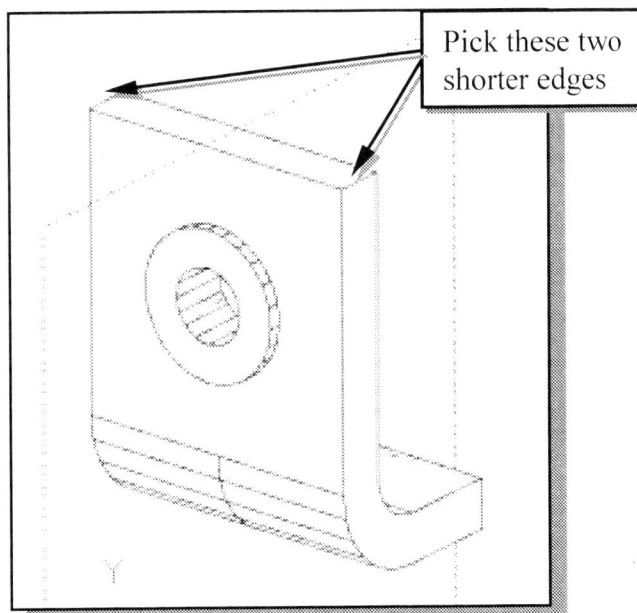

Pick these two shorter edges

3. Press the **ENTER** key or click the middle-mouse-button to accept the selected entities.

4. The message "*Enter Radius*" is displayed in the prompt window. Enter **1.5** for the radius.

5. Press the **ENTER** key to accept the settings and create the *Fillets*. Your screen should appear as shown on the next page.

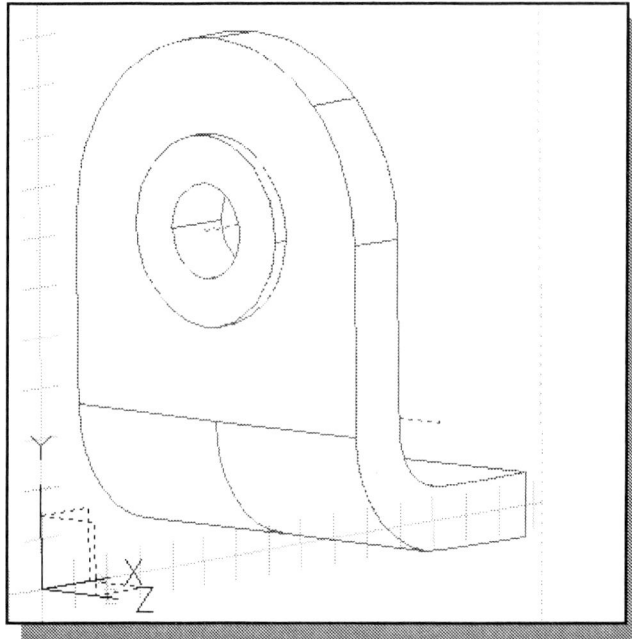

10.11 Setup of an FEA Model

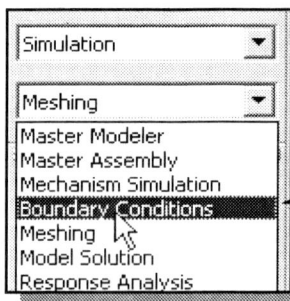

1. Switch to the **Boundary Conditions** task by selecting it in the task menu as shown.

    Pick *Boundary Conditions*

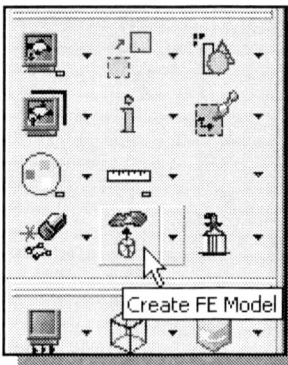

2. Choose **Create FE Model** in the icon panel. (The icon is located in the last row of the application specific icon panel.) The *FE Model Create* window appears.

FE Model Create

Bin | Part or Assembly
Main | ? | Bracket | Get..

Part Material

GENERIC_ISOTROPIC_S | Select..

FE Model Name | FE Model Part #

Bracket |

☐ All Sections   (SC)

☐ P Element Analysis

☑ Geometry Based Analysis Only

☐ Auto Create Dynamic Groups

☑ 1D Elements   ☑ Include Related Nodes

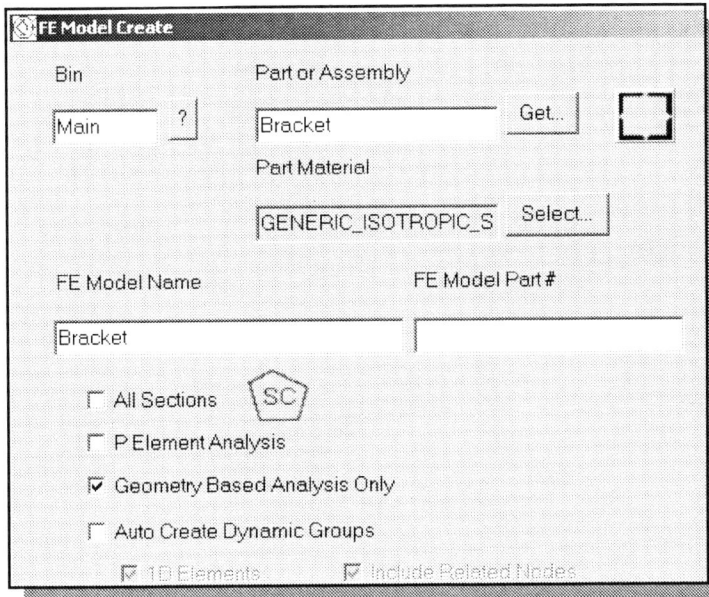

3. In the *FE Model Create* window, enter **Bracket** in the *FE Model Name* box.

4. Switch *on* the *Geometric Based Analysis Only* option.

5. Click on the **OK** button to accept the settings.

10.12 Apply Boundary Conditions

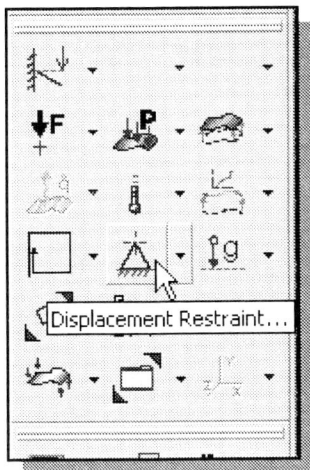

1. Choose ***Displacement Restraint*** in the icon panel. (The icon is located in the fourth row of the task icon panel.)

2. The message "*Pick Nodes/Centerpoints/Vertices*" is displayed in the prompt window. Pick the bottom face of the model as shown. Note the displayed **Fxx** symbol as the cursor is moved on top of the surface.

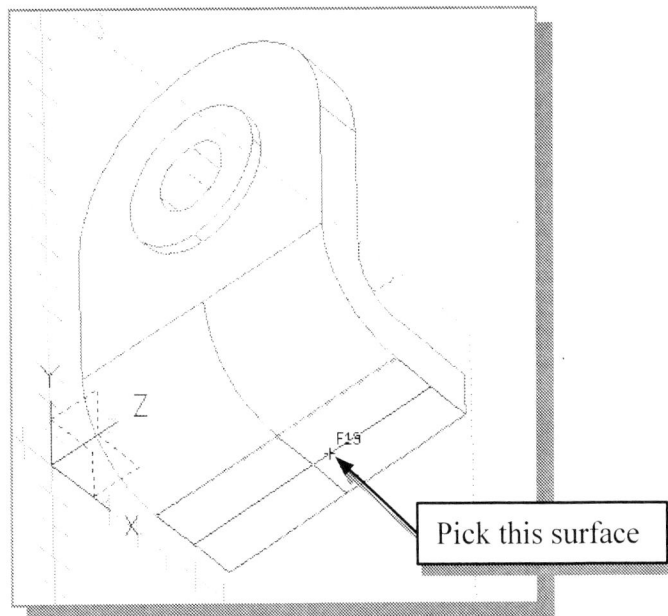

Displacement Restraint...

Pick this surface

3. Press the **ENTER** key or click the middle-mouse-button to continue. The *Displacement Restraint on Surface* window appears.

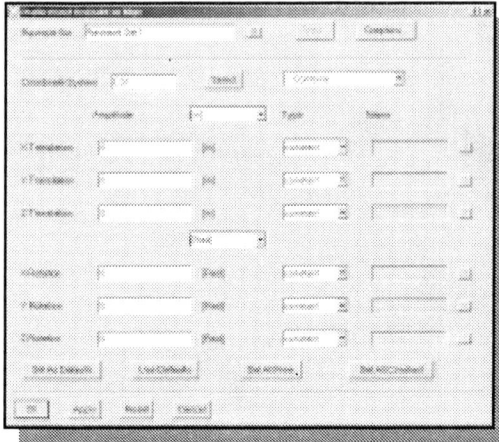

4. We will restrict any displacement of the bottom of the **Bracket**. Click on the **OK** button to accept the default settings and exit the *Displacement Restraint on Surface* window.

10.13 Apply the External Loads

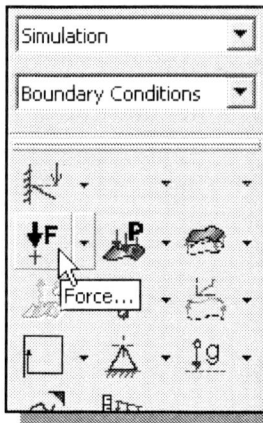

1. Choose **Force...** in the icon panel. (The icon is located in the second row of the task icon panel.)

2. The message "*Pick entities*" is displayed in the prompt window.

3. Pick the circular surface as shown below.

4. Press the **ENTER** key or click the middle-mouse-button to accept the selection. The *Traction on Surface* window appears.

Pick this surface

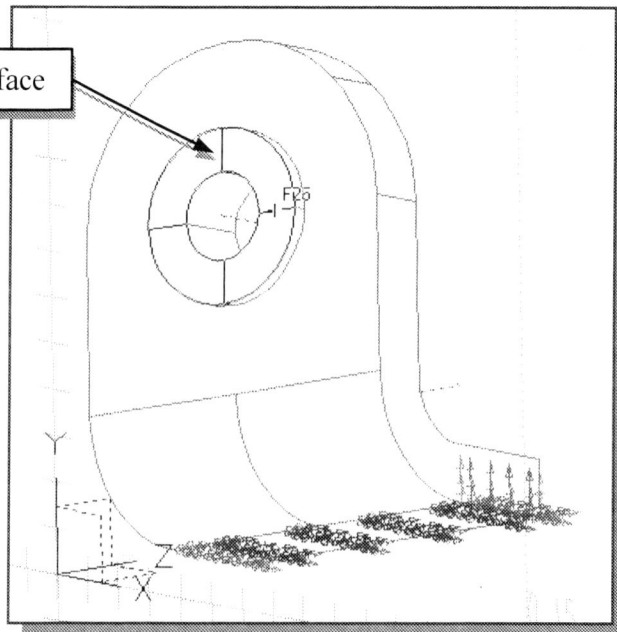

5. In the *Traction on Surface* window, enter **300** in the *Amplitude* box.

6. Set the force option to **Total Force**.

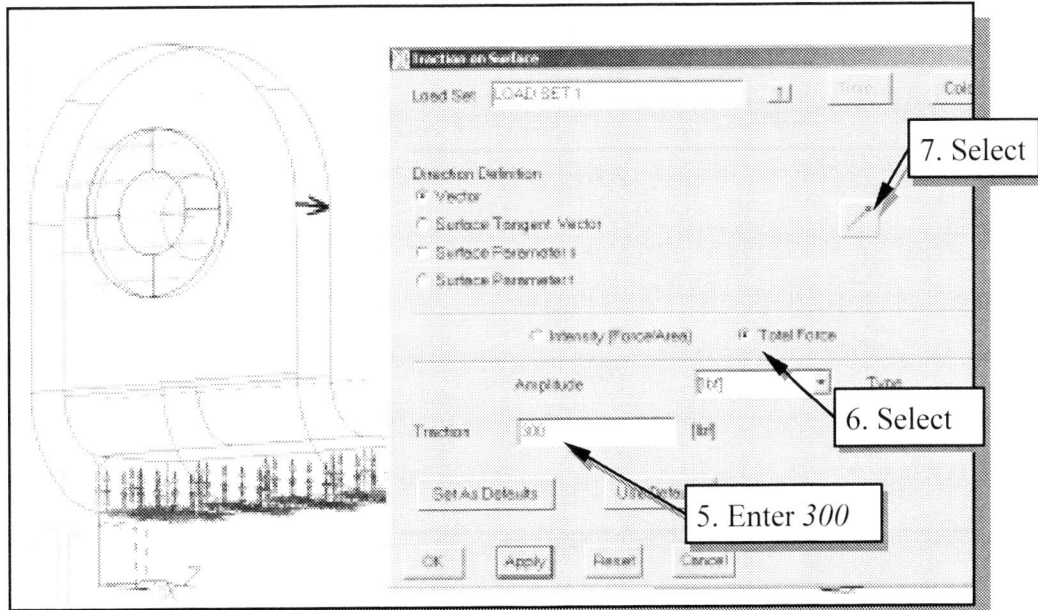

7. Click on the **Select Vector** icon to set the direction of the load.

8. Select one of the edges that is perpendicular to the selected face (the direction parallel to the center hole direction).

9. Click on the **Apply** button to see a preview of the applied load. If necessary, make additional modifications so that the settings are as shown.

10. Click on the **OK** button to accept the settings and exit the *Traction on Surface* window. Your screen should appear as shown below.

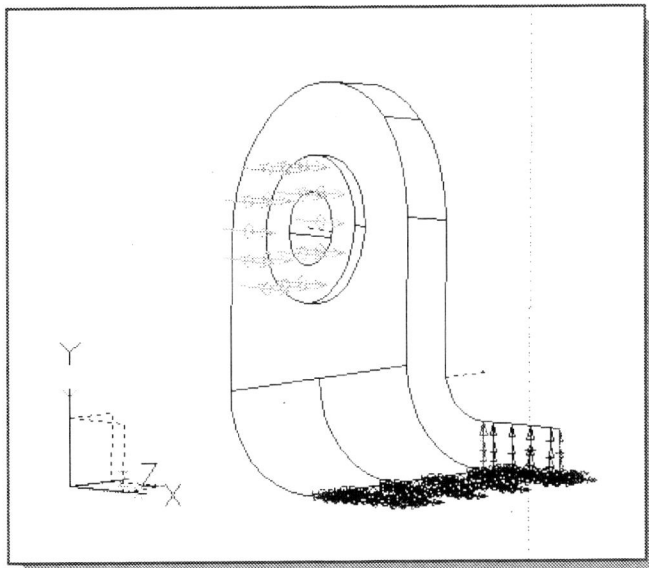

10.14 Boundary Condition Set

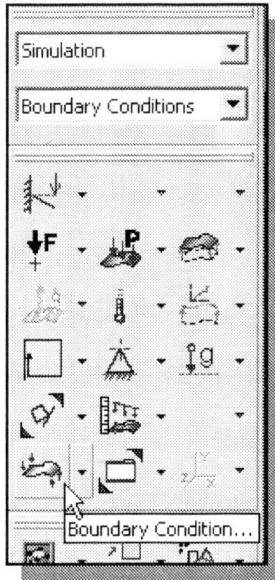

1. Choose **Boundary Condition Set Management** in the icon panel. (The icon is located in the sixth row of the task icon panel.) The *Boundary Condition Set Management* window appears.

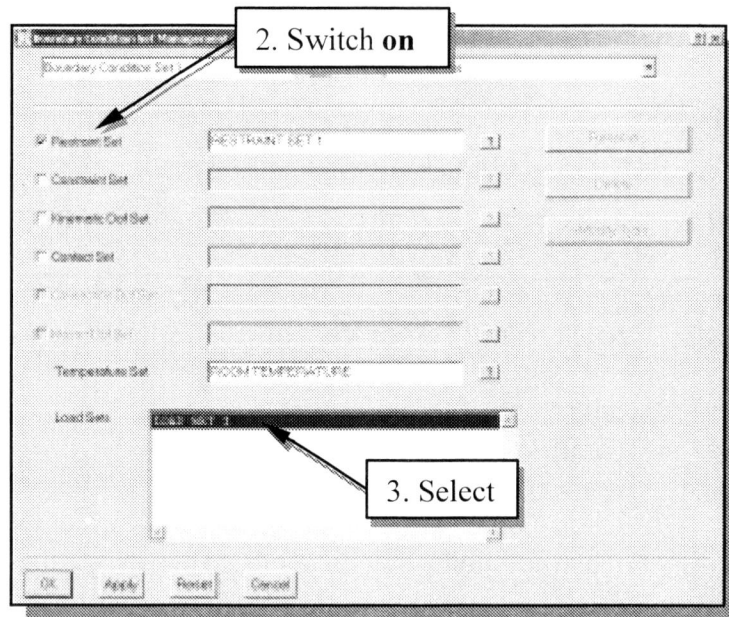

2. Switch *on* the *Restraint Set.*

3. Select **Load Set 1** in the *Load Sets* list.

4. Click on the **OK** button to accept the settings and exit the *Boundary Condition Set Management* window.

10.15 Material Property Table

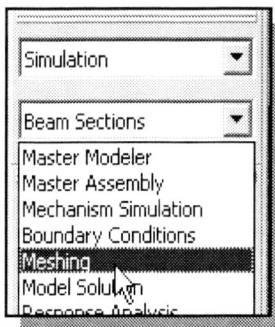

1. Switch back to the **Meshing** task by selecting the Meshing task in the task menu as shown.

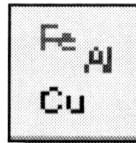

2. Choose **Materials** in the icon panel. (The icon is located in the fourth row of the task icon panel.) The *Materials* window appears.

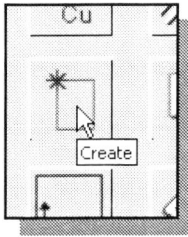

3. Choose **Quick Create** in the *Materials* window. The *Quick Create* window appears.

4. Enter **Aluminum Alloy 6061 T6** in the *Material Name* box.

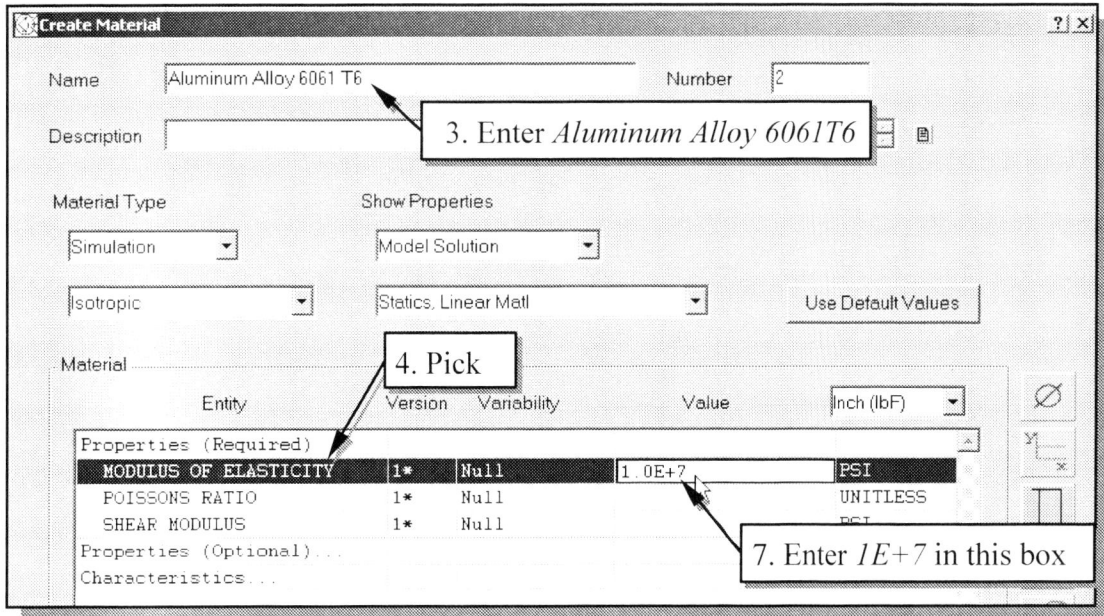

5. Pick **Modulus of Elasticity** in the *Properties* list.

6. Click in the *Value* box to activate the modification mode.

7. Type **1E+7** in the *Properties Value* box.

8. On your own, set the following values in the *Properties* list:

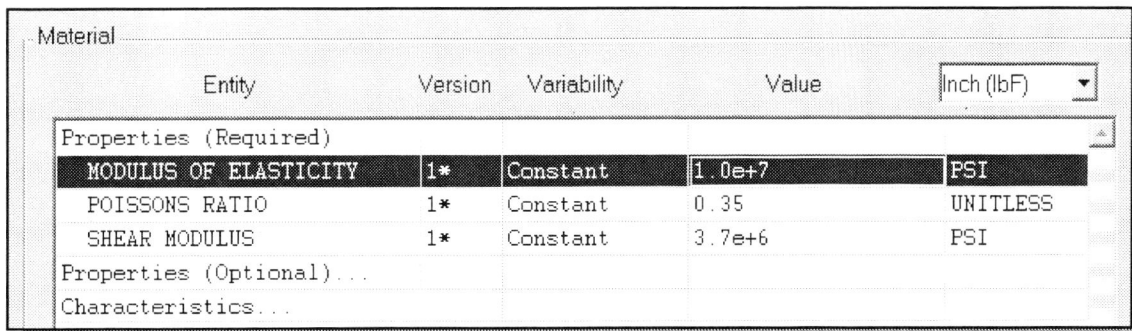

9. Click on the **OK** button to exit the *Create* window.

10. Click on the **OK** button to exit the *Materials* window.

10.16 Define Auto-Mesh

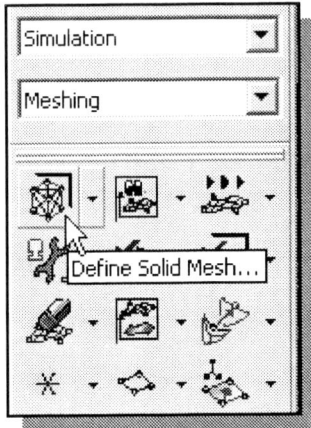

1. Choose **Define Solid Mesh** in the icon panel. The icon is located in the first row of the task icon panel.

2. The message "*Pick Volumes*" is displayed in the prompt window. Pick any edge of the part.

3. Press the **ENTER** key or click the middle-mouse-button to accept the selection. The *Define Mesh* window appears.

4. Enter **0.25** for *Element Length*.

5. Set **Other** for the *Material Property* option.

6. Set *Material* to **Aluminum Alloy 6061 T6**.

7. Confirm the *Element Type* is set to the higher order element type (additional node points on the edges of the tetrahedral element).

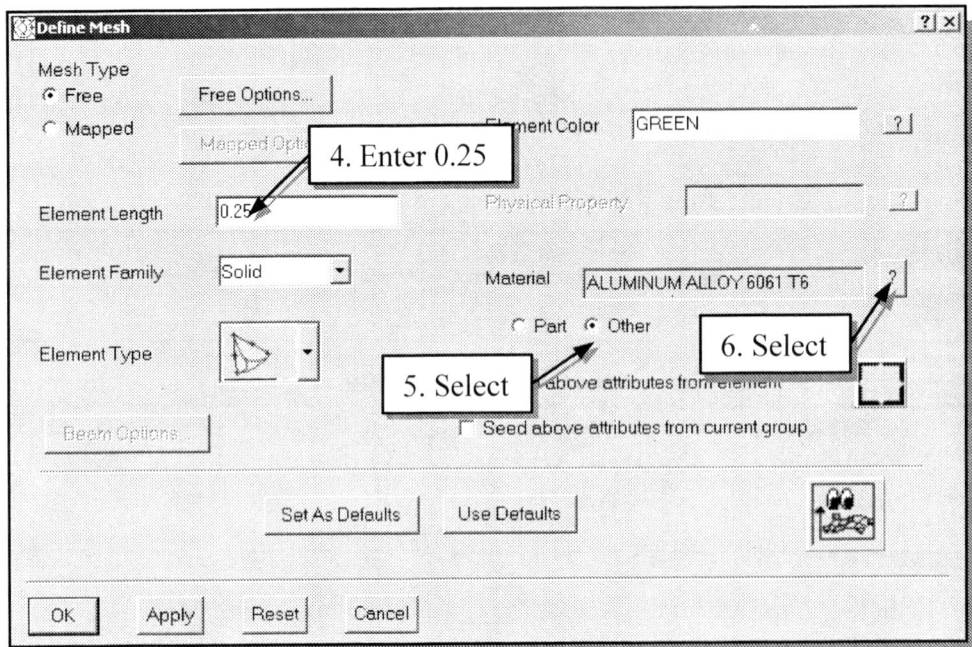

8. Click on the **OK** button to accept the settings and exit the *Define Mesh* window.

10.17 Create the Mesh

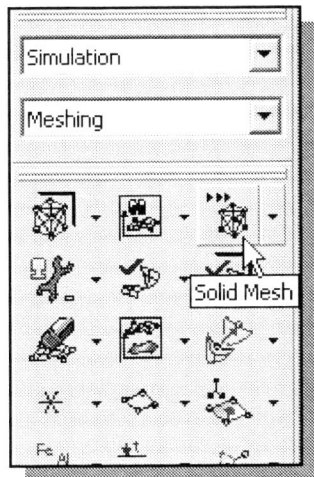

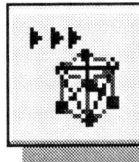

1. Choose **Create Solid Mesh** in the icon panel. The icon is located in the first row of the task icon panel.

2. The message "*Pick Volume*" is displayed in the prompt window. Pick any edge of the plate.

3. Press the **ENTER** key or click the middle-mouse-button to accept the selection. New nodes and elements are created.

   ❖ How many nodes and elements are created for this model? How does it compare to the 2D solid elements we did in Chapter 9?

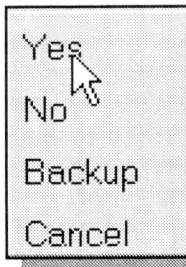

4. Select **Yes** in the popup menu or press the **ENTER** key or click the middle-mouse-button to accept the additions.

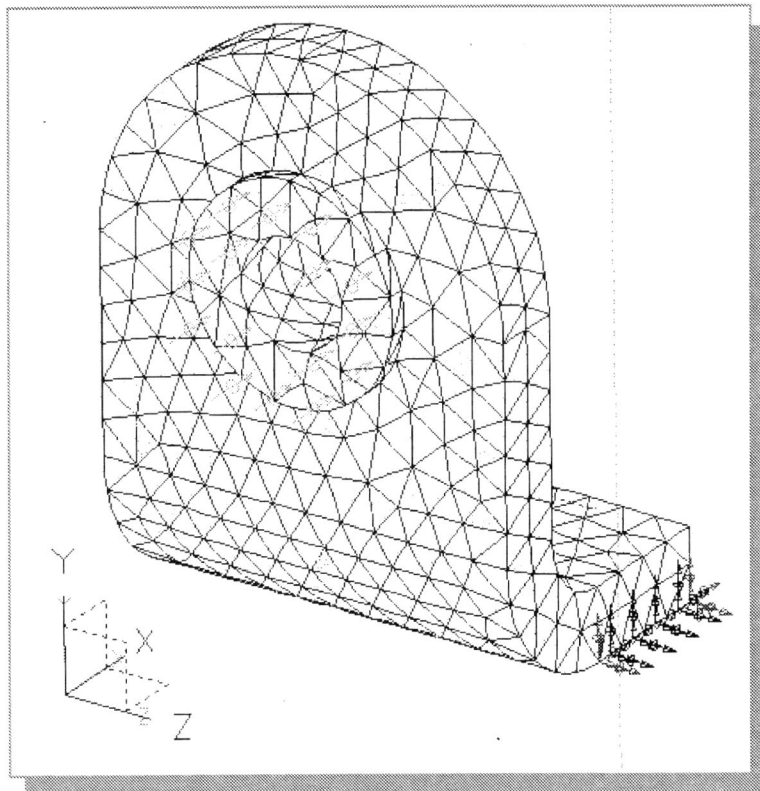

10.18 Run the Solver

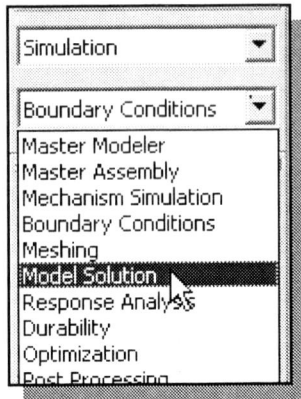

1. Switch to the **Model Solution** task by selecting the task in the task menu as shown.

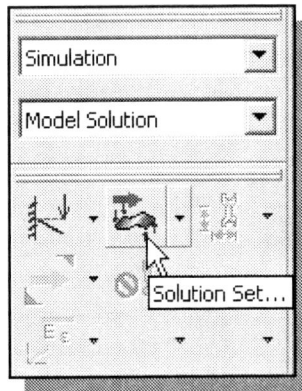

2. Choose **Solution Set** in the icon panel. (The icon is located in the first row of the task icon panel.) The *Manage Solution Sets* window appears.

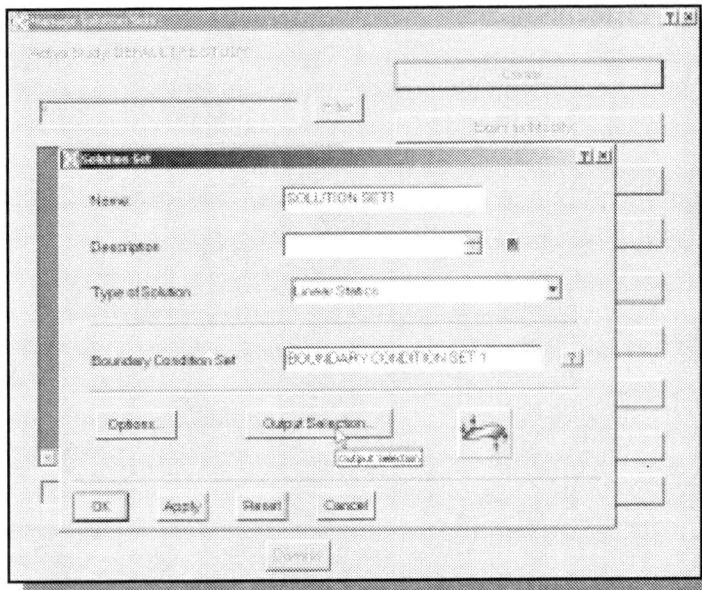

3. Click on the **Create** button. The *Solution Set* window appears.

4. We will use the default settings; click the **OK** button to exit the *Solution Set* window.

5. Pick **Dismiss** to exit the *Manage Solution Sets* window.

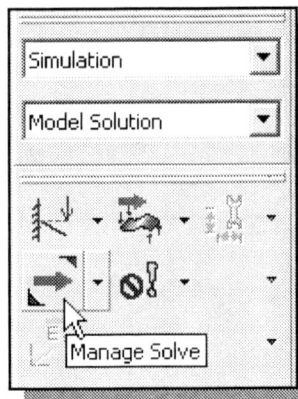

6. Choose **Manage Solve** in the icon panel. (The icon is located in the second row of the task icon panel.) The *Solve* window appears.

7. In the *Solve* window, pick the **Solve** icon to find the solutions.

➢ I-DEAS will begin the solving process. <u>DO NOT</u> close any windows. Any errors or warnings are displayed in the list window.

10.19 View the Results

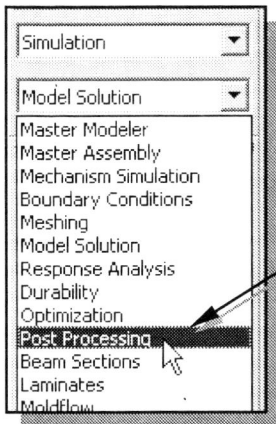

1. Switch to the **Post Processing** task by selecting in the task menu as shown.

Pick *Post Processing*

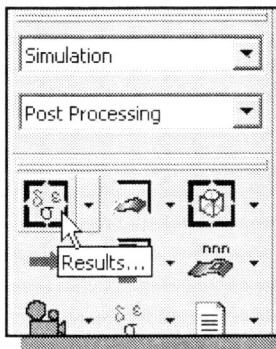

2. Choose **Results Selection** in the icon panel. (The icon is located in the first row of the task icon panel.) The *Results Selection* window appears.

3. Confirm the *Display Component* is set to **Von Mises**.

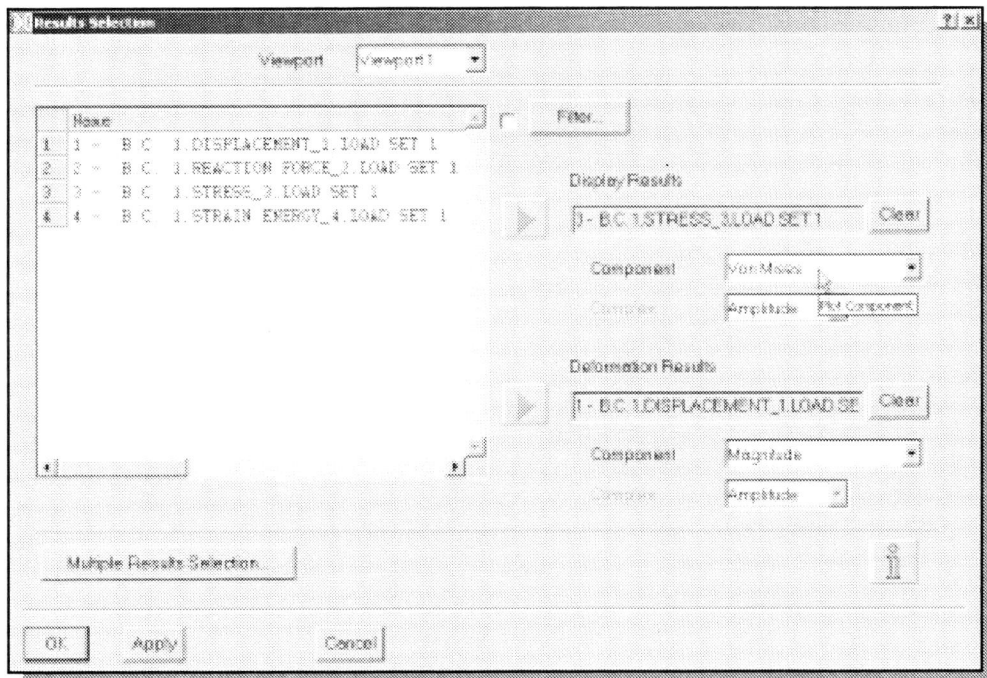

4. Click on the **OK** button to accept the settings and exit the *Results Selection* window.

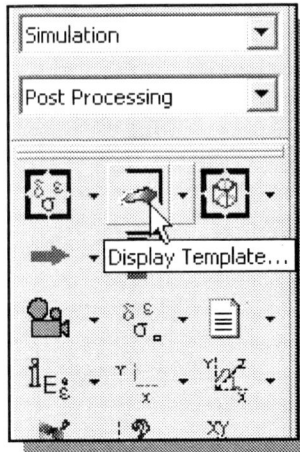

5. Choose **Display Template** in the icon panel. The icon is located in the first row of the task icon panel. The *Display Template* window appears.

6. In the *Display Template* window, select **Smooth Shaded** and toggle **on** the *Lighting Effects.*

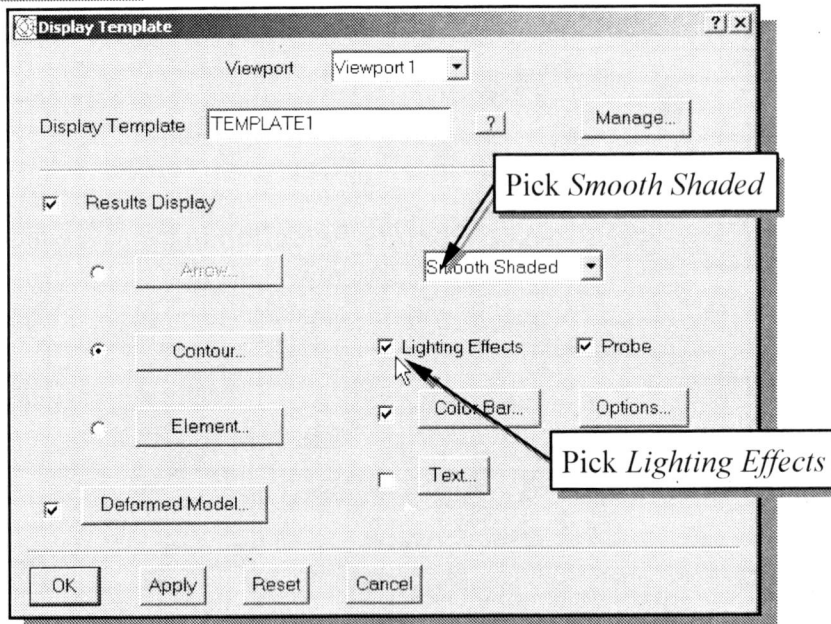

7. Click on the **OK** button to exit the *Display Template* window.

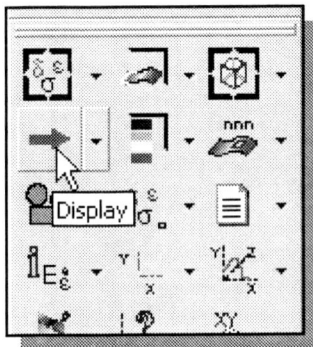

8. Choose **Results Display** in the icon panel. (The icon is located in the second row of the task icon panel.)

9. The message "*Pick elements*" is displayed in the prompt window. Press the **ENTER** key to continue. (Stresses on all elements will be processed.)

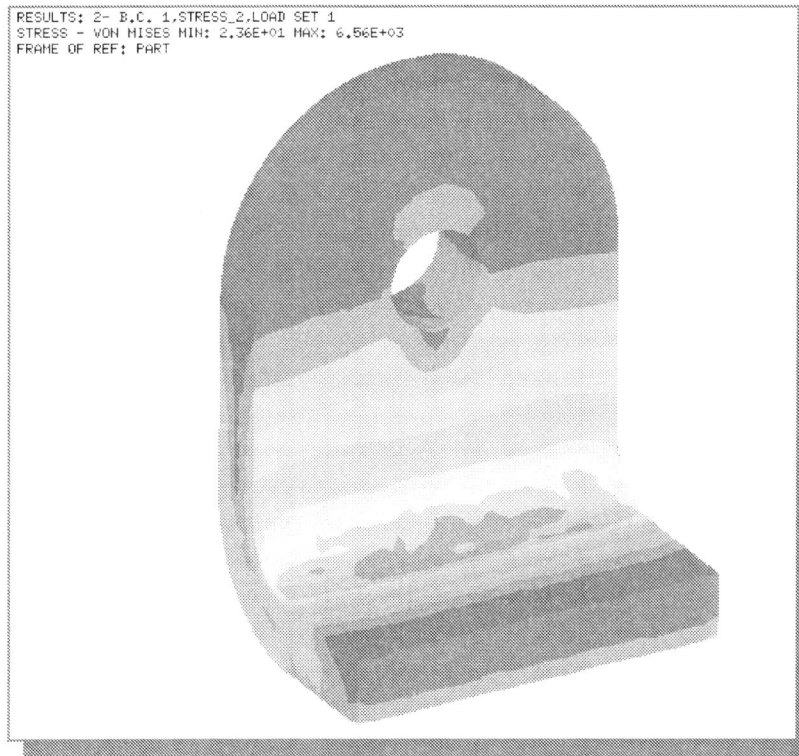

```
RESULTS: 2- B.C. 1,STRESS_2,LOAD SET 1
STRESS - VON MISES MIN: 2.36E+01 MAX: 6.56E+03
FRAME OF REF: PART
```

❖ On your own, perform a convergence study by using the *Free* options and the *Define Local* options to refine the mesh of the current FE model. (Hint: concentrate on the regions that displayed higher stresses in the original analysis. Refer to page 9-22 through 9-26 for the procedures to refine the mesh.) Record the number of elements generated and compare the runtime of the solver between each analysis.

10.20 Conclusion

In examining the FEA results, one should first examine the deformed shape to check for proper placement of boundary conditions and to determine if the calculated deformation of the model is reasonable. For the example illustrated in this chapter, it is also possible to consider analyzing the problem using simple beam theory. Although the part is short and wide, the flexural stress can still give us some indication of the reasonable stress range of the FEA results. It is always important to perform a *convergence study* and refine the mesh in high-stress regions to obtain more accurate results.

It should be emphasized that, when performing FEA analysis, besides confirming that the systems remain in the elastic regions for the applied loading, other considerations are also important. For example, large displacements and buckling of beams, which can also invalidate the *linear statics* analysis results. In performing finite element analysis, it is also necessary to acquire some knowledge of the theory behind the method and understand the restrictions and limitations of the software. There is no substitution for experience.

Finite element analysis has rapidly become a vital tool for design engineers. However, use of this tool does not guarantee correct results. The design engineer is still responsible for doing approximate calculations, using good design practice, and applying good engineering judgment to the problem. It is hoped that FEA will supplement these skills to ensure that the best design is obtained.

Throughout this text, various techniques have been presented. The goals are to make use of the tools provided by *I-DEAS* and to successfully enhance the DESIGN capabilities. We have gone over the fundamentals of FEA procedures. In many instances, only a single approach to the stress analysis tasks was presented; you are encouraged to repeat any of the chapters and develop different ways of thinking in order to accomplish the goal of making better designs. We have only scratched the surface of *I-DEAS'* functionality. The more time you spend using the system, the easier it will be to perform *computer aided engineering* with *I-DEAS*.

Questions:

1. What are the most widely used failure criteria?

2. What is the main objective of finite element analysis?

3. What will happen to the system if it is stressed beyond the elastic limit?

4. What is the purpose of doing a convergence study?

5. Identify and describe the following commands:

(a)

(b)

(c)

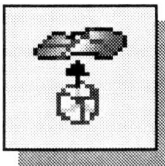

(d)

Exercises:

1. For the steel cantilever beam (5" X 1" X 4"), perform a FEA analysis using 3D solid elements.

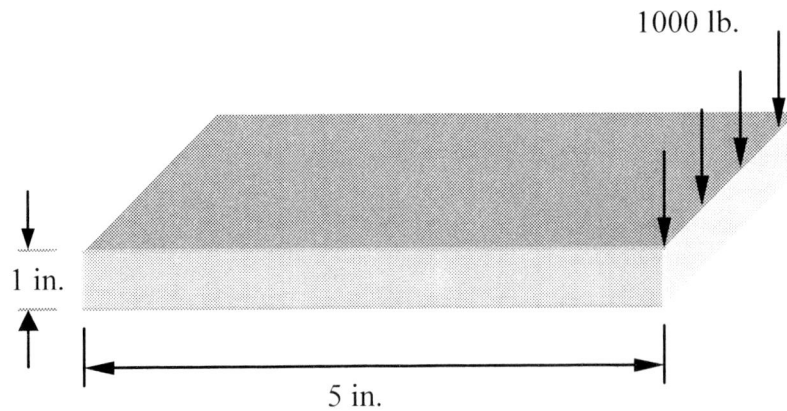

1000 lb.

1 in.

5 in.

2. For the steel brake-pedal, perform the FEA analysis to examine the stress distribution and determine the maximum deflection developed.

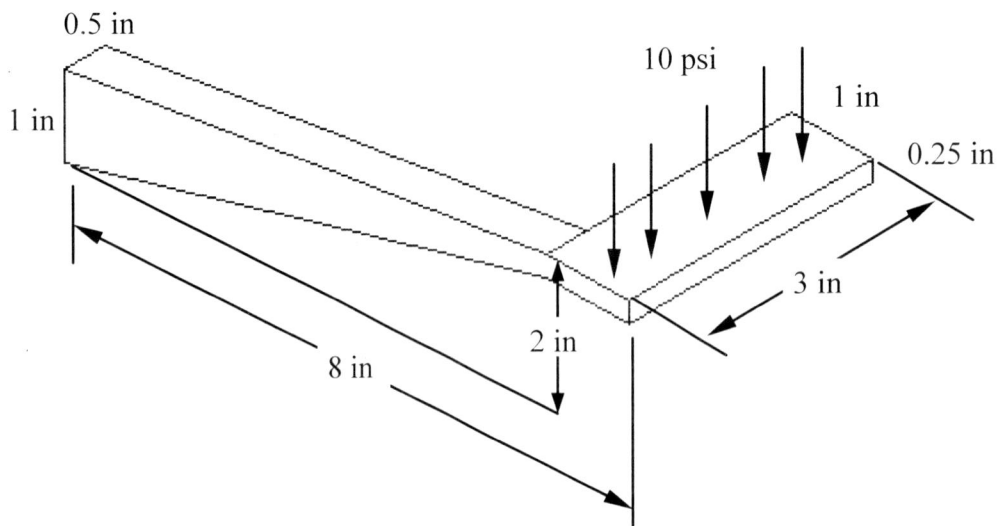

0.5 in

1 in

10 psi

1 in

0.25 in

3 in

2 in

8 in

INDEX